Brock/Springer Series in Contemporary Bioscience

The Search for Bioactive Compounds from Microorganisms

Brock/Springer Series in Contemporary Bioscience

Series Editor: Thomas D. Brock
   University of Wisconsin-Madison

Tom Fenchel
ECOLOGY OF PROTOZOA:
The Biology of Free-living Phagotrophic Protists

Johanna Döbereiner and Fábio O. Pedrosa
NITROGEN-FIXING BACTERIA IN NONLEGUMINOUS
CROP PLANTS

Tsutomu Hattori
THE VIABLE COUNT: Quantitative and Environmental Aspects

Roman Saliwanchik
PROTECTING BIOTECHNOLOGY INVENTIONS:
A Guide for Scientists

Hans G. Schlegel and Botho Bowien (Editors)
AUTOTROPHIC BACTERIA

Barbara Javor
HYPERSALINE ENVIRONMENTS: Microbiology and Biogeochemistry

Ulrich Sommer (Editor)
PLANKTON ECOLOGY: Succession in Plankton Communities

Stephen R. Rayburn
THE FOUNDATIONS OF LABORATORY SAFETY:
A Guide for the Biomedical Laboratory

Gordon A. McFeters (Editor)
DRINKING WATER MICROBIOLOGY:
Progress and Recent Developments

Mary Helen Briscoe
A RESEARCHER'S GUIDE TO SCIENTIFIC AND
MEDICAL ILLUSTRATIONS

Max M. Tilzer and Colette Serruya (Editors)
LARGE LAKES: Ecological Structure and Function

Jürgen Overbeck and Ryszard J. Chróst (Editors)
AQUATIC MICROBIAL ECOLOGY:
Biochemical and Molecular Approaches

(Continued after index)

Satoshi Ōmura
Editor

The Search for Bioactive Compounds from Microorganisms

With 74 Figures

Springer-Verlag
New York Berlin Heidelberg London Paris
Tokyo Hong Kong Barcelona Budapest

Satoshi Ōmura, Ph.D.
Research Center for Biological Function
The Kitasato Institute
and
School of Pharmaceutical Sciences
Kitasato University
Minato-ku, Tokyo 108, Japan

Cover: Colonies of *Streptomyces* sp. SK-1894, which produces the acyl-CoA synthetase inhibitors triacsins discovered by S. Ōmura and colleagues.

Library of Congress Cataloging-in-Publication Data
The search for bioactive compounds from microorganisms / [editor], Satoshi Ōmura.
    p.   cm. — (Brock/Springer series in contemporary bioscience)
  Includes bibliographical references and index.
  ISBN 0-387-97755-4. — ISBN 3-540-97755-4
  1. Pharmaceutical microbiology. I. Ōmura, Satoshi, 1935–
II. Series
  [DNLM: 1. Biological Products. 2. Biotechnology.
3. Microbiology. QW 800 S439]
QR46.5.S43 1992
660'.62—dc20
DNLM/DLC
for Library of Congress   91-5204

Printed on acid-free paper.

© 1992 Spring-Verlag New York, Inc.
All rights reserved. This work may not be translated or copied in whole or in part without the written permission of the publisher (Springer-Verlag New York, Inc., 175 Fifth Avenue, New York, NY 10010, USA), except for brief excerpts in connection with reviews or scholarly analysis. Use in connection with any form of information storage and retrieval, electronic adaptation, computer software, or by similar or dissimilar methodology now known or hereafter developed is forbidden.
The use of general descriptive names, trade names, trademarks, etc., in this publication, even if the former are not especially identified, is not to be taken as a sign that such names, as understood by the Trade Marks and Merchandise Marks Act, may accordingly be used freely by anyone.

Production managed by Terry Kornak; manufacturing supervised by Jacqui Ashri.
Typeset by Impressions, Madison, WI, a division of Edwards Brothers, Inc.
Printed and bound by Edwards Brothers, Inc., Ann Arbor, MI.
Printed in the United States of America.

9 8 7 6 5 4 3 2 1

ISBN 0-387-97755-4 Springer-Verlag New York Berlin Heidelberg
ISBN 3-540-97755-4 Springer-Verlag Berlin Heidelberg New York

*This book is dedicated to the memory of
Professor Max Tishler, deceased March 18, 1989*

Preface

In the days in 1973 when I had my own research group at the Kitasato Institute, it was said that most of the useful antibiotics had already been discovered and that the approaches, which are still in use today, had been exhausted. Some researchers who were engaged in the development of antibiotics were rather pessimistic about the discovery of any further new antibiotic. Some industrial research groups decided to withdraw from screening for new antibiotics. Actually, however, important antibiotics such as the avermectins, monobactams, and lactivicin were discovered later. This was the result of enthusiastic efforts aimed at development of new methods for the isolation of producing strains, modification of production conditions, and improvement of assay methods using unique test organisms such as hypersensitive mutants.

As of today more than 6000 antibiotics have been discovered. Although many antibiotics have unpleasant side effects, some show interesting modes of action on the basis of which they can be utilized clinically or as agrochemical agents.

For example, J.F. Borel and his co-workers observed an interesting side effect of cyclosporin A, an antifungal compound found by G. Thiel and his co-workers in 1970. Cyclosporin A showed specific toxicity to lymphocytes. Borel et al. then developed the antibiotic as an immunosuppressant. The new application of this antibiotic, in combination with other broad-spectrum antibiotics such as semisynthetic cephalosporins and penicillins, showed a good prophylactic effect and helped to make organ transplantation possible. In another study concerning side effects of erythromycin, Z. Itoh found that this antibiotic acts to mimic motilin, a gastrointestinal peptide hormone. Recently, we obtained an erythromycin derivative, EM-536, that exhibits gastrointestinal motor-stimulating activity 2860-fold more active than erythromycin and which has no antimicrobial activity. The motilides, named for this series of macrolide compounds with gastrointestinal motor-stimulating activity, are expected to be very useful not only in therapy for digestive disorders but also as reagents in studies of gastrointestinal motility. In another example, bialaphos, originally discovered as an antifungal agent, was later found to be a potent herbicide. These examples indicate that reevaluation and redevelopment of known antibiotics as pharmacological or agrobiological drugs may be highly promising.

It was the late Dr. H. Umezawa and his co-workers who initiated research and development of microbial enzyme inhibitors as therapeutic agents for the control of abnormalities of homeostasis. This approach led his and other research groups to discover a number of interesting pharmacologically active microbial metabolites such as bestatin (aminopeptidase inhibitor, immunostimulant), FK-506 (immunosuppressant), mutastein (glycosyltransferase inhibitor), and pravastatin (HMG-CoA reductase inhibitor). There is the greatest possibility of finding many more substances with interesting bioactivities (such as immunomodulative, antiulcer, and hypotensive) among microbial metabolites.

It is an enjoyable and rewarding task to find new physiologically active substances. Toward this goal, it is very important to set the direction and the methods of searching in proper paths to avoid any waste of time and money. The first step toward the successful discovery of new substances requires both natural science and technology. In this book the principles, processes, and examples of finding physiologically active substances, including antibiotics, are described with emphasis on the strategy and the methods of research.

I have been engaged for a long time in educating and encouraging the researchers of my group at Kitasato. From our laboratory, 31 researchers have obtained the Ph.D. degree to date. They have learned their ways of thinking and details of proceeding studies with us, and today these people participate actively in research at various places in Japan and overseas. In commemoration of the twentieth anniversary of my laboratory, those who belong to my group have helped me with the preparation of this book. I sincerely hope that it will be helpful not only to researchers but also to students who are interested in bioactive compounds at their microbial origins.

I wish to thank the late Dr. Shigenobu Okuda for his helpful advice and assistance in editing this book, for which he was also responsible for preparing some chapters.

Satoshi Ōmura, Ph.D.

Contents

Preface   vii
Contributors   xiii

Part 1   Antimicrobial Substances

1   **Antibacterial Agents**   1
    Ruiko Ōiwa

2   **Antifungal Agents**   30
    Yoshitake Tanaka

3   **Antiviral Agents**   45
    Hideo Takeshima

4   **Antiparasitic Agents**   63
    Kazuhiko Otoguro and Haruo Tanaka

Part 2   Antitumor Substances

5   **Antitumor Agents**   79
    Kanki Komiyama and Shinji Funayama

6   **Cell Differentiation Inducers**   104
    Haruki Yamada

Part 3 Enzyme Inhibitors

7 General Screening of Enzyme Inhibitors 117
Haruo Tanaka, Kazuhito Kawakita, Nobutaka Imamura, Kazuo Tsuzuki, and Kazuro Shiomi

8 New Strategy for Search of Enzyme Inhibitors 161
Hiroshi Tomoda and Satoshi Ōmura

Part 4 Pharmacologically Active Substances

9 Immunomodulators 171
Haruki Yamada

10 Vasoactive Substances 198
Akira Nakagawa

Part 5 Agrochemicals

11 Fungicides and Antibacterial Agents 213
Shigenobu Okuda and Yoshitake Tanaka

12 Herbicides 224
Shigenobu Okuda

13 Insecticides, Acaricides, and Anticoccidial Agents 237
Yoshitake Tanaka and Shigenobu Okuda

Part 6 Chemical Screening

14 Chemical Screening 263
Akira Nakagawa

Part 7 Sources, Fermentation, and Improvement of Producing Microorganisms

15 Selection of Microbial Sources of Bioactive Compounds 281
Yuzuru Iwai and Yōko Takahashi

16 Fermentation Processes in Screening for New Bioactive Substances 303
Yoshitake Tanaka

17 Genetic Engineering of Antibiotic-Producing Microorganisms 327
Haruo Ikeda

Index 337

Contributors

Shinji Funayama Research Center for Biological Function, The Kitasato Institute, Minato-ku, Tokyo 108, Japan. Present address: Faculty of Pharmacy, Tohoku University, Sendai 980, Japan.

Haruo Ikeda School of Pharmaceutical Sciences, Kitasato University, Minato-ku, Tokyo 108, Japan

Nobutaka Imamura Research Center for Biological Function, The Kitasato Institute, Minato-ku, Tokyo 108, Japan. Present address: Marine Biotechnology Institute Ltd., Sodeshi-cho, Shimizu-shi, Shizuoka 424, Japan

Yuzuru Iwai Research Center for Biological Function, The Kitasato Institute, Minato-ku, Tokyo 108, Japan

Kazuhito Kawakita School of Pharmaceutical Sciences, Kitasato University, Minato-ku, Tokyo 108, Japan. Present address: Faculty of Agriculture, Nagoya University, Nagoya 464-01, Japan

Kanki Komiyama Research Center for Biological Function, The Kitasato Institute, Minato-ku, Tokyo 108, Japan

Akira Nakagawa School of Pharmaceutical Sciences, Kitasato University, Minato-ku, Tokyo 108, Japan. Present address: School of Science and Engineering, Teikyo University, Utsunomiya 320, Japan

Ruiko Ōiwa Research Center for Biological Function, The Kitasato Institute, Minato-ku, Tokyo 108, Japan

Shigenobu Okuda (Deceased March, 1991) Formerly at Research Center for Biological Function, The Kitasato Institute, Minato-ku, Tokyo 108, Japan

Satoshi Ōmura Research Center for Biological Function, The Kitasato Institute, and School of Pharmaceutical Sciences, Kitasato University, Minato-ku, Tokyo 108, Japan

Kazuhiko Otoguro School of Pharmaceutical Sciences, Kitasato University, Minato-ku, Tokyo 108, Japan. Present address: Bioiatric Center, The Kitasato Institute, Minato-ku, Tokyo 108, Japan

Kazuro Shiomi Research Center for Biological Function, The Kitasato Institute, Minato-ku, Tokyo 108, Japan

Yōko Takahashi Research Center for Biological Function, The Kitasato Institute, Minato-ku, Tokyo 108, Japan

Hideo Takeshima Research Center for Biological Function, The Kitasato Institute, Minato-ku, Tokyo 108, Japan

Haruo Tanaka School of Pharmaceutical Sciences, Kitasato University, and Research Center for Biological Function, The Kitasato Institute, Minato-ku, Tokyo 108, Japan

Yoshitake Tanaka Research Center for Biological Function, The Kitasato Institute, Minato-ku, Tokyo 108, Japan

Hiroshi Tomoda Research Center for Biological Function, The Kitasato Institute, Minato-ku, Tokyo 108, Japan

Kazuo Tsuzuki Research Center for Biological Function, The Kitasato Institute, Minato-ku, Tokyo 108, Japan. Present address: Tsukuba Research Laboratories, Upjohn Pharmaceutical Ltd., Tsukuba 300-42, Japan

Haruki Yamada Research Center for Biological Function and Institute of Oriental Medicine, The Kitasato Institute, Minato-ku, Tokyo 108, Japan

Part 1
Antimicrobial Substances

1

Antibacterial Agents

Ruiko Ōiwa

1.1 Introduction

Penicillin, discovered by Fleming in 1928, was rediscovered by Chain et al. (1940) as a chemotherapeutic agent. In the same year, Waksman started the screening of antibacterial substances produced by a soil actinomycete, and discovered the actinomycins, some of which are used as antitumor agents. After many clinically useful antibiotics—streptomycin, chloramphenicol (chloromycetin), chlortetracycline (aureomycin), neomycin, oxytetracycline (terramycin), erythromycin, etc.—were discovered, most bacterial infections seemed to be conquered. However, about 10 years after the spread of antibiotic therapy, a number of species of *Staphylococcus*, *Mycobacterium*, and Gram-negative enteric bacteria had developed resistance to antibiotics. Then, the resistant strains of bacteria were used as test organisms to obtain new useful antibiotics. After the mechanisms of resistance to antibiotics in bacteria, such as enzymatic transformation, decrease in cell permeability, or decrease in ribosomal affinity had been elucidated, many derivatives of antibiotics were prepared chemically or biologically on the basis of information about the resistance mechanisms.

Strategies of screening methods for useful antibiotics were also modified. Results of the basic studies of mechanisms of action of antibiotics were applied to screening methods. The selective toxicity of a therapeutic agent should be superior to avoid the problem of side effects. From this point of view, some types of β-lactam antibiotic and inhibitors of bacterial cell wall biosynthesis were detected by various screening methods. A penicillinase inhibitor itself was also detected.

With antibiotic therapy, various septicemia caused by *Streptococcus* or *Staphylococcus* infection or violent enteric infections caused by Gram-negative bacteria decreased remarkably, and the average span of a human life has been gradually increased.

However, in hospitals, there is a problem of opportunistic infections of patients after operations, partly because of a decrease of immunological activity against various pathogens, even against weak ones. These infectious diseases have been mainly caused by various Gram-negative bacteria such as *Pseudomonas aeruginosa, Escherichia coli, Klebsiella pneumoniae, Proteus* sp., and *Serratia marcescens*. Nowadays, some major efforts have been made to obtain new types of antibacterial antibiotics, which are superior in selective toxicity, in pharmacokinetic properties, and in treating complex infections including opportunistic ones.

To discover new effective antibiotics, there are two subjects that cannot be considered separately. One is "production" of metabolites and the other is "detection" of activities of the metabolites. The former includes how to discover novel microorganisms and how to have the microorganisms produce various new metabolites. The latter problem includes how to sensitively and effectively detect activities of the metabolites. The latter subject, the "detection," is mainly described in this chapter, while the "production" is mentioned in Chapter 16.

Screening methods for antibacterial antibiotics have been modified by the changes of target in pathogens and on the basis of the elucidation of mechanisms of action of antibiotics. Traditionally, antibacterial activities were detected by the agar diffusion method using some Gram-positive and Gram-negative bacteria as test organisms. From the 1970s, these methods have been successfully replaced by detection of morphological changes of the test organisms, use of supersensitive mutants, inhibition of enzyme activities of bacteria, or application of monoclonal antibodies to an enzyme-linked immunosorbent assay.

In this section, some traditional methods and screening processes are described in the former part, and some target directed methods are shown in the latter part.

1.2 Conventional Assay Methods

The agar diffusion assay methods using some bacteria as test organisms is simple and easy to use in finding antibacterial activity in cultured broths or test materials. This method gives only limited information, for example, "one or more substances contained in the test sample inhibit the growth of the test organism." This method, however, is still useful in the improved screening procedures.

Use of antibiotic-resistant mutants After various pathogens that acquired resistance to antibiotics appeared in the clinical field, efforts to find new types of antibiotics active against such resistant bacteria were accelerated. Then resistant mutants of the assay organisms were also employed in the screening procedures, and mechanisms of resistance such as enzyme inhibition or blocking of transport systems also have been taken into consideration.

In recent screening programs to find new antibiotics active against the bacterial cell surface, an antibiotic selectively active against antibiotic resistant mutant *Staphylococcus aureus* 4R was sought (Higashide et al., 1985). The mutant strain which is resistant to tetracycline, erythromycin, chloramphenicol, and streptomycin was derived from *S. aureus* FDA 209P in vitro. In this screening process, about 2,000 actinomycete strains were tested, and a new dipeptide antibiotic named alahopcin was found. The antibiotic was assumed to be an inhibitor of bacterial cell wall synthesis, since it inhibited the incorporation of [^{3}H]diaminopimelic acid into *Escherichia coli*. Alahopcin also showed synergistic activity with erythromycin against drug-resistant bacteria, especially constitutive types of macrolide resistant strains.

In another screening program to detect oligopeptide antibiotics (Yoshida et al., 1986), a resistant mutant *Bacillus subtilis* SRY-7, was used as an assay organism. The mutant strain, which was selected by its resistance to a tripeptide antibiotic SF-1293, seemed to be an oligopeptide transport deficient mutant.

SF-1293 (L-phosphinothricyl-alanyl-alanine) is known to be taken up by *E. coli* K-12 *via* an oligopeptide transport system. On the other hand, a triornithine resistant mutant of *E. coli* K-12 was known to be an oligopeptide transport deficient strain. Therefore, a mutant of *B. subtilis* resistant to SF-1293 was isolated from *B. subtilis* M8193 (met$^-$, trp$^-$) as an oligopeptide transport deficient mutant. Growth response of the mutant and the parent strain to di- and tripeptides was tested. When the di- and tripeptides contained methionine or methionine plus alanine the mutant did not grow, although the parent utilized both types.

In this screening system, antibiotics that are active against the parent strain M8193 and inactive against the mutant strain SRY-Y were selected. A new antibiotic SF-2339, 2-(valylvalylaminoxy)malic acid, was discovered.

Use of mutant supersensitive to antibiotics To detect directly small amounts of antibiotic produced in a culture broth by the traditional agar diffusion method, a supersensitive mutant of a conventional assay organism was used. Sometimes a multiple drug sensitive mutant was used, and sometimes a supersensitive mutant to a specific antibiotic was used. It was also useful to employ a larger paper disc and a thin-layered agar medium for an assay plate. In the screening programs to find new β-lactam antibiotics, supersensitive mutants of various test organisms were found to be useful, as mentioned in the latter part in this section.

In a recent example of a screening program using a multiple drug sensitive strain of *Micrococcus flavus*, a new pyrrole-amidine antibiotic TAN-868 A, which is active against bacteria, fungi, and protozoa, was found to be produced by *Streptomyces idiomorphus* (Takizawa et al., 1987).

Assay under anaerobic condition It was found that the antibacterial activity of fosfomycin is more potent under anaerobic condition than under aerobic condition. Based on this fact, screening for antibiotics that are more active under anaerobic than under aerobic conditions was carried out using *Escherichia coli* NIHJ as assay organism.

To examine the screening system, whether the phenomenon is specific or not, inhibitory activities of various known antibiotics against *E. coli* NIHJ were assayed under aerobic and anaerobic conditions. As a result, inhibitory activities of most types of antibiotics—except for penicillin G and cycloserine—were not found to be enhanced under anaerobic conditions.

Among several antibiotics selected in this screening system, a new antibiotic SF-2312, 1,5-dihydroxy-2-oxopyrrolidin-3-yl-phosphonic acid, was found (Watanabe et al., 1986).

Use of anaerobic bacteria Opportunistic infections caused by some anaerobes are sometimes a clinical problem with associated high mortality rates. For instance, pseudomembranous colitis, which tends to occur in association with antibiotic therapy, is one of such serious infectious diseases.

To search for new antianaerobic antibiotics, *Bacteroides, Clostridium, Fusobacterium, Peptococcus*, and other anaerobic bacteria are used as assay organisms. One of them, *Clostridium difficile*, the major causative pathogen of pseudomembranous colitis, is extremely sensitive to oxygen. Even after being exposed to air once, it cannot grow any further. Strict anaerobic conditions are necessary for its growth. Such anaerobic bacteria are too difficult to use as assay organisms for a routine process in a conventional assay system.

Masuma et al. (1987) reported a simple and convenient agar medium for the growth of *C. difficile* as assay organism. It can be handled under air for several hours before the assay of test materials, and incubation in an anaerobic chamber.

To protect *C. difficile* cells from exposure to oxygen, a double-layered agar plate was prepared. *C. difficile* was seeded in the bottom agar layer and an upper agar layer was used as an oxygen barrier and nutrient reservoir. This is the reverse phase of the double layer of the conventional agar plate for paper disc or cup assay method.

Using this assay technique, thiotetromycin (Ōmura et al., 1983), luminamicin (Ōmura et al., 1985), lustromycin (Tomoda et al., 1986), and clostomicin A, B_1, B_2, C, and D (Ōmura et al., 1986) were found.

In other recent screening systems from 1987 to 1988, although the methods were not described, discovery of abbeymycin, coloradocin (identified with luminamicin), and tirandalydigin were reported.

Detection of synergistic activity In the clinical field, the use of two or more antimicrobial agents in combination for the treatment of serious bacterial infections is practical. Such an approach is founded on the premise that the agents may act synergistically in vivo. To improve an antimicrobial spectrum or activity of a clinically useful antibiotic that acts as a protein or cell wall synthesis inhibitor etc., a bacterial membrane affecting antibiotic may be useful.

Ichimura et al. (1987) tried to screen for new antibiotics that showed antibacterial activity against *E. coli* when tested together with spiramycin, which itself is not effective against Gram-negative bacteria. Antibiotic activity in the fermentation broth was determined by the paper disc agar diffusion method using *E. coli* ATCC 26 as assay organism in the presence of 40 μg/ml of spiramycin.

Using this screening system, a new antibiotic, CV-1, was discovered. Although the antibacterial activity of CV-1 is very weak, it showed a cooperative effect with spiramycin against *E. coli*. The mode of action of CV-1, which is 1,2-diamino-1,2-*N*,*N*'-carbonyl-1,2-dideoxy-α-D-glucose hydrate, seemed to be the inhibition of lipopolysaccharide synthesis. The lipopolysaccharide of the outer membrane seems to have an important role as a barrier to transport. The cooperative effect of CV-1 seems due to the increased entry of spiramycin into cells of *E. coli*.

Radioimmune and enzyme-linked immunosorbent assays   Immunoassays of various types have been widely used in clinical laboratories for both the detection and the quantitative analysis of antibodies, hormones, and drugs in body fluids. In microbiological or agricultural laboratories, enzyme-linked immunosorbent assays (ELISA) have been applied to the rapid detection and quantification of mycotoxins (Pestka et al., 1981). Yao and Mahoney (1984) developed an ELISA for detecting aminoglycoside antibiotics in screening for novel fermentation products. In other studies, radioimmune assays and the ELISA technique were successfully applied to erythromycin derivatives (Tanaka et al., 1988) and antichlamydial agents (Cervenini et al., 1987).

Identification of known antibiotics   Since the discovery of new antibiotics has become more and more difficult, it is necessary to increase assay sensitivity with shorter assay times. Moreover, rapid and accurate identification of new antibiotics is an absolute necessity to prevent wasteful duplication of effort. In most laboratories, accumulated personal data about antimicrobial spectra and paper or thin-layer chromatographic properties of known antibiotics have been utilized for initial antibiotic identification and classification in crude culture broths.

Although there are few reviews mentioning early identification of antibiotics, Mitscher and Omoto (1977) published a brief report with 158 references about physical methods, X-ray methods, mass-spectrometry, nuclear magnetic resonance, and circular dichroism and their application to identification of antibiotics.

To know whether a newly detected antibiotic is new or identified, it is necessary to compare the physical and chemical characteristics of the unknown compound with those of known compounds. When the number of antibiotics to be compared is not large, edge-punched cards were used. Today, however, more than 6,000 antibiotics and a similar number of inactive metabolites have been reported to be produced by microorganisms. Two computer systems have been designed to assist in the identification of known antibiotics. One is by NIC, Frederick Cancer Research Center in Hungary (Bostian et al., 1977), and the other is by the Microbial Chemistry Laboratories of the Kitasato Institute in Japan (unpublished).

The former, BERDY data base, was reported to contain physical, chemical, and biological data except for structural and IR spectra. The latter, KMC (Kitasato Microbial Chemistry) data base, contains similar items of physical, chemical, and biological characteristics of microbial products including low molecular weight metabolites. In the near future, structures of the compounds will also be covered by this data base system.

1.3 Target-Directed Screening

The screening target is usually focused on a mechanism of action of an antibiotic that shows highly selective toxicity. Usually, on the basis of various findings of mechanisms of action of useful known antibiotics, some congeners or alternative structures of these are sought. Sometimes rationally selected enzymes or receptor sites, for which there are no known inhibitors, are noted and used as the screening target.

Inhibitors of bacterial cell wall synthesis Many efforts were made in searching for inhibitors of bacterial cell wall synthesis, because their specific inhibitory activity on the biosynthesis of bacterial cell wall, which animal cells lack, provides a guarantee of their low toxicity to human cells. Therefore, the bacterial cell wall received great emphasis as the target of antibiotic action, and many inhibitors have been sought. However, only a few inhibitors of cell wall synthesis were found for clinical use, because some of them lack activity *in vivo*, and some of them show cross-resistance with the clinically used antibiotics. However, various inhibitors acting on the different sites of the pathway of cell wall peptidoglycan synthesis were used as important tools for elucidating details of bacterial cell wall biosynthesis. Inhibitors of cell wall biosynthesis which were found in recent target-directed screening are shown in Table 1.1.

Observation of morphological changes Spheroplast or bulge formation by bacterial cells is one proof that an inhibitor of cell wall biosynthesis is present in the test material. For the purpose of screening for inhibitors of bacterial cell wall synthesis, spheroplast or bulge formation was observed microscopically. Sometimes the screening procedure is effectively carried out in combination with other detection methods, such as the use of supersensitive mutants and the observation of synergistic activity with other inhibitors of bacterial cell wall synthesis.

In recent work, L-cycloserine, which does not show antimicrobial activity, was detected by spheroplast formation of *E. coli* LS-1, a supersensitive mutant

Table 1.1 Screening methods for inhibitors of bacterial cell wall formation

Name of antibiotic	Group	Producing organism	Screening method	Reference
Penicillin N (identified) Metabolite 2 Metabolite 3 Metabolite 4 (= cephamycin C)	β-Lactam β-Lactam β-Lactam	*Streptomyces* sp. *Streptomyces lipmanii* *S. clavuligerus*		Nagarajan, 1971
Cephamycins A and B	Cephem	*S. griseus* and other *Streptomyces* strains	Using *Proteus vulgaris* and *Vibrio percolans*	Stapley, 1972
Cephamycin C	Cephem	*S. lactamdurans*		
Thienamycin Epithienamycin	Carbapenem Carbapenem	*S. cattleya* *S. flavogriseus*	Using large size paper disc and thin layer agar medium for assay plate	Kahan, 1979 Stapley, 1981
Penicillins and cephalosporines (identified)		80 fungal and 30 actinomycete strains out of 30,000 strains	Using *Pseudomonas aeruginosa* PsC^{ss} (hypersensitive to β-lactams)	Kitano, 1975b
C-19393 S$_2$ and H$_2$ (= Carpetimycins B and A)	Carbapenem	*S. griseus*		Imada, 1980
Sulfazecin	Monocyclic β-lactam (monobactam)	*Pseudomonas acidophilla*	Using *P. aeruginosa* PsC^{ss} and *Escherichia coli* PG8 (supersensitive to β-lactams, observation of spheroplast formation, and inactivation by β-lactamase	Imada, 1981
Isosulfazecin		*P. mesoacidophilla*		Imada, 1982
Bulgecins A, B, and C	Sugar	*P. acidophilla* and *P. mesoacidophilla*		
Cephabacins F$_{1-9}$ and H$_{1-6}$	Cephem	*Lysobacter lactamgenus* and *Xanthomonas lactamgena*	Using *P. aeruginosa* C141 and *E. coli* (sensitive to β-lactams) and observation of spheroplast formation	Ono, 1984

Continued next page

Table 1.1 (cont.)

Name of antibiotic	Group	Producing organism	Screening method	Reference
Formadicins A, B, C, and D	Monocyclic β-lactam	*Flexibacter alginoliquefaciens*		Katayama, 1985
Nocardicins A and B	Monocyclic β-lactam	*Nocardia uniformis*	Using *E. coli* Es-11 (supersensitive to β-lactams)	Aoki, 1976
FR-900098	Phosphonic acid	*S. rubellomurinus*	Using *P. aeruginosa* Ps-III (supersensitive to β-lactams) and observation of spheroplast formation	Okuhara, 1980a
Fosmidomycin (FR-31564) and FR-32863	Phosphonic acid	*S. lavendulae*		Okuhara, 1980b
FR-33289	Phosponic acid	*S. rubellomurinus*		
FR-900137	Amino acid containing	*S. unzenensis*	Using *P. aeruginosa* Ps-IV (supersensitive to β-lactams) and observation of spheroplast formation	Kuroda, 1980c,
FR-900148	Amino acid containing	*S. xanthocidicus*		Kuroda, 1980b
FR-900130	Amino acid	*S. catenulae*	Observation of synergism with cycloserine and spheroplast formation	Kuroda, 1980a
FR-900318	Penam	*Aspergillus candidus*	Using *E. coli* 8S-1 (supersensitive to β-lactams) and detection of penicillinase inhibitory activity	Yamashita, 1983
Amphomycin (identified) 3-Amino-3-deoxy-D-glucose (identified)	Peptide Amino sugar	*Streptomyces* sp. *Bacillus cereus*	Difference assay using *Bacillus* and *Acholeplasma*, and inhibition of incorporation of bacterial cell wall precursor	Ōmura, 1975 Iwai, 1977
Azureomycin Izupeptins A and B	Sugar, peptide Glycopeptide	*Pseudonocardia azurea* *Nocardia* sp.		Ōmura, 1979 Spiri-Nakagawa, 1986

Table 1.1 (cont.)

Name of antibiotic	Group	Producing organism	Screening method	Reference
Clavulanic acid (MM 14151)	Bicyclic β-lactam	S. clavuligerus	Detection of β-lactamase inhibitory activity	Brown, 1976
Olivanic acids MM 4550, MM 13902 and MM 17880	Carbapenem	S. olivaceus		Box, 1979
Olivanic acids MM 22380, MM 22381, MM 22382 and MM 22383	Carbapenem	S. olivaceus		Box, 1982
Olivanic acid MM 27696	Carbapenem	S. olivaceus		Box, 1988
MM 42842	Monobactam	Pseudomonas cocovenenans	Using Acinetobacter lwoffii LSS7 or P. aeruginosa ES48 (hypersensitive to β-lactams)	Inukai, 1978
Globomycin	Peptide	Streptoverticillium cinnamoneum and three Streptomyces strains	Specific activity against E. coli and observation of spheroplast formation	Okamura, 1978
PS-5	Carbapenem	S. cremeus	Combination assay using Comamonas terrigena B-996 (sensitive to β-lactam) and C. terrigena B-996R (resistant to β-lactams) with and without β-lactamase	Shibamoto, 1980
PS-6 abd PS-7	Carbapenem	S. cremeus		
PS-8	Carbapenem	S. cremeus		Shibamoto, 1982
OA-6129 A, B₁, B₂ and C	Carbapenem	Streptomyces sp.		Okabe, 1982
SQ 26, 445 (=sulfazecin)	Monobactam	Gluconobacter sp.	Using Bacillus licheniformis SC 9262 (supersensitive to β-lactams)	Sykes, 1981
SQ 27, 860	Carbapenem	Erwinia carotovora E. herbicola and Serratia sp.		Parker, 1982a
SQ 26, 180	Monobactam	Chromobacterium violaceum		Wells, 1982a

Continued next page

Table 1.1 (cont.)

Name of antibiotic	Group	Producing organism	Screening method	Reference
SQ 26, 517	Lactone	Bacillus sp.		Parker, 1982b
Deacetoxycephalosporin C (identified)	Cephem	Flavobacterium sp. and Xanthomonas sp.		Singh, 1982
SQ 26, 823, SQ 26, 875, SQ 26, 700, SQ 26, 970, and SQ 26, 812	Monobactam	Agrobacterium radiobacter		Wells, 1982b
SQ 28, 332	Monobactam	Flexibacter sp.		Singh, 1983
SQ 28, 502 and SQ 28, 503	Monobactam	Flexibacter sp.		Cooper, 1983
SQ 28, 516 and SQ 28, 517	Cephem	Flavobacterium sp.		Singh, 1984
SF-2103 A	Carbapenem	Streptomyces sulfonofaciens	Detection of β-lactamase inhibitory activity	Ito, 1982
SF-1623, SF-1623 B and SF-1623 C (= deacetylcephalosporin C)	Cephem	S. chartreus	By HPLC technique after treatment with β-lactamases	Inouye, 1983
SF-2050 B	Carbapenem	Streptomyces sp.	(Not described, but showing β-lactamase inhibitory activity)	Ohba, 1986
Asparenomycins A, B, and C	Carbapenem	S. tokunonensis and S. argenteorus	Using E. coli JC-2 (supersensitive to β-lactams) and inactivation with β-lactamase	Shoji, 1982
L-Cycloserine (identified)		Erwinia uredovora	Using E. coli LS-1 (supersensitive to β-lactams) and observation of spheroplast formation	Shoji, 1984a
Chitinovorins A, B, and C	Cephem	Flavobacterium chitinovorum		Shoji, 1984b
Forsfomycin (identified)		Pseudomonas syringae		Shoji, 1986
PB-5266 A (=SQ 28, 332) PB-5266 B and C	Monobactam	Cytophaga johnsonea		Kato, 1987

Table 1.1 (cont.)

Name of antibiotic	Group	Producing organism	Screening method	Reference
Pluracidomycins A, B, and C	Carbapenem	S. pluracidomyceticus	Detection of β-lactamase inhibitory activity	Tsuji, 1982
Pluracidomycins A_1, A_2, C_2, C_3, and D	Carbapenem	S. pluracidomyceticus		Tsuji, 1985
Antibiotic TA		Myxococcus xanchis	Using E. coli ESS (sensitive to β-lactams)	Rosenberg, 1982
Oganomycins F, G, H, and I	Cephem	S. oganoensis	Produced by precursor fermentation	Osono, 1980
Carpetimycins A and B (= C-19393 H_2 and S_2)	Carbapenem	Streptomyces sp.	(Not described, but showing β-lactamase inhibitory activity)	Nakayama, 1980
Carpetimycins C and D	Carbapenem			Nakayama, 1983
Chlrocardicin	Monocyclic β-lactam	Streptomyces sp.		Nisbet, 1985
Aridicins A, B, and C	Glycopeptide	Kibdelsporangium aridum	Using Staphylococcus aureus VAN^R (resistant to vancomycin) and antagonism test for glycopeptide receptor	Sitrin, 1985
Kibdelins (AAD-609) A, B, C_1, C_2, and D	Glycopeptide	K. aridum		Shearer, 1986
Actinoidin A_2	Glycopeptide	Nocardia sp.		Dingerdissen, 1987
Parvodicin	Glycopeptide	Actinomadula parvosata		Christensen, 1987
U-68, 204	Thiolactone	S. thiolactonus	Using P. aeruginosa UC 6513 (supersensitive to β-lactams)	Dolak, 1986

to β-lactam antibiotics (Shoji et al., 1984a). By the same screening system, Shoji et al. (1986) isolated a cell wall inhibitor, which was identified as fosfomycin, from a bacterial producer *Pseudomonas syringae*, although all of the known fosfomycin-producing organisms had been streptomycete strains.

Production of β-lactam antibiotics from actinomycete strains For a long time, β-lactam antibiotics were thought to be classified into two types, penicillins and cephalosporins, which were produced exclusively by filamentous fungi. However in 1971, the production of β-lactam antibiotics from *Streptomyces* was reported (Nagarajan et al., 1971). In the next year, it was reported that cephamycins A and B (Fig. 1.1) were produced by *S. griseus* and eight other different species of *Streptomyces*. Cephamycin C was produced from *S. lactamdurance* and *S. clavuligerus* (Stapley et al., 1972). The finding of the production of cephamycins by streptomycete strains stimulated the screening efforts for β-lactam antibiotics.

Use of thinner-layer agar medium for an assay plate In further screening procedures, the same group of researchers mentioned above reported the discovery of thienamycin (Fig. 1.2) produced by *S. cattieya* (Kahan et al., 1979) and epithienamycins A, B, C, D, E, and F, which are structurally related to N-acetylthienamycin (Stapley et al., 1981). They isolated a great many strains of *S. flavogriseus* and found that 43 of them produced members of the epithienamycin family. In this screening program, they employed a larger paper

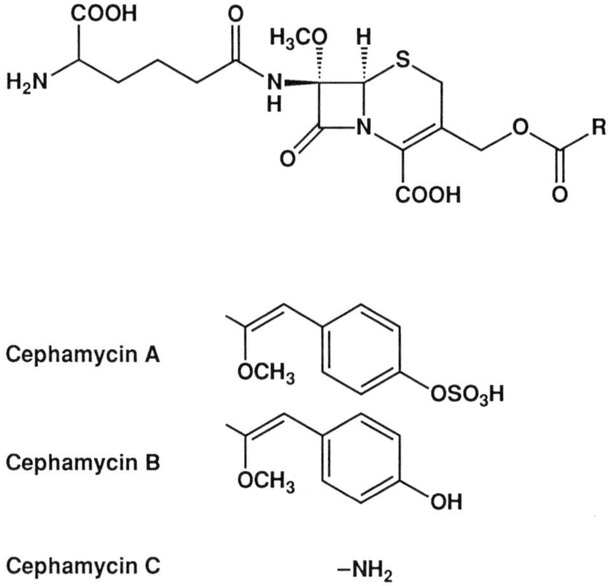

Figure 1.1 Structures of cephamycins A, B, and C.

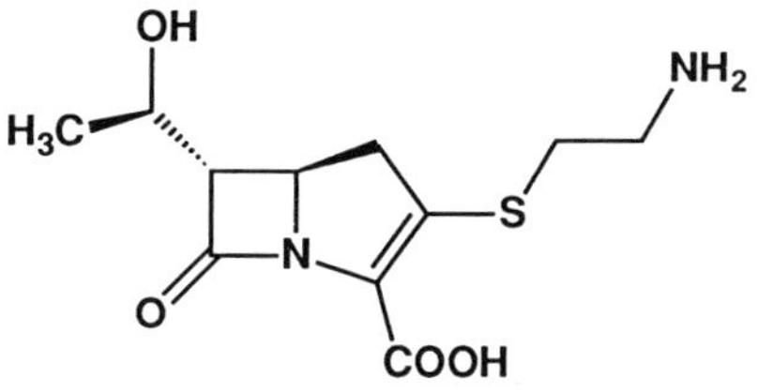

Thienamycin

Figure 1.2 Structure of thienamycin.

disc (14 mm in diameter) and a thin agar layer (2 mm in depth) plate seeded with *Staphylococcus aureus* ATCC 6538P.

Use of sensitive mutants to detect β-lactam antibiotics When an antibiotic shows the same selective inhibitory activity against supersensitive bacteria as close a β-lactam antibiotic, together with its ability to induce the formation of spheroplasts, it is presumed to be a β-lactam antibiotic. Therefore, some supersensitive mutants were prepared to help discover novel β-lactam antibiotics.

In the screening program using *E. coli* Es-11, which is specifically sensitive to β-lactam antibiotics such as penicillin G (MIC: 0.8 μg/ml) and cephalosporin C (MIC: 0.4 μg/ml), a novel β-lactam antibiotic named nocardicin A was discovered. Nocardicin A is a novel monocyclic β-lactam antibiotic (Fig. 1.3) produced by *Nocardia uniformis* which induced protoplast or spheroplast formation by *Pseudomonas aeruginosa* (Aoki et al., 1976).

Use of hypersensitive mutant, Pseudomonas aeruginosa PsC^{ss} A supersensitive mutant to β-lactam antibiotics was derived from *P. aeruginosa* IFO 3080 (Ps), through three-step mutagenesis with N-methyl-N'-nitro-N-nitrosoguanidine (Kitano et al., 1977). The parent strain of *P. aeruginosa* (Ps), was insensitive to cepharosporin C (MIC: >1,000 μg/ml). The first-step mutant Ps53 was selected to be sensitive to cepharosporine C (MIC: 28.5 μg/ml), the second-step mutant PsC^s was more sensitive (MIC: 1.8 μg/ml), and the final selected strain PsC^{ss} was obtained as a hypersensitive mutant (MIC: 0.05 μg/ml).

The sensitivity of the mutant strain PsC^{ss} to various β-lactam antibiotics, such as penicillins, cephalosporins, nocardicin A, and semisynthetic β-lactam antibiotics, increased greatly. On the other hand, the sensitivity to antibiotics other than β-lactams showed little or no increase.

Detection of active antibiotics was carried out by using four agar plates. In the first plate, the parent strain *P. aeruginosa* Ps was grown. The other plates were seeded with the mutant strain PsC^{ss}. The third and the fourth plates contained penicillinase and cephalosporinase, respectively.

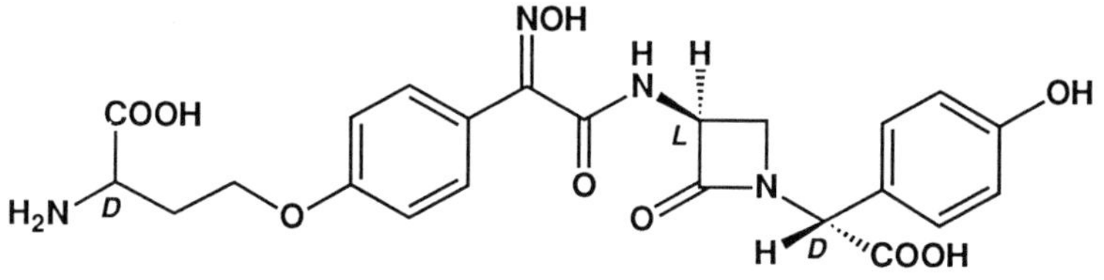

Nocardicin A

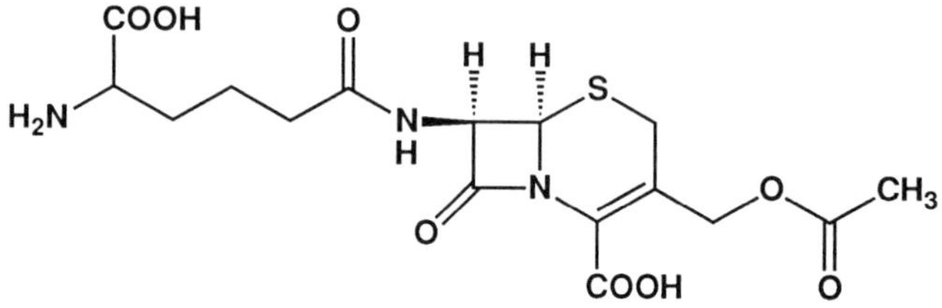

Cephalosporin C

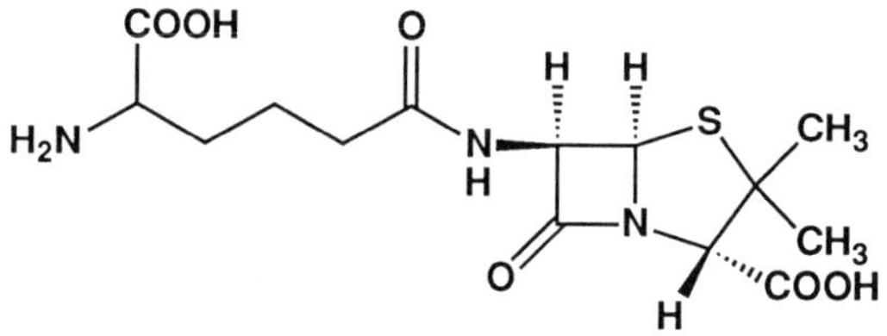

Penicillin N

Figure 1.3 Structures of nocardicin A and other β-lactam antibiotics.

As shown in Table 1.2, active principles are classified into five groups according to the pattern of the inhibitory zone. Cephalosporins and cephamycins are found to be included in group I, penicillins in group II, 6-aminopenicillanic acid (6-APA) in group III, clavulanic acid and nocardicins in group IV, and non-β-lactam antibiotics in group V.

In this screening system, up to 1975, about 30,000 strains of fungi, yeasts, bacteria, and actinomycetes were examined and 116 strains (80 fungal and 36 actinomycete) were found to produce both cephalosporins and penicillin N or only penicillin N, and 25 fungal strains produced penicillin G type antibiotics. Actinomycete strains producing cephalosporins were found to belong to *Streptomyces*. The fungal strains *Emericellopsis glabra* and *E. microspora* had been reported to produce cephalosporins, but in this screening system it was newly found that *E. terricola, E. minima, E. synematicola, E. salmosynne-*

Table 1.2 Four plates system for screening β-lactam antibiotics

Test organisms	Ps	PsC^{ss}	PsC^{ss}	PsC^{ss}		
β-Lactamase			PCase	CPase		
Inhibitory zone	−	+	+	−	I	Cephalopsorins Cephamycins
	−	+	−	−	II	Penicillins
	−	+	−	+	III	6-APA
	−	+	+	+	IV	Clavulanic acid Nocardicins
	+	+	+	+	V	Non β-lactam antibiotics

Ps, *Pseudomonas aeruginosa* IFO 3080; PsC^{ss}, a cephalosporin supersensitive mutant obtained from Ps.

mata, E. micrabilis, Arachnomyces minimus, Anixiopsis peruviana, and *Spiroidium fascum* also produced cephalosporin antibiotics. Fungal strains belong to *Penicillium, Aspergillus, Trichophyton, Epidermophyton*, and *Malbranchea* had already been reported to produce G-type penicillins, but they were newly detected in cultured broths of *Thermoascus crustaceus, Gymnoasus ucinatus, Polypaecilum insolitum, Eupenicillium altaceum, Talaromyces avellaneus*, and *Malbranchea* sp. (Kitano et al., 1975b).

Detection of 6-APA was also possible using this screening system, and it was found that all strains tested—*Thermoascus crustaceus, Gymnoascus uncinatus, Polypaecilum* sp., *Polypaecilum insolitum, Penicillium chrysogenu* and *Malbranchia* sp.—produced 6-APA-like substances (Kitano et al., 1975a).

Use of supersensitive mutants with observation of morphological changes
Imada et al. (1982) looked at three features to indicate the presence of β-lactam antibiotics: induction of bulge formation, activity against supersensitive mutants to β-lactam antibiotics, and inactivation by β-lactamase. In this screening system using *P. aeruginosa* PsC^{ss} and *E. coli* PG8 as test organisms, they found the new antibiotics sulfazecin and isosulfazecin (Imada et al., 1981) and bulgecins A, B, and C (Imada et al., 1982) which were produced from two bacterial strains, *Pseudomonas acidophilia* and *P. mesoacidophilia*.

Cultured broth containing sulfazecin and the bulgecins showed bulge formation on assay organisms such as *E. coli*. However, after purification, the β-lactam antibiotic sulfazecin was found to have no bulge forming activity, and it caused only filamentation on bacteria. The bulgecins having bulge forming activity, however, were found to show no antibacterial activities against any bacteria tested, including supersensitive mutants to β-lactam antibiotics. Instead, they showed strong synergistic activity with β-lactam antibiotics.

Use of hypersensitive mutant Bacillus licheniformis Sykes et al. (1981), using a hypersensitive mutant strain of *Bacillus licheniformis* that is sensitive to β-lactam antibiotics <0.1 μg/ml, reported that over 10,000 bacterial isolates were screened for their ability to produce β-lactam antibiotics. Among antibiotics found, the monocyclic β-lactam antibiotics SQ 26,445 produced from *Gluconobacter* sp. proved to be identical with sulfazecin. For this new family of monocyclic β-lactams, characterized by the 2-oxoazetidine-1-sulfonic acid nucleus as shown in Fig. 1.4, Sykes suggested the class name "monobactam" (Sykes et al., 1981).

Selection of monobactams After the discovery of sulfazecin, screening for β-lactam antibiotics including monobactams was carried out actively. To select a new monobactam, Cooper (1983) reported a rapid and simple method to identify monobactams and to distinguish them from other families of β-lactam antibiotics. The method was accomplished by spectroscopic studies and analyses of infrared and fast atom bombardment (FAB) mass spectral data. The detection of a sulfamic acid group ($RNHSO_3^-$) in the monobactam was done by treatment with $BaCl_2$/2 N HCl followed by the addition of $NaNO_2$.

Combined use of β-lactamase and sensitive and resistant mutant Two strains of *Commamonas terrigena*, strain B-996 which is highly sensitive to cephalotin (MIC: >0.1 μg/ml) and strain B-996R which is resistant to cephalotin (MIC: 400 μg/ml) were used as assay organisms. Antibacterial activities of broth supernatants and the concentrated solutions were compared using five kinds of assay plates with or without addition of penicillinase (1 unit/ml) or cephalosporinase (1 unit/ml) seeded with *C. terrigana* B-996 or B-996R (Okamura et al., 1979).

By this assay system, penicillins, cephalosporins, or other non-β-lactam antibiotics were found to be distinguishable from each other. Penicillins were inactivated by both of the enzymes. Cephalosporins were not inactivated by the penicillinase. The 7-methoxy type of cephalosporins were differentiated from other β-lactam antibiotics, because *C. terrigena* B-996R was specifically susceptible to the 7-methoxy type of cephalosporins.

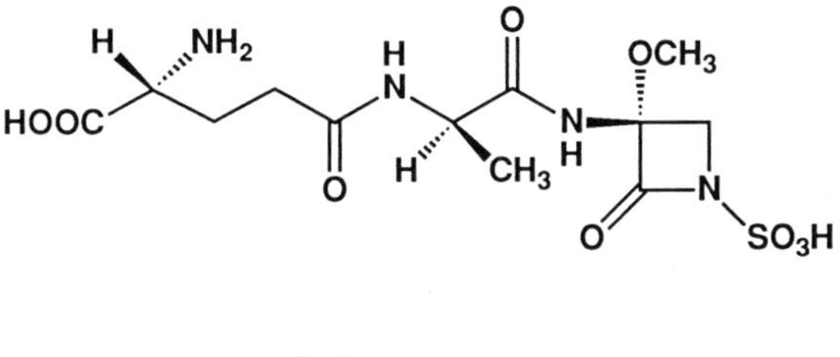

Sulfazecin

Figure 1.4 Structure of sulfazecin.

As a result, nine *Streptomyces* strains out of about 2,000 strains tested, were found to produce new β-lactam antibiotics—PS-5, PS-6, PS-7, and PS-8. In the continued screening for new β-lactam antibiotics using this system, four new carbapenem antibiotics—OA-6129 A, B_1, B_2, and C—were found (Okabe et al., 1982).

Screening for β-lactamase inhibitor It was known that there are some pathogens which are resistant to penicillins and cephalosporins owing to their production of β-lactamase. If the activity of the β-lactamase could be inhibited adequately in vivo, penicillins and cephalosporins could be useful in treating infectious disease caused by such pathogens. Thus, the screening of inhibitors of β-lactamase was started (Hata et al., 1972).

Penicillinase solution, prepared from a penicillinase producing strain of *Staphylococcus aureus*, was mixed to react with test materials such as culture broths of actinomycete strains, followed by the addition of penicillin G solution to bind with the residual enzyme. Therefore, the enzyme inhibitory activity was estimated by the determination of residual penicillin content in the reaction mixture assayed by the conventional agar diffusion method using *S. aureus* FDA 209P. In this screening system, a novel penicillinase inhibitor KA-107, a high molecular weight protein, was found to be produced from *Streptomyces gedanensis*.

Detection of a β-lactamase inhibitor by biological activity To search for a novel compound containing the β-lactam ring, determination of decreased β-lactamase activity can be used as an index of the presence of a β-lactam ring. Therefore, the screening of β-lactamase inhibitors that show antibacterial activity was carried out by Butterworth et al. (1979).

Klebsiella aerogenes ATCC 15380, which owes its penicillin resistance to the production of β-lactamase, was used as a test organism. Cultured broths of soil actinomycetes were placed in holes (8 mm in diameter) cut in two agar plates; a test plate and a control plate. The test plate contained sodium benzylpenicillin at a concentration of 10 µg/ml. The control plate was prepared identically except for the addition of the penicillin. A difference in zone diameter between the test and control plates was indicative of the presence of a β-lactamase inhibitor. The plates were incubated overnight at 28°C. When a sample contained a diffusible β-lactamase inhibitor, an inhibitory zone of the test organism appeared, resulting from the protection of the penicillin contained in the agar. In the absence of β-lactamase inhibitor, bacterial growth was observed, resulting from the inactivation of the penicillin by the β-lactamase produced by the test organism. When meticilin, which has inhibitory activity against the β-lactamase of *K. aerogenes*, was used at a concentration of 200 µg/ml as a positive control, a clear zone of inhibition was obtained on the test plate, whereas no zone occurred on the control plate.

In this system, clavulanic acid produced from *Streptomyces clavuligerus* and the olivanic acid complex including MM4550, MM13902, and MM17880 produced from S. olivaceus were discovered by the screening for β-lactamase inhibitors (Brown et al., 1976). In the same screening system, four olivanic acids, MM22380, MM22381, MM22382, and MM22383 produced from S. olivaceus were found (Box et al., 1979).

Detection of β-lactamase inhibitors by color change Uri et al. (1978) reported a rapid and simple method for detection of β-lactamase inhibitors. A unique property of nitrocefin, a chromogenic cephalosporin, was utilized to search for β-lactamase inhibitors of natural origin. Upon β-lactamase hydrolysis of the amide bond of the β-lactam ring, the color of nitrocefin changes from light yellow to red. The color change is a result of an electron shift along the molecule that contains a highly conjugated 2,4-dinitrostyryl moiety at the 3'-position. In the test system, it was the most sensitive to use an 18-hour culture of K. pneumoniae 1200 grown in a colorless semisynthetic peptone-glucose-buffered medium allowing for optimal visualization of the color change. The bacterial culture and a test sample were mixed for 5 minutes and a nitrocefin solution was added. The test solution containing an inhibitor remained colorless, whereas in the control tube, a purple-red color developed within a few minutes. This test appeared to be as sensitive as the biological test for detection of clavulanic acid which was used as a standard control material in this test procedure.

Induction of β-lactamase Sykes and Wells (1985) described a simple screening method for β-lactam antibiotics, measuring the induction of β-lactamase from Bacillus licheniformis with the use of a chromogenic cephalosporin. This bacteria produces minimal levels of β-lactamase in the absence of β-lactam antibioties, and it produces a large amount of the enzyme in the presence of β-lactams. The β-lactamase activity induced was detected by the appearance of a red color proportional to hydrolysis of the amide bond of the β-lactam ring. This method proved to be highly sensitive and specific for β-lactams and β-lactone-containing compounds.

Aliquots of fermentation broths or test materials were placed onto agar plates seeded with B. licheniformis. After incubation for 2 or 3 hours, the test plates were overlaid with a solution of a chromogenic cephalosporin. When the production of β-lactamase was induced, a red zone around the discs appeared as a result of hydrolysis of the chromogenic cephalosporin. The sensitivity of the induction assay was compared to that of the screening methods dependent on observation of morphological changes of test organisms or on the detection of activity against β-lactam supersensitive mutants. As a result, the amount of β-lactam compound detected by the induction assay was 1/15 of that by the agar diffusion method using a β-lactam supersensitive mutant, E. coli SC 12155, and was 1/200 of that by the observation of morphological changes of test organisms.

Detection by HPLC using β-lactamase Two novel cephamycins, SF-1623 and SF-1623 B, were discovered by a detection method using the HPLC technique after treatment with β-lactamases (Inouye et al., 1983). Cultured broths of soil isolates were analyzed by HPLC with a Zipax SAX 0.2 × 100 cm column developed with 0.1M NaH_2PO_4 solution containing 0.002 M $NaClO_4$ (pH 4.5). The peaks assigned to cephamycin antibiotics were stable against the penicillinase of *Bacillus cereus*, but disappeared on treatment with the cephalosporinase of *Citrobacter freundii*. The new cephamycins, SF-1623 and SF-1623B, were differentiated from those of other known cephamycin group antibiotics by their retention times.

Inhibition of DD-carboxypeptidase It was known that β-lactam antibiotics inhibit bacterial peptidoglycan biosynthesis at its final step. The peptide cross-links are introduced by a transpeptidase, accompanied by the release of the ultimate D-Ala residue. In addition to the transpeptidation reaction, the terminal D-Ala is also removed by DD-carboxypeptidase. The activity of this enzyme can be assayed by using diacetyl-L-lysyl-D-alanyl-D-alanine as substrate, monitoring the liberated D-alanine by quantitative analysis.

Fleming et al. (1982) reported a detection method for β-lactam antibiotics using DD-carboxypeptidase prepared from *Actinomadura* R 39, which is quite sensitive to β-lactam antibiotics. However, this method was not suitable for high-throughput screening for β-lactam compounds in crude microbial broths, since there was some interference by various broth constituents. To overcome this difficulty, Fleming improved the method, using the cell wall precursor UDP-MurNAc-L-Ala-D-Glu-Dap-D-Ala-D-Ala as substrate. In this method, the enzyme activity was assayed by the HPLC technique. Unreacted substrate (UDP pentapeptide) and the product UDP-tetrapeptide were monitored by the UV absorbance of uridine nucleotides at 254 nm. Bioactive metabolites of 10,047 actinomycete isolates were screened, and 394 isolates were found to be positive. Out of these isolates, 382 were found to produce olivanic acid, penicillin N, and cephamycin/cephalosporins.

The screening method mentioned above was improved again by Schindler et al. (1986). Since the isolation of UDP-MurNac-pentapeptide from bacteria is rather laborious, a suitable chromophore was introduced into synthetic α-acetyl-L-Lys-D-Ala-D-Ala which was used as substrate in this assay. Thus synthetic Nα-acetyl-Nε-4-(7-nitro-benzofurazanyl)-L-Lys-D-Ala-D-Ala ($ANLA_2$) is converted by DD-carboxypeptidase into the corresponding dipeptide ($ANLA_1$) with only one D-alanine residue. Both compounds are yellow and highly fluorescent, and can be separated by thin-layer chromatography. Using this substrate, high-throughput screening for β-lactam antibiotics was possible using a simple visual inspection of chromatograms.

DD-Carboxypeptidase and a sample in reaction buffer were mixed and incubated for 10 minutes at room temperature. The substrate $ANLA_2$ was added and the assay sample further incubated at 37°C for 20 minutes. The reaction was stopped by applying the reaction mixture onto a silica gel plate.

The plate was developed in a solvent system, butanol–acetic acid–n-heptane (10:8:7). The separated spots on the plate are viewed under UV light at 366 nm. $ANLA_1$ was detected as a yellow-orange spot above that of $ANLA_2$. In this screening system, 35 cultures out of 6,200 strains were found to be positive. The compounds identified were penicillin N, 7-methoxy-deacetylcephalosporin C, cephamycin A, N-acetylthienamycin, epithienamycin D, and sulfazecin.

Detection of synergistic activity with cycloserine Substances that show synergy with cycloserine are considered to be cell wall synthesis inhibitors, according to the theory of sequential blockade of enzymatic reaction in cell wall biosynthesis (Dulaney et al., 1970).

Kuroda et al. (1980a) undertook a screening program using a cycloserine-containing agar plate, and discovered an antibiotic, FR-900130, which caused spheroplast formation of *Staphylococcus aureus*. The activity of the antibiotic against *S. aureus* 279 is 16-fold potentiated by the addition of 5 μg/ml of D-cycloserine. The antibiotic was found to inhibit the alanine racemase of *S. aureus*.

Using Bacillus *and* Acholeplasma *as test organisms* This screening method is based on lack of activity against *Mycoplasma* or *Acholeplasma*, and selective inhibition of incorporation of *meso*-diaminopimelic acid (Dpm) into the acid-insoluble macromolecular fraction of growing cells of *Bacillus* (Ōmura et al., 1979a).

In the primary screening test, culture broths of soil microorganisms were selected simply by differential microbial activity, having activity against *B. subtilis* and lacking activity against *A. laidlawii*. Inhibitors of bacterial cell wall biosynthesis are inactive against mycoplasmas that lack a cell wall. Antibiotics other than the cell wall inhibitors, which were inactive against mycoplasmas, were eliminated by the following procedure.

In the secondary test, inhibitors of bacterial cell wall synthesis were selected by their ability to prevent the incorporation of *meso*-[^{3}H]Dpm but not that of the incorporation of L-[^{14}C]leucine into the acid-insoluble macromolecular fraction of growing cells of a Dpm-requiring strains of *Bacillus* sp. Dpm is an amino acid present only in the peptidoglycan of Dpm-containing bacteria such as bacilli. To examine whether or not the secondary procedure was suitable for the object of this screening, various known antibiotics of differing modes of action were tested. As a result, it was found that this procedure is useful to eliminate antibiotics having a mode of action other than that of inhibition of cell wall synthesis.

Broth filtrates of 10,200 strains, including fungal, bacterial, and actinomycete soil isolates, were submitted to this screening program. Two new antibiotics, azureomycins (Ōmura et al., 1979b) and izupeptines (Spiri-Nakagawa et al., 1986), were found, and six known antibiotics, amphomycin, 3-

amino-3-deoxy-D-glucose, D-cycloserine, penicillin G, and ristocetins A and B were identified, respectively.

Screening for glycopeptide antibiotics   Glycopeptide antibiotics, which are subgrouped into three types—vancomycin, ristocetin, and avoparcin—are inhibitors of bacterial cell wall synthesis. They show antibacterial activity due to a high affinity for cell wall receptors that bind with the D-Ala-D-Ala terminus of an intermediate of the growing peptidoglycan. Vancomycin is effectively used for the treatment of severe infections caused by methicillin-resistant *S. aureus*. Because of the increasing importance of vancomycin, a search for novel members of this class of glycopeptide antibiotics was carried out (Rake et al., 1986).

In the primary screening, differences of inhibitory activity against a vancomycin-resistant strain and its susceptible parent strain were measured to find new antibiotics of the vancomycin type. The resistant strain, *S. aureus* 209P VAN^R, was at least 60-fold more resistant to vancomycin than the parent strain. In the secondary screening, deacetyl-L-Lys-D-Ala-D-Ala, a tripeptide analogue of the glycopeptide receptor, was used to reverse the activity of vancomycin against the susceptible test organism. In this assay system *B. subtilis* was used, since it was 2- to 32-fold more sensitive to different glycopeptide antibiotics than was *S. aureus* 209P. For routine screening a disc loaded with 100 μg/ml of the tripeptide was used, since this amount completely reversed the activity of 3.1 μg/ml of vancomycin against the susceptible test organism.

During the first period of screening using this system, a total of 2,457 cultures were tested, and 344 cultures were selected primarily. Finally, 5 cultures out of 27, which were selected in the secondary test using the tripeptide antagonist, were found to produce glycopeptide antibiotics. During the second period, the screening was enhanced, by the use of improved strain selection and improved production conditions, to find 42 glycopeptide producers. Aridicins came from four cultures, LL-AM-347 from three cultures, A47934 from two cultures, actinoidin B from one culture, ristocetin from six cultures, vancomycin from 13 cultures, and a novel glycopeptide from 14 cultures. Antibiotics from four other cultures were not identified. Novel glycopeptide antibiotics found in this screening system were aridicins (Sitrin et al., 1985), kibdelins (AAD-609) A, B, C_1, C_2, and D (Folena-Wasserman et al., 1986), actinoidin A_2 (Dingerdissen et al., 1987), and parvodicin (Christensen et al., 1987). They were purified using Diaion HP-20 resin followed by a specific glycopeptide affinity column, Affigel-10-D-Ala-D-Ala.

Inhibitors of fatty acid synthesis   Recently several inhibitors of fatty acid synthesis that show selective toxicity against procaryotes have been noted. Thiolactomycin, a thiolactone antibiotic, which was discovered in the screening system using a β-lactam antibiotic supersensitive mutant of *Pseu-*

domonas aeruginosa, selectively inhibits the fatty acid synthetase of *E. coli* (type II) but has little effect on that of mammalian tissue (type I).

Mochizuki et al. (1986) reported a screening system used to find a new antibiotic that inhibits the fatty acid synthetase of *E. coli* (type II). The activity of the fatty acid synthetase was determined by the radioactive assay method, which measured the incorporation of [2-^{14}C]malonyl-CoA into the fatty acid fraction in the presence of acetyl-CoA and NADPH. By this screening method, new ansamycin antibiotics—naphthoquinomycins A, B, and C—were discovered. However, they showed only weak antibacterial activity by the conventional agar diffusion method.

Aminoglycoside antibiotics   Aminoglycoside antibiotics also have received attention; many of them are clinically important, e.g., streptomycin, kanamycin, gentamicin, sisomicin, tobramycin, and their derivatives. Most aminoglycoside antibiotics were obtained as products of actinomycete strains, but various bacterial soil isolates were also examined as aminoglycoside antibiotic producers and some were found: Bu-1709 A_1 and A_2 (identified with butirosins A and B, respectively), Bu-1975 (4'-deoxybutirosin) and sorbistin (Tsukiura et al., 1976).

Use of hypersensitive mutant   In a further effort to discover new types of aminoglycoside antibiotics, it was attempted to select a sorbistin-hypersensitive mutant that might have specific sensitivity to certain groups of aminoglycoside antibiotics. Sorbistin (Fig. 1.5) has a unique chemical structure, being a non-aminocyclitol-aminoglycoside antibiotic (Numata et al., 1986).

Klebsiella pneumoniae type No. 22 (Kp-8), which was a clinical isolate and highly resistant to various antibiotics, was selected as a parent strain.

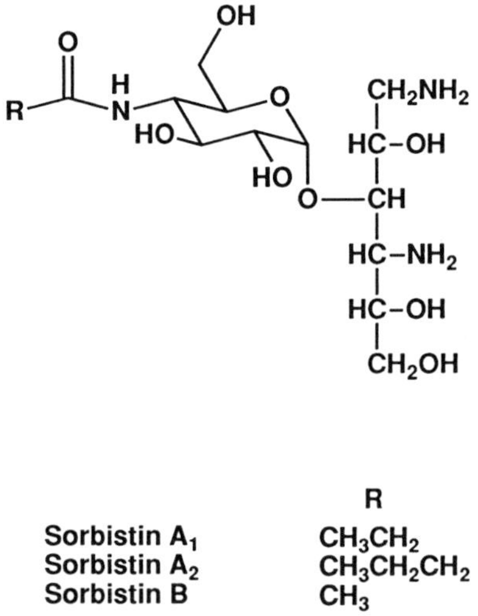

	R
Sorbistin A_1	CH_3CH_2
Sorbistin A_2	$CH_3CH_2CH_2$
Sorbistin B	CH_3

Figure 1.5   Structures of sorbistins A and B.

After a series of treatments with N-methyl-N'-nitro-N-nitrosoguanidine (NTG), a sorbistin-sensitive mutant, Kp-126 was obtained. The parent strain Kp-8 showed resistance to most aminoglycoside antibiotics tested; MICs were over 1,000 µg/ml, except for 4'-deoxybutirosin A and amikacin. The mutant strain Kp-126 was sensitive to all aminoglycoside antibiotics tested; MICs were under 0.2 µg/ml.

In this screening system, about 20,000 soil isolates were tested, and 10 strains were found to produce butirosins, sorbistins, BMY-28160 (peptide), capreomycin, streptothricin group antibiotics, and a new antibiotic, BMY-28251. The antibiotic BMY-28251, 3,3'-neotrehalosadiamine, exhibited hazy inhibition zones on nutrient agar plate seeded with most test organisms except the hypersensitive mutant of *K. pneumoniae* Kp-126 (Tsuno et al., 1986).

Detection by enzyme-linked immunosorbent assay   In the clinical field, rapid and susceptible antibiotic assays, without interference from coadministrated antibiotics, have been requested. The most frequently requested antibiotic assays are those for the aminoglycoside antibiotics, chiefly gentamicin. To confirm attainment of therapeutically effective peak serum levels and to avoid excess accumulation of drug with concomitant risk of ototoxicity and nephrotoxicity, the monitoring of serum levels of aminoglycoside antibiotics is important. Therefore, rapid assay methods for gentamicin have been developed including immunoassays, enzyme assays, gas–liquid chromatography, etc. Immunoassay techniques were found to be sensitive, accurate, and precise on application to the assay of low-molecular-weight compounds including aminoglycoside antibiotics. The enzyme-linked immunosorbent assay technique was developed and applied to the screening for novel aminoglycoside antibiotics in fermentation broth (Yao et al., 1984).

A gentamicin antibody was purified from goat antiserum to gentamicin and coated onto the surface of the wells of microtiter plates and incubated with gentamicin–alkaline phosphatase conjugate. After incubation, the plates were reacted with *p*-nitrophenylphosphate substrate solution. The total bound conjugate was determined spectrophotometrically at 410 nm. For the competitive assay, antibiotic solution or fermentation broth was added to the wells first followed by the addition of enzyme conjugate.

As a result, it was shown that the antibody probe used in this assay cross-reacted with all aminoglycosides tested, except neomycins B and C. No cross-reaction was detected with nonaminoglycoside antibiotics. This assay method was so sensitive that known aminoglycoside antibiotics produced in culture broths were easily detected at levels below the concentrations required for antimicrobial activity, and without interference by complex fermentation broths. Yao et al. (1984) wrote that this technique is amenable for high volume screening and detection of fermentation metabolites.

1.4 Conclusions

Detection of antibacterial activities of various metabolites of microorganisms have been mostly carried out by the traditional agar diffusion method. The method is simple and convenient for assaying a great many test materials, except for it taking several hours or more to grow the assay organisms. In target directed screens placing the focus on the mechanism of action or the structure of antibiotics, various cell free assays including inhibition of enzymes, radioimmuno assay, enzyme-linked immunosorbent assay, chemical assay, etc. were employed in combination with the traditional method.

A large number of antibacterial compounds have been discovered by now, but the need for discovering new types of antibiotics and the necessity for new types of therapeutic agents will not soon be exhausted. The naturally occurring mutation of microorganisms including both pathogenic bacteria and antibiotic producing organisms will continue forever.

A newly discovered metabolite whose mechanism of action is specific, even if its antibiotic activity is not enough to make it a therapeutic agent, will be useful as a biochemical tool and moreover often as a starting material for chemical synthesis.

References

Aoki, H., Sakai, H., Kohsaka, M., Konomi, T., Hosoda, J., Kubochi, Y., Iguchi, E. and Imanaka, H. 1976. Nocardicin A, a new monocyclic β-lactam antibiotic. *Journal of Antibiotics* (Tokyo) 29:492–500.

Bostian, M., McNitt, K., Aszalos, A. and Berdy, J. 1977. Antibiotic identification: A computerized data base system. *Journal of Antibiotics* (Tokyo) 30:633–634.

Box, S. J., Hood, J. D. and Spear, S. R. 1979. Four further antibiotics related to olivanic acid produced by *Streptomyces olivaceus*. *Journal of Antibiotics* (Tokyo) 32:1239–1247.

Box, S. J., Corbett, D. F., Robins, D. G., Spear, S. R. and Verrall, M. S. 1982. A new olivanic acid derivative produced by *Streptomyces olivaceus*. *Journal of Antibiotics* (Tokyo) 35:1394–1396.

Box, S. J., Brown, A. G., Gilpin, M. L., Gwynn, M. N. and Spear, S. R. 1988. MM 42842, a new member of the monobactam family produced by *Pseudomonas cocovenenans*. *Journal of Antibiotics* (Tokyo) 41:7–12.

Brown, A. G., Butterworth, D., Cole, M., Hanscomb, G., Hood, J. D. and Reading, C. 1976. Naturally-occurring β-lactamase inhibitors with antibacterial activity. *Journal of Antibiotics* (Tokyo) 29:668–669.

Butterworth, D., Cole, M., Hanscomb, G., and Rolinson, G. N. 1979. Olivanic acids, a family of β-lactam antibiotics with β-lactamase inhibitory properties produced by *Streptomyces* species. I. *Journal of Antibiotics* (Tokyo) 32:287–294.

Cevenini, R., Donati, M., Sambri, V., Rimpianesi, F. and LaPlaca, M. 1987. Enzyme-linked immunosorbent assay for the *in-vitro* detection of sensitivity of *Chlamydia trachomatis* to antimicrobial drugs. *Journal of Antimicrobial Chemotherapy* 20:677–684.

Chain, E. B., Florey, H. W., Gardner, A. D., Heatley, N. G., Jerrings, M. A., Orr-Ewing, J. and Sanders, A. G. 1940. Penicillin as a chemotherapeutic agent. *Lancet* ii:226–228.

Christensen, S. B., Allaudeen, H. S., Burke, M. R., Carr, S. A., Chung, S. K., DePhillips, P., Dingerdissen, J. J., DiPaolo, M., Giovenella, A. J., Heald, S. L., Killmer, L. B., Mico, B. A., Mueller, L., Pan, C. H., Poehland, B. L., Rake, J. B., Roberts, G. D., Shearer, M. C., Sitrin, R. D., Nisbet, L. J. and Jeffs, P. W. 1987. Parvodicin, a novel

glycopeptide from a new species, *Actinomadura parvosata. Journal of Antibiotics* (Tokyo) 40:970–990.
Cooper, R. 1983a. Chemical and spectroscopic characterization of monobactams. *Journal of Antibiotics* (Tokyo) 36:1258–1262.
Cooper, R., Bush, K., Principe, P. A., Trejo, W. H., Wells, J. S. and Sykes, R. B. 1983b. Two new monobactam antibiotics produced by a *Flexibacter* sp. *Journal of Antibiotics* (Tokyo) 36:1252–1257.
Dingerdissen, J. J., Sitrin, R. D., DePhillips, P. A., Giovenella, A. J., Grappel, S. F., Mehta, R. J., Oh, Y. K., Pan, C. K., Roberts, G. D., Shearer, M. C. and Nisbet, L. J. 1987. Actinoidin A_2, a novel glycopeptide. *Journal of Antibiotics* (Tokyo) 40:165–172.
Dolak, L. A., Castle, T. M., Truesdell, S. E. and Sebek, O. K. 1986. Isolation and structure of antibiotic U-68,204, a new thiolactone. *Journal of Antibiotics* (Tokyo) 39:26–31.
Dulaney, E. L. 1970. 1-Aminoethylphosphonic acid, an inhibitor of bacterial cell wall synthesis. *Journal of Antibiotics* (Tokyo) 23:567–568.
Fleming, I. D., Nisbet, L. J. and Brewer, S. J. 1982. Target directed antimicrobial screens. In: *Bioactive Microbial Products. Search and Discovery* (J. D. BuLock et al., eds), pp. 107–130. Academic Press, London.
Folena-Wasserman, G., Poehland, B. L., Yeung, E. W. K., Staiger, D., Killmer, L. B., Snader, K., Dingerdissen, J. J. and Jeffs, P. W. 1986. Kibdelins (AAD-609), novel glycopeptide antibiotics. *Journal of Antibiotics* (Tokyo) 39:1395–1406.
Hata, T., Ōmura, S., Iwai, Y., Ohno, H., Takeshima, H. and Yamaguchi, N. 1972. Studies on penicillinase inhibitors produced by microorganisms. *Journal of Antibiotics* (Tokyo) 25:473–474.
Higashide, E., Horii, S., Ono, H., Mizokami, N., Yamazaki, T., Shibata, M. and Yoneda, M. 1985. Alahopcin, a new dipeptide antibiotic produced by *Streptomyces albulus* subsp. *ochragenes* subsp. nov. *Journal of Antibiotics* (Tokyo) 38:285–295.
Ichimura, M., Koguchi, T., Yasuzawa, T. and Tomita, F. 1987. CV-1, a new antibiotic produced by a strain of *Streptomyces* sp. *Journal of Antibiotics* (Tokyo) 40:723–726.
Imada, A., Nozaki, Y., Kintaka, K., Okonogi, K., Kitano, K. and Harada, S. 1980. C-19393 S_2 and H_2, new carbapenem antibiotics. I. *Journal of Antibiotics* (Tokyo) 33:1417–1424.
Imada, A., Kitano, K., Kintaka, K., Muroi, M. and Asai, M. 1981. Sulfazecin and isosulfazecin, novel β-lactam antibiotics of bacterial origin. *Nature* 289:590–591.
Imada, A., Kintaka, K., Nakao, M. and Shinagawa, S. 1982. Bulgecin, a bacterial metabolite which in concert with β-lactam antibiotics causes bulge formation. *Journal of Antibiotics* (Tokyo) 35:1400–1403.
Inouye, S., Kojima, M., Shomura, T., Iwamatsu, K., Niwa, T., Kondo, Y., Niida, T., Ogawa, Y. and Kusama, K. 1983. Discovery, isolation and structure of novel cephamycins of *Streptomyces chartreusis. Journal of Antibiotics* (Tokyo) 36:115–124.
Inukai, M., Enokita, R., Torikata, A., Nakahara, M., Iwado, S. and Arai, M. 1978. Globomycin, a new peptide antibiotic with spheloplast-forming activity. I. *Journal of Antibiotics* (Tokyo) 31:410–420.
Ito, T., Ezaki, N., Ohba, K., Amano, S., Kondo, Y., Miyadoh, S., Shomura, T., Sezaki, M., Niwa, T., Kozima, M., Inouye, S., Yamada, Y. and Niida, T. 1982. A novel β-lactamase inhibitor, SF-2103 A produced by a *Streptomyces. Journal of Antibiotics* (Tokyo) 35:533–535.
Iwai, Y., Tanaka, H., Ōiwa, R., Shimizu, S. and Ōmura, S. 1977. Studies on bacterial cell wall inhibitors. 3-Amino-3-deoxy-D-glucose, an inhibitor of bacterial cell wall synthesis. *Biochimica et Biophysica Acta* 498:223–228.
Kahan, J. S., Kahan, F. M., Goegelman, R., Currie, S. A., Jackson, M., Stapley, E. O., Miller, T. W., Miller, A. K., Hendlin, D., Mochales, S., Hernandez, S., Woodruff,

H. B. and Birnbaum, J. 1979. Thienamycin, a new β-lactam antibiotic. *Journal of Antibiotics* (Tokyo) 32:1–12.
Katayama, N., Nozaki, Y., Okonogi, K., Ono, H., Harada, S. and Okazaki, H. 1985. Formadicins, new monocyclic β-lactam antibiotics of bacterial origin. I. *Journal of Antibiotics* (Tokyo) 38:1117–1127.
Kato, T., Hinoo, H., Shoji, J., Matsumoto, K., Tanimoto, T., Hattori, T., Hirooka, K. and Kondo, E. 1987. PB-5266 A, B and C, new monobactams. I. *Journal of Antibiotics* (Tokyo) 40:135–138.
Kitano, K., Kintaka, K., Katamoto, K., Nara, K. and Nakao, Y. 1975a. Occurrence of 6-aminopenicillanic acid in culture broths of strains belonging to the genera *Thermoascus, Gymnoascus, Polypaecilum* and *Malbranchea. Journal of Fermentation Technology* 53:339–346.
Kitano, K., Kintaka, K., Suzuki, S., Katamoto, K., Nara, K. and Nakao, Y. 1975b. Screening of microorganisms capable of producing β-lactam antibiotics. *Journal of Fermentation Technology* 53:327–338.
Kitano, K., Nara, K. and Nakao, Y. 1977. Screening for β-lactam antibiotics using a mutant of *Pseudomonas aeruginosa. Journal of Antibiotics* (Tokyo) 30(Suppl.):s239–s245.
Kuroda, Y., Okuhara, M., Goto, T., Iguchi, E., Kohsaka, M., Aoki, H. and Imanaka, H. 1980a. FR-900130, a novel amino acid antibiotic. I. *Journal of Antibiotics* (Tokyo) 33:125–131.
Kuroda, Y., Okuhara, M., Goto, T., Yamashita, M., Iguchi, E., Kohsaka, M., Aoki, H. and Imanaka, H. 1980b. FR-900148, a new antibiotic. I. *Journal of Antibiotics* (Tokyo) 33:259–266.
Kuroda, Y., Goto, T., Okamoto, M., Yamashita, M., Iguchi, E., Kohsaka, M., Aoki, H. and Imanaka, H. 1980c. FR-900137, a new antibiotic. I. *Journal of Antibiotics* (Tokyo) 33:272–279.
Masuma, R., Okuyama, K., Tanaka, Y., Hirose, R., Tomoda, H., Iwai, Y. and Ōmura, S. 1987. A new agar medium suitable for screening of anti-clostridium agents. *Journal of Antibiotics* (Tokyo) 40:1773–1775.
Mitscher, L. A. and Omoto, S. 1977. Modern instrumental methods for identification of antibiotics. *Journal of Antibiotics* (Tokyo) 30(suppl.):s-246–261.
Mochizuki, J., Kobayashi, E., Furihata, K., Kawaguchi, A., Seto, H. and Otake, N. 1986. New ansamycin antibiotics, naphthoquinomycins A and B, inhibitors of fatty acid synthesis in *Escherichia coli. Journal of Antibiotics* (Tokyo) 39:157–161.
Nagarajan, R., Boeck, L. D., Gorman, M., Hamill, R. L., Higgens, C. E., Hoehn, M. M., Stark, W. M. and Whitney, J. G. 1971. β-Lactam antibiotics from *Streptomyces. Journal of the American Chemical Society* 93:2308–2310.
Nakayama, M., Iwasaki, A., Kimura, S., Mizoguchi, T., Tanabe, S., Murakami, A., Watanabe, I., Okuchi, M., Itoh, H., Saina, Y., Kobayashi, F. and Mori, T. 1980. Carpetimycins A and B, new β-lactam antibiotics. *Journal of Antibiotics* (Tokyo) 33:1380–1389.
Nakayama, M., Kimura, S., Mizoguchi, T., Tanabe, S., Iwasaki, A., Murakami, A., Okuchi, M., Itoh, H. and Mori, T. 1983. New β-lactam antibiotic, carpetimycins C and D. *Journal of Antibiotics* (Tokyo) 36:943–949.
Nisbet, L. J., Mehta, R. J., Oh, Y., Pan, C. H., Phelen, C. G., Polansky, M. J., Shearer, M. C., Giovenella, A. J. and Grappel, S. F. 1985. Chlorocardicin, a monocyclic β-lactam from a *Streptomyces* sp. 1. *Journal of Antibiotics* (Tokyo) 38:133–138.
Numata, K., Yamamoto, H., Hatori, M., Miyaki, T. and Kawaguchi, H. 1986. Isolation of an aminoglycoside hypersensitive mutant and its application in screening. *Journal of Antibiotics* (Tokyo) 39:994–1000.
Ohba, K., Nojiri, C., Itoh, J., Yoshida, T., Ogawa, Y., Shomura, T., Niwa, T., Inouye, S., Yamada, Y. and Ito, T. 1986. Isolation and biological activity of SF-2050 B, a

novel carbapenem antibiotic of *Streptomyces* origin. *Science Reports of Meiji Seika Kaisha* No. 25:1–11.
Oishi, H., Noto, T., Sasaki, H., Suzuki, K., Hayashi, T., Okazaki, H., Ando, K. and Sawada, M. 1982. Thiolactomycin, a new antibiotic. *Journal of Antibiotics* (Tokyo) 35:391–395.
Okabe, M., Azuma, S., Kojima, I., Kouno, K., Okamoto, R., Fukagawa, Y. and Ishikura, T. 1982. Studies on the OA-6129 group of antibiotics, new carbapenem compounds. 1. *Journal of Antibiotics* (Tokyo) 35:1255–1263.
Okamura, K., Hirata, S., Okumura, Y., Fukagawa, Y., Shimauchi, Y., Kouno, K., Ishikura, T. and Lein, J. 1978. PS-5, a new β-lactam antibiotic from *Streptomyces*. *Journal of Antibiotics* (Tokyo) 31:480–482.
Okamura, K., Koki, A., Sakamoto, M., Kubo, K., Mutoh, Y., Fukagawa, Y., Kouno, K., Shimauchi, Y., Ishikura, T. and Lein, J. 1979. Microorganisms producing a new β-lactam antibiotic. *Journal of Fermentation Technology* 57:265–272.
Okuhara, M., Kuroda, Y., Goto, T., Okamoto, M., Terano, H., Kohsaka, M., Aoki, H. and Imanaka, H. 1980a. Studies on new phosphonic acid antibiotics. I. FR-900098, isolation and characterization. *Journal of Antibiotics* (Tokyo) 33:13–17.
Okuhara, M., Kuroda, Y., Goto, T., Okamoto, M., Terano, H., Kohsaka, M., Aoki, H. and Imanaka, H. 1980b. Studies on new phosphonic acid antibiotics. III. Isolation and characterization of FR-31564, FR-32863 and FR-33289. *Journal of Antibiotics* (Tokyo) 33:24–28.
Ōmura, S., Tanaka, H., Shinohara, M., Ōiwa, R. and Hata, T. 1975. Inhibition of bacterial cell wall synthesis by amphomycin. In: *Chemotherapy (Proceeding of 9th International Congress of Chemotherapy)* 5:365–369.
Ōmura, S., Tanaka, H., Ōiwa, R., Nagai, T., Koyama, Y., and Tanaka, Y. 1979a. Studies on bacterial cell wall inhibitors. II. Screening method for the specific inhibitors of peptidoglycan synthesis. *Journal of Antibiotics* (Tokyo) 32:978–984.
Ōmura, S., Tanaka, H., Tanaka, Y., Spiri-Nakagawa, P., Ōiwa, R., Takahashi, Y., Matsuyama, K. and Iwai, Y. 1979b. Studies on bacterial cell wall inhibitors. VII. Azureomycins A and B, new antibiotics produced by *Pseudonocardia azurea* nov. sp. Taxonomy of the production organism, isolation, characterization and biological properties. *Journal of Antibiotics* (Tokyo) 32:985–994.
Ōmura, S., Iwai, Y., Nakagawa, A., Iwata, R., Takahashi, Y., Shimizu, H. and Tanaka, H. 1983. Thiotetromycin, a new antibiotic. *Journal of Antibiotics* (Tokyo) 36:109–114.
Ōmura, S., Iwata, R., Iwai, Y., Taga, S., Tanaka, Y. and Tomoda, H. 1985. Luminamicin, a new antibiotic. *Journal of Antibiotics* (Tokyo) 38:1322–1326.
Ōmura, S., Imamura, N., Ōiwa, R., Kuga, H., Iwata, R., Masuma, R. and Iwai, Y. 1986. Clostomicins, new antibiotics produced by *Micromonospora echinospora* subsp. *armeniaca* subsp. nov. 1. *Journal of Antibiotics* (Tokyo) 39:1407–1412.
Rosenberg, E., Fytlovitch, S., Carmeli, S. and Kashman, Y. 1982. Chemical properties of *Myxococcus xanchus* antibiotic TA. *Journal of Antibiotics* (Tokyo) 35:788–793.
Schindler, P. W., Koenig, W., Chatterjee, S. and Ganguli, N. 1986. Improved screening for β-lactam antibiotics. A sensitive, high-throughput assay using DD-carboxypeptidase and a novel chromophore-labeled substrate. *Journal of Antibiotics* (Tokyo) 39:53–57.
Shearer, M. C., Grovenella, A. J., Grappel, S. F., Hedde, R. D., Mehta, R. J., Oh, Y. K., Pan, C. H., Pitkin, D. H. and Nisbet, L. J. 1986. Kibdelines, novel glycopeptide antibiotics. I. *Journal of Antibiotics* (Tokyo) 39:1386–1394.
Shibamoto, N., Koki, A., Nishino, M., Nakamura, K., Kiyoshima, K., Okuhara, K., Okabe, M., Okamoto, R., Fukagawa, Y., Shimauchi, Y., Ishikura, T. and Lein, J. 1980. PS-6 and PS-7, new β-lactam antibiotics, isolation, physicochemical properties and structures. *Journal of Antibiotics* (Tokyo) 33:1128–1137.

Shibamoto, N., Nishino, M., Okamura, K., Fukagawa, Y. and Ishikura, T. 1982. PS-8, a minor carbapenem antibiotic. *Journal of Antibiotics* (Tokyo) 35:763–765.

Shoji, J., Hinoo, H., Sakazaki, R., Tsuji, N., Nagashima, K., Matsumoto, K., Takahashi, Y., Kozuki, S., Hattori, T., Kondo, E. and Tanaka, K. 1982. Asparenomycins A, B and C, new carbapenem antibiotics. *Journal of Antibiotics* (Tokyo) 35:15–23.

Shoji, J., Hinoo, H., Masunaga, R., Hattori, T., Wakisaka, Y. and Kondo, E. 1984a. Isolation of L-cycloserine from *Erwinia uredovora*. *Journal of Antibiotics* (Tokyo) 37:1198–1203.

Shoji, J., Kato, T., Sakazaki, R., Nagata, W., Terui, Y., Nakagawa, Y., Shiro, M., Matsumoto, K., Hattori, T., Yoshida, T. and Kondo, E. 1984b. Chitinovorins A, B and C, novel β-lactam antibiotics of bacterial origin. *Journal of Antibiotics* (Tokyo) 37:1486–1490.

Shoji, J., Kato, T., Hinoo, H., Hattori, T., Hirooka, K., Matsumoto, K., Tanimoto, T. and Kondo, E. 1986. Production of fosfomycin (phosphonomycin) by *Pseudomonas syringae*. *Journal of Antibiotics* (Tokyo) 39:1011–1012.

Singh, P. D., Ward, P. C., Wells, J. S., Ricca, C. M., Trejo, W. H., Principe, P. A. and Sykes, R. B. 1982. Bacterial production of deacetoxycephalosporin C. *Journal of Antibiotics* (Tokyo) 35:1397–1399.

Singh, P. D., Johnson, J. H., Ward, P. C., Wells, J. S., Trejo, W. H. and Sykes, R. B. 1983. SQ 28,332, a new monobactam produced by a *Flexibacter* sp. *Journal of Antibiotics* (Tokyo) 36:1245–1251.

Singh, P. D., Young, M. G., Johnson, J. H., Cimarusti, C. M. and Sykes, R. B. 1984. Bacterial production of 7-formamidocephalosporins isolated and structure determination. *Journal of Antibiotics* (Tokyo) 37:773–780.

Sitrin, R. D., Chan, G. W., Dingerdissen, J. J., Holl, W., Hoover, J. R. E., Valenta, J. R., Webb, L. and Snader, K. M. 1985. Aridicins, novel glycopeptide antibiotics. *Journal of Antibiotics* (Tokyo) 38:561–571.

Spiri-Nakagawa, P., Fukushi, Y., Maebashi, K., Imamura, N., Takahashi, Y., Tanaka, Y., Tanaka, H. and Omura, S. 1986. Izupeptin A and B, new glycopeptide antibiotics produced by an actinomycete. *Journal of Antibiotics* (Tokyo) 39:1719–1723.

Stapley, E. O., Jackson, M., Hernandez, S., Zimmerman, S. B., Currie, S. A., Mochales, S., Mata, J. M., Woodruff, H. B. and Hendlin, D. 1972. Cephamycins, a new family of β-lactam antibiotics. I. *Antimicrobial Agents and Chemotherapy* 2:122–131.

Stapley, E. O., Cassidy, P. J., Tunac, J., Monaghan, R. L., Jackson, M., Hernandez, S., Zimmerman, S. B., Mata, J. M., Currie, S. A., Daoust, D. and Hendlin, D. 1981. Epithienamycins, novel β-lactams related to thienamycin. I. *Journal of Antibiotics* (Tokyo) 34:628–636.

Sykes, R. B., Cimarsuti, C. M., Bonner, D. P., Bush, K., Floyd, D. M., Georgopapadakou, N. H., Koster, W. H., Liu, W. C., Parker, W. L., Principe, P. A., Rathnum, M. L., Slusarchyk, W. A., Trejo, W. H. and Wells, J. S. 1981. Monocyclic β-lactam antibiotics produced by bacteria. *Nature* 291:489–491.

Sykes, R. B. and Wells, J. S. 1985. Screening for β-lactam antibiotics in nature. *Journal of Antibiotics* (Tokyo) 38:119–121.

Takizawa, M., Tsubotani, S., Tanida, S., Harada, S. and Hasegawa, T. 1987. A new pyrrole-amidine antibiotic TAN-868 A. *Journal of Antibiotics* (Tokyo) 40:1220–1230.

Tanaka, Y., Kimura, K., Komagata, Y., Tsuzuki, K., Tomoda, H. and Ōmura, S. 1988. Radioimmuno assay for erythromycin derivatives. *Journal of Antibiotics* (Tokyo) 41:258–260.

Tomoda, H., Iwata, R., Takahashi, Y., Iwai, Y., Ōiwa, R. and Ōmura, S. 1986. Lustromycin, a new antibiotic produced by *Streptomyces* sp. *Journal of Antibiotics* (Tokyo) 39:1205–1210.

Tsuji, N., Nagashima, K., Kobayashi, M., Terui, Y., Matsumoto, K. and Kondo, E. 1982. The structures of pluracidomycins, new carbapenem antibiotics. *Journal of Antibiotics* (Tokyo) 35:536–540.

Tsuji, N., Kobayashi, M., Terui, Y., Matsumoto, K., Takahashi, Y. and Kondo, E. 1985. Pluracidomycin A_2, a new carbapenem bearing a sulfinic acid, and other minor pluracidomycins. *Journal of Antibiotics* (Tokyo) 38:270–274.

Tsukiura, H., Hanada, M., Saito, K., Fujisawa, K., Miyaki, T., Koshiyama, H. and Kawaguchi, H. 1976. Sorbistin, a new aminoglycoside antibiotic complex of bacterial origin. 1. *Journal of Antibiotics* (Tokyo) 29:1137–1146.

Tsuno, T., Ikeda, C., Numata, K., Tomita, K., Konishi, M. and Kawaguchi, H. 1986. 3,3'-Neotrehalosadiamine (BMY-28251), a new aminosugar antibiotic. *Journal of Antibiotics* (Tokyo) 39:1001–1003.

Uri, J. V., Actor, P. and Weisbach, J. A. 1978. A rapid and simple method for detection of β-lactamase inhibitors. *Journal of Antibiotics* (Tokyo) 31:789–791.

Watanabe, H., Yoshida, J., Tanaka, E., Ito, M., Miyadoh, S. and Shomura, T. 1986. Studies on a new phosphonic acid antibiotic, SF-2312. *Science Reports of Meiji Seika Kaisha* No. 25:12–17.

Wells, J. S., Trejo, W. H., Principe, P. A., Bush, K., Georgopapadakou, N., Bonner, D. P. and Sykes, R. B. 1982a. SQ 26,180, a novel monolactam. I. *Journal of Antibiotics* (Tokyo) 35:184–188.

Wells, J. S., Trejo, W. H., Principe, P. A., Bush, K., Georgopapadakou, N., Bonner, D. P. and Sykes, R. B. 1982b. EM5400, a family of monobactam antibiotics produced by *Agrobacterium radiobacter*. I. *Journal of Antibiotics* (Tokyo) 35:295–299.

Yamashita, M., Hashimoto, S., Ezaki, M., Iwami, M., Komori, T., Kohsaka, M. and Imanaka, H. 1983. FR-900318, a novel penicillin with β-lactamase inhibitory activity. *Journal of Antibiotics* (Tokyo) 36:1774–1776.

Yao, R. C. and Mahoney, D. F. 1984. Enzyme-linked immunosorbent assay for the detection of fermentation metabolites, aminoglycoside antibiotics. *Journal of Antibiotics* (Tokyo) 37:1462–1468.

Yoshida, S., Watanabe, H., Miyadoh, S., Ohba, K., Moriyama, C., Shomura, T. and Sezaki, M. 1986. A microbiological method for oligopeptide screening: Discovery of a new tripeptide antibiotic, SF-2339. *Science Reports of Meiji Seika Kaisha* No. 25:23–30.

2
Antifungal Agents

Yoshitake Tanaka

2.1 Introduction

Fungal infection in man and animals is divided into two types, dermatomycosis and deep or systemic mycosis. Systemic mycosis is often lethal. Table 2.1 lists the principal pathogenic fungi. Systemic mycosis has attracted attention in recent studies of mycology and antifungal chemotherapy for the following reasons. In association with antibacterial chemotherapy with broadspectrum antibiotics, etiological changes have occurred. Although potent bacterial pathogens have decreased markedly, opportunistic infections have increased gradually, with *Candida* and *Pseudomonas* as the major fungal and bacterial pathogens, respectively. In addition, fungal diseases are increasing in immunocompromised hosts. Such patients have a defect in the self-defense system, party because of medication with immunosuppressive drugs such as anticancer agents.

Current factors that also increase opportunistic fungal infections include aging and acquired immunodeficiency syndrome (AIDS). Although tumors take the top rank among diseases with high mortality in adults and the elderly, simultaneous fungal infections should not be overlooked. AIDS and the related immunodeficiency diseases caused by human immunodeficiency virus (HIV) and other immunodeficiency viruses are also associated with fungal infections involving opportunistic fungi such as *Candida* and *Aspergillus*.

Table 2.1 Mycoses and pathogenic fungi

Mycosis	Pathogenic fungi
Aspergillosis	*Aspergillus fumigatus, A. flavus,*
Candidiasis	*Candida albicans, C. krusei, C. tropicalis*
Coccidiomycosis	*C. immitis*
Cryptococcosis	*C. neoformans*
Mycotic keratitis	*Fusarium* spp., *Aspergillus* spp.
Onychomycosis	*Fusarium* spp., *Aspergillus* spp.
Paracocccidiomycosis	*Paracoccidioides brasiliensis*
Dermatophytoses	*Trichophyton* spp., *Microsporum* spp., *Histoplasma* spp.

In many of these cases, the actual cause of patient death is fungal growth in the organs and blood, rather than tumors or AIDS itself. If effective, safe, and long-acting antifungal agents become available, the mortality rates in patients with tumors and AIDS are expected to decrease considerably.

Garlic and Bordeaux mixture are among the oldest antifungal agents so far recorded. Garlic did not find practical application because of its characteristic unpleasant odor. The Bordeaux mixture, which is composed of cupric sulfate and calcium hydroxide, has been used for more than a century to control fungal infections on fruit trees. Other early medicinal antifungals include sulfur, iodine, mercury, and organic acids such as salicylic acid and undecylenic acid.

In 1935, griseofulvin was discovered as a metabolite of *Penicillium griseofulvus*. The compound was again identified in 1945 as a potent antifungal agent suitable for controlling plant pathogens. The antidermatophytic effect of griseofulvin in humans was discovered in 1958 (Gentile). Since then, griseofulvin in oral dose form has become an important antifungal chemotherapeutic. Griseofulvin was followed by other antifungal agents of microbial origin such as the polyene macrolide antibiotics (nystatin and amphotericin B), pyrrolnitrin (Arima et al., 1964), and other newer antifungal antibiotics. In 1969 the synthetic antifungal compounds, clotrimazole (Buchel et al., 1972) and miconazole (Van Cutsem and Thienpont, 1972), were introduced, followed by a series of azole derivatives.

Today, many antifungal drugs of clinical usefulness are synthetic chemicals. Amphotericin B is the sole compound of microbial origin now available for systemic use.

This limited state of available antifungal antibiotics contrasts greatly with the successful and rapid progress leading to antibacterial antibiotics, primarily for the following reasons.

1. Fungi are eucaryotes, as are animal and human cells. Therefore, selective toxicity is less likely to occur with antifungal than with antibacterial agents.

2. Methods in antifungal screening have not progressed very rapidly. These include the lack of standard methods for in vitro evaluation of antifungal activity such as minimal inhibitory concentration (MIC), the lack of reliable animal models for evaluation of in vivo efficacy, and the lack of sensitive and selective techniques for diagnosis of fungal infections.

3. Although potent fungicidal effects are required, almost all the antifungal drugs now available for clinical use are fungistatic, and those potent fungicides picked up to date by in vitro screening show serious cytotoxicity.

4. The host defense system and drug action are the two major factors in successful antimicrobial chemotherapy. However, the former is not expected to function well in fungal diseases, because fungal infection is mostly associated with a depression in host immune activity. This fact was often ignored in the past during screening for antifungal drugs.

As outlined, there is an increasing need for effective and safe antifungal agents for human as well as veterinary and agricultural uses. This chapter covers screening methods for detecting antifungal antibiotics, mainly those for human use.

2.2 Screening Methods

Paper disk method   The agar diffusion method, or paper disk method, is most often employed in antifungal screening. A paper disk, after being dipped into a sample solution and dried, is placed on the surface of an agar medium that has been seeded with indicator fungal cells. The agar medium is incubated at an appropriate temperature (usually 25°–40°C) for 2 to 4 days. The antifungal activity of a test sample is indicated by a growth-inhibition zone around the disk. Table 2.2 lists some antifungal antibiotics with clinical utility. These were probably discovered by the paper disk method, although the method was not reported in detail. The chemical structures of these drugs are shown in Figure 2.1

The advantage of the paper disk method is that a large number of test samples can be assayed within a small space; also, organic solvents that are used to dissolve the test substances can be evaporated.

Cylinder cups, which are used in the standard assay of antibacterial activity, also are employed in antifungal screening. The results with cylinder cups and paper disks are almost identical, although the former gives larger inhibition zones for compounds with higher molecular weight. One disadvantage of the cup assay is that organic solvents in the sample solution that are harmful to the test fungi must be evaporated before assay.

Table 2.2 Antifungal antibiotics of medical uses

Antibiotic[1]	Producing organism
Griseofulvin	*Penicillium griseofulvus*
Echinocandin B	*Aspergillus nidulans* subsp. *echinulatus*
Nystatin	*Streptomyces noursei*
Amphotericin B	*S. nodosus*
Trichomycin	*S. hachijoensis*
Pimaricin	*S. natalensis*
Azaromycin	*S. hygroscopicus* subsp. *azalomyceticus*
Variotin	*Paecilomyces varioti*
Siccanin	*Helminthosporium siccans*
Pyrrolnitrin	*Pseudomonas pyrrocinia*

[1]For structures of these antibiotics, see Figure 2.1.

Figure 2.1   Antifungal antibiotics of clinical uses.

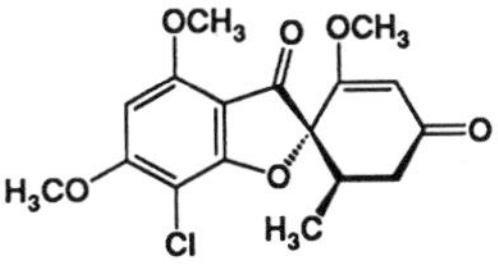

Chapter 2 Antifungal Agents 33

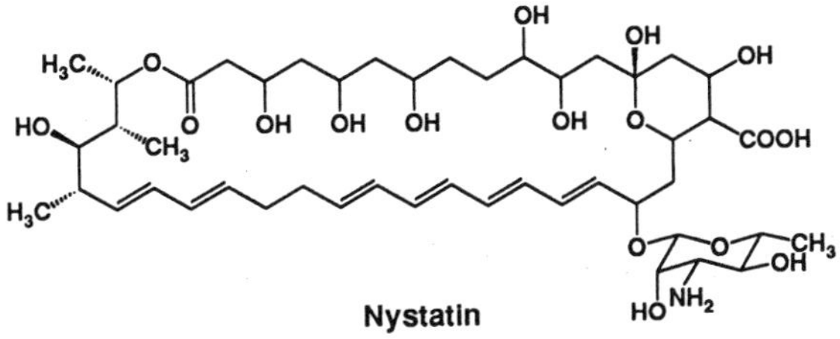

Griseofulvin

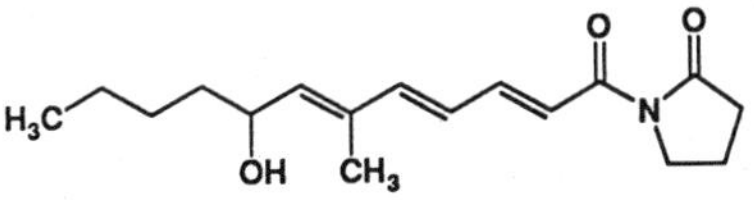

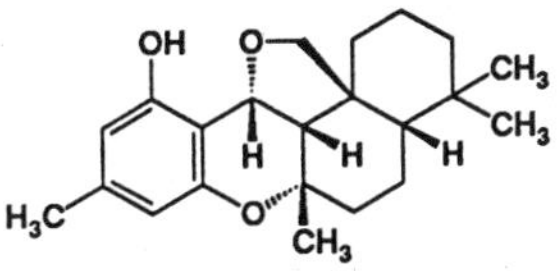

Nystatin

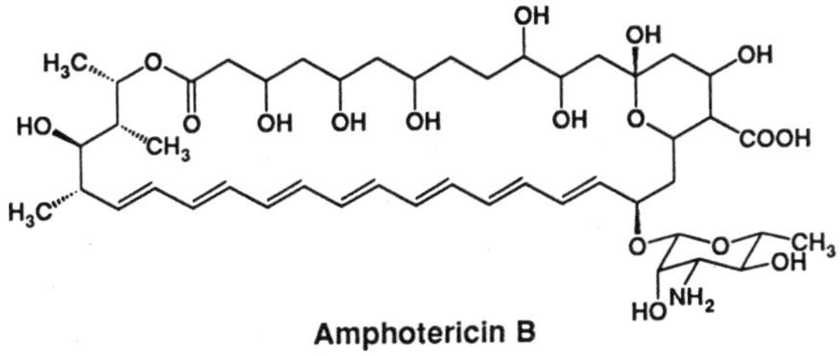

Amphotericin B

Trichomycin

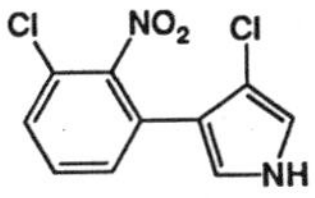

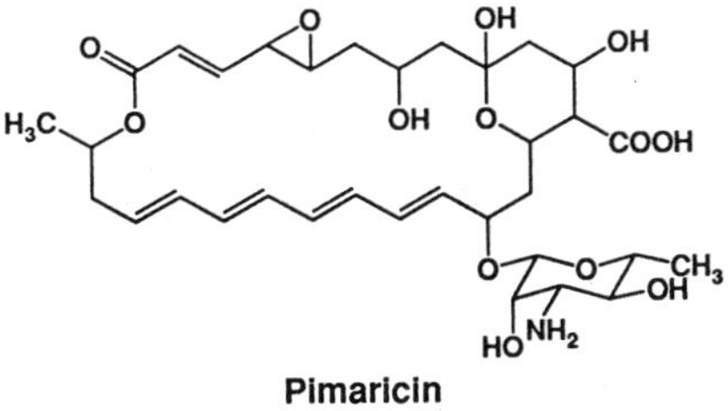

Pimaricin

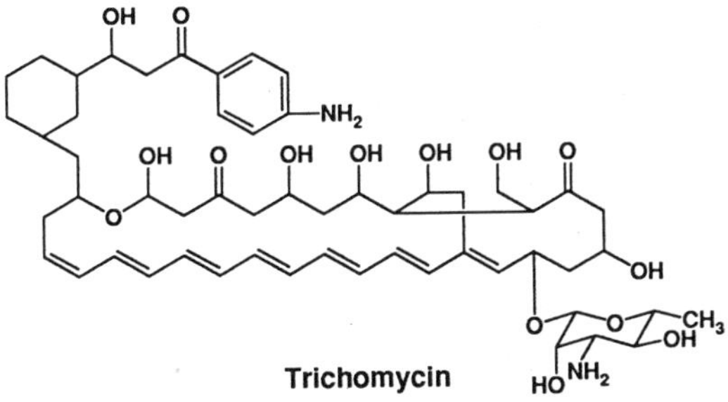

Variotin

Siccanin

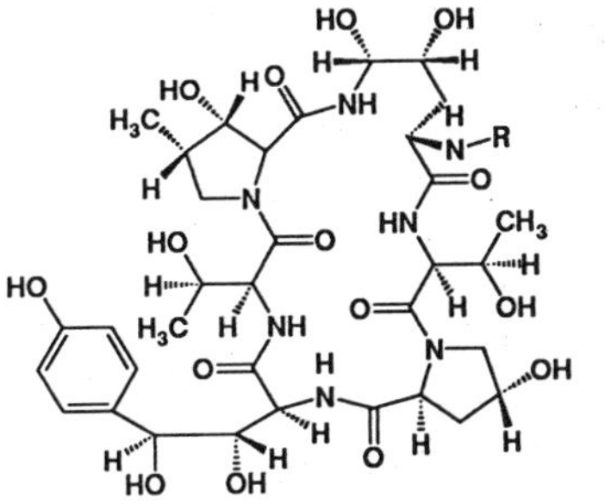

Pyrrolnitrin

Echinocandin B
Aculeacin A

R
linoleoyl
palmitoyl

In the paper disk assay, several factors affect the sensitivity of the test fungi to the antifungal antibiotics. The composition and pH of the agar media and the concentration of seeded fungal cells have the greatest influence on the diameter of the inhibition zone (Mikami and Uno, 1988).

Media   Potato-glucose medium and Czapek medium are conventionally employed in the paper disk assay. However, Sabaourand medium and sensitivity disk medium support good or better growth of many kinds of test fungi and are useful for assay of the MIC.

Yeast nitrogen base (Difco) supplemented with a carbon source (mostly glucose) and amino acids is a good chemically defined medium. Asparagine, aspartic acid, glutamine, glutamic acid, arginine, and lysine are often employed as nitrogen sources.

Cell concentration   Fungi in yeast form, such as *Candida* and *Saccharomyces*, are easy to use when controlling seeded cell concentrations. When filamentous fungi are seeded, their spores, reserved in suspension in saline phosphate buffer or 20%–30% glycerol, are used. When spores are not readily available, mycelia are mechanically broken before inoculation or are seeded without control of concentration.

pH   The pH is adjusted to around 6 because fungal growth is often abundant at this pH. However, the diameters of inhibition zones versus *Candida albicans* are greater with griseofluvin, nystatin, and ketoconazole when the pH of the agar medium is in the region of 7–8.

Antifungal activity of antibiotics can also be assayed in liquid media by photometric measurement of fungal growth. Automatic density recording machines are now available commercially. They give good results for yeast form of fungi but not for filamentous fungi.

Inhibition of cell wall synthesis   The cell wall of *Candida albicans* and *Saccharomyces cerevisiae* is composed of a network of three polysaccharides; β-1,3-glucan, chitin, and mannan. β-1,3-Glucan serves as the backbone giving rigidity and shape to the cell. Chitin and mannan are attached to this polymer, probably by chemical bonding (Farkas, 1985). Because a cell wall is indispensable for fungal growth, but is lacking in animal cells, the inhibitors of fungal cell wall synthesis are expected to be safe and effective antifungal agents, similar to the antibacterial penicillins and cephalosporins, which inhibit bacterial cell wall synthesis.

In the search for antibiotics effective against plant pathogenic fungi, polyoxins were discovered (Isono and Suzuki, 1979). Studies of the mode of action revealed that the polyoxins are inhibitors of chitin synthesis (Endo et al., 1970). The polyoxins have found agricultural uses. Unfortunately, they are almost inactive against human pathogens such as *Candida albicans* in vivo. Isono's group directed their efforts toward finding other inhibitors of fungal cell wall synthesis suitable for medical uses. They established enzyme systems for the assay of the synthesis of the three cell wall polysaccharides, β-1,3-glucan, chitin, and mannan.

These assay systems were used to find several new chitin synthesis inhibitors, neopolyoxins A, B, and C, active against *Candida* (Kobinata et al., 1980), and β-1,3-glucan synthesis inhibitors, lipopeptins A, B, and C (Tsuda et al., 1980), and neopeptins A, B, and C (Ubukata et al., 1986) (Figure 2.2). The latter two antibiotics inhibited mannan synthesis, as well. Microscopic observations showed that these compounds induced swollen, enlarged cell morphology, or pseudospheroplasts, of sensitive fungal cells when in contact with the drugs. This morphological abnormality of fungi was also employed as a marker in the screening of cell wall synthesis inhibitors.

Methods

Assay of chitin synthesis (Hori et al., 1974)   Chitin synthesis is measured with a particulate fraction of *Pyricularia oryzae* as enzyme source and UDP-[^{14}C]N-acetylglucosamine as substrate. The reaction mixture contains, in a total volume of 60 μl: 0.6 mM UDP-[U-^{14}C]N-acetylglucosamine (22,000 dpm), 16.6 mM Tris-maleate buffer (pH 7.2), 3.2 mM $MgCl_2$, 0.07 mM EDTA, 0.7 mM dithiothreitol, 16.6 mM N-acetylglucosamine, the test sample in 10 μl, and 10 μl of enzyme solution. After a 2-h incubation at 25°C, the reaction mixture is subjected to paper chromatography, and the radioactivity of a polymerization product located at the origin of the chromatogram is counted.

Assay of β-1,3-glucan synthesis (Satomi et al., 1982)   β-1,3-Glucan synthesis is measured with a particulate fraction of *Saccharomyces cerevisiae* cells as enzyme source and UDP[^{14}C]-glucose as substrate. Incubation is at 30°C for 2 h. The reaction mixture contains, in a final volume of 50 μl: 2.5 mM UDP-[U-^{14}C]glucose (27,500 dpm), 3.80 mM Tris-HCl buffer (pH 8.0), 0.25 mM EDTA, the test sample in 10 μl, and 10 μl of enzyme solution. Enzyme activity is measured by paper chromatography and radioactivity assay.

The particulate fraction of a wild-type strain of *S. cerevisiae* can catalyze both β-1,3-glucan synthesis and glycogen (a glucose polymer with α-1,4 linkage) synthesis with UDP[^{14}C] glucose as common starting material. The two polymers, glycogen and β-1,3-glucan, are not distinguishable by paper chromatographic behavior. A mutant of *S. cerevisiae* defective in glycogen synthesis was useful in the β-1,3-glucan assay (private communication from K. Isono).

Assay of mannan synthesis (Satomi et al., 1982)   Mannan synthesis is assayed with a particulate fraction of *S. cerevisiae* as the enzyme source and [^{14}C]GDP-mannose as substrate. The reaction mixture contains, in a total volume of 50 μl: 12.5 μM GDP[U-^{14}C]mannose (25,500 dpm), 20 mM cackodylate buffer (pH 6.5), 10 mM $MnCl_2$, 1 mM dithiothreitol, the test sample in 10 μl, and 10 μl of enzyme solution. Incubation is carried out at 25°C for 30 min. After incubation, the radioactive polymer formed is assayed after paper chromatography.

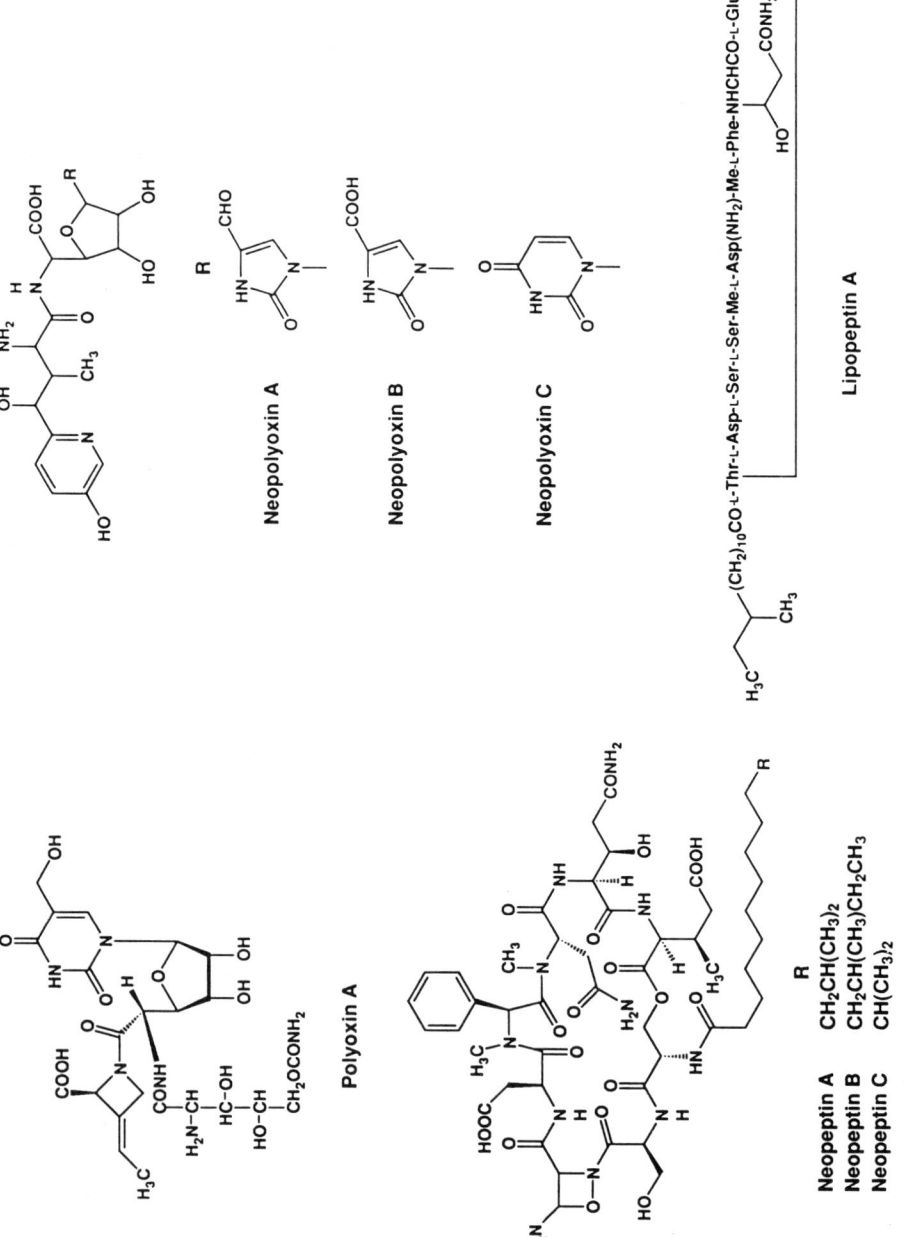

Figure 2.2 Structures of cell wall-active antifungal antibiotics.

Morphological changes as indicator Inhibitors of fungal cell wall synthesis such as the polyoxins, neopolyoxins, nikkomycins, aculeacins, and anticapsins are all known to induce round swollen pseudospheroplasts in sensitive fungal cells. This abnormal cell morphology can be observed around or within an inhibition zone on an agar plate using a microscope at low magnification. When a cultured broth shows antifungal activity and simultaneously the ability to induce pseudospheroplasts of sensitive fungi, it may be an indication that the culture broth contains inhibitors of cell wall synthesis. On the basis of this consideration, Tanaka et al. (1987) surveyed actinomycetes for their ability to produce inhibitors of cell wall synthesis. They discovered globopeptin, a peptide antibiotic. Studies of the mode of action of globopeptin suggested that it inhibited chitin synthesis in *S. cerevisiae* cells treated with toluene.

King et al. (1986) employed morphological changes as a marker together with other criteria to screen antifungal compounds from microorganisms. They discovered clavamycins A–F (Figure 2.3). These compounds were antifungal β-lactams active against *Candida*, *Trichophyton*, and *Neurospora* but inactive against *Aspergillus*. Clavamycin did not inhibit chitin synthesis. The anti-*Candida* activity of the clavamycins was antagonized by alanylalanine but not by ergosterol. These compounds were stable to β-lactamase. The mechanism of action of the clavamycins is not known.

Candida and *Mucor* are typical fungal pathogens showing dimorphism. They grow in a yeast form under certain conditions but in filamentous form under other conditions. *Candida albicans* cells growing in yeast form are less

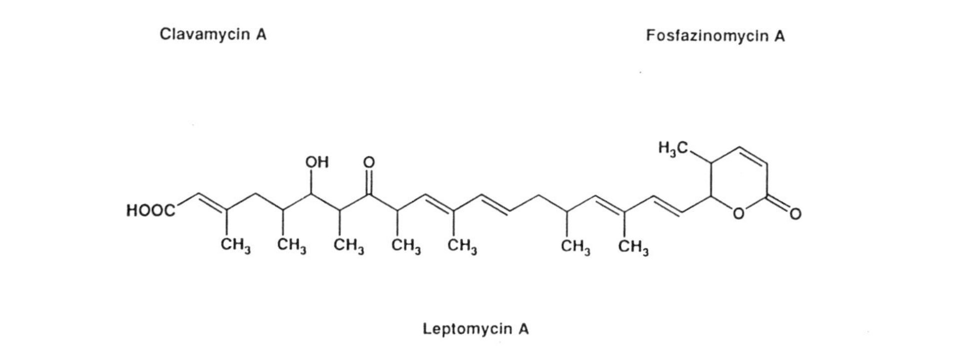

Clavamycin A

Fosfazinomycin A

Leptomycin A

Figure 2.3 Structures of antifungal antibiotics inducing morphological abnormality.

pathogenic because they are susceptible to attack by macrophages. However, when self-defense activity of the host decreases, *Candida* cells begin to grow in a filamentous form and penetrate into various tissues in the body. This form of *Candida* cells is less susceptible to attack by macrophages. Opportunistic infection by *Candida* is thus established. Dimorphic conversion can be observed under a microscope.

Gunji et al. (1983) selected an agar medium in which *Candida albicans* growth was sensitive to various antibiotics and morphological changes were easily observed. They surveyed microbial metabolites that induced abnormal morphology in *Candida albicans* and other test fungal cells and discovered leptomycin and fosphazinomycins A and B (Ogita et al., 1983) (see Figure 2.3). Leptomycin is highly active against *Schizosaccharomyces pombe* only and is highly cytotoxic (Hamamoto et al., 1983). Fosphazinomycin induced abnormal morphology against *Candida*, *Aspergillus*, and several other fungal cells tested (Gunji et al., 1983; Ogita et al., 1983). These authors failed to find any inhibitors of dimorphic interconversion.

Care has to be taken in interpreting the results of morphological observations, because many antifungal compounds can induce abnormal morphology. Swollen or enlarged cells are often observed when *Candida* or *Aspergillus* strains are exposed to antifungal agents other than inhibitors of cell wall synthesis. These include the polyene antibiotics (nystatin, trichomycin, eurocidin), antimycins, nanaomycins, DNA-active antibiotics such as pyrrolnitrin, etc.

Methods

Morphology as screen (Tanaka et al., 1987)

Step 1. In vitro antifungal activity against *Candida albicans*. Those cultures, which give clear inhibition zones versus both *Candida albicans* and *Aspergillus niger*, are given the higher priority and reserved for further study. The two fungal strains are grown in a complex agar medium (glucose, 1%, yeast extract, 0.5%, agar, 1%, pH 5.5)

Step 2. Agar plates of selected cultured broths are examined under a microscope to observe pseudospheroplasts microscopically (x100 magnification). The cultures showing the ability to induce swollen *Candida* cells are selected.

Step 3. To select interesting cultures, additional criteria are employed: no haemolytic activity in vitro versus horse red blood cells, low molecular weight as estimated using ultrafiltration membranes.

Inhibition of dimorphic conversion of Candida albicans *(Gunji et al., 1983)* An agar medium (sucrose, 0.3%, malt extract, 1%; yeast extract, 0.1%, agar, 0.7%) is used in which morphological changes of *Candida* and *Mucor* cells are more significant than in others. The morphological changes are observed around a paper disk under a light microscope.

Synergy with polyene antibiotics The polyene antibiotics are potent antifungal agents with broad antifungal spectra. However, their clinical application is limited because of severe toxicity from hemolytic activity. Amphotericin B is the only polyene now used in systemic dose form. Many attempts aimed at the discovery of less toxic polyenes were unsuccessful. A compound showing synergy with polyenes is expected to be useful because it may reduce the dose of (and thus the toxicity of) amphotericin B when used in combination, and also because it may be a novel type of antifungal agent.

A screening system was established for compounds showing synergy with a polyene antibiotic, trichomycin (Fukuda et al., 1973). Using this method, these authors found a series of new antibiotics possessing hydroxamate groups (enniatins). The enniatins are cyclic depsi-peptides showing broad antifungal spectra (Zocher et al., 1983).

Method The screening method (Fukuda et al. 1973) is based on the fact that ergosterol antagonizes the antifungal activity of the polyenes. A *Candida* strain was grown in two agar media at 37°C overnight. Streptomycete metabolites were examined for their synergy with polyenes or antagonism to ergosterol (or cholesterol) according to the inhibitory activity on the two media, as shown in Table 2.3.

Chemical screening Chemical screening is another route to new antifungal antibiotics. A general description of chemical screening appears in Chapter 7.

Several groups established efficient chemical screening systems for new microbial metabolites. Briefly, cultured broths were extracted with easily evaporable organic solvents, and the organic layers, combined and evaporated, were subjected to thin-layer chromatography (TLC). The chromatograms were visualized with several color reagents, such as Ehrlich's regent, tetrazolium bromide, 2,6-dichloroindophenol, periodate etc.

Compounds of chemical interest were purified and characterized, and biological activity was then assayed. The nikkomycins (Hagenmaier et al.,

Table 2.3 Screening methods for compounds showing synergy with polyenes or antagonism to sterol

Compound	Anti-*Candida* activity	
	S[1]	S + T + E[1]
Polyene	+	−
Nonpolyene	+	+
Synergist for polyene	+	+ +
Antagonist for sterol	−	+

[1]Symbols: S, sample tested; T, trichomycin, a polyene; E, ergosterol or cholesterol.

1976) are antifungal compounds, inhibitng cell wall synthesis, which were discovered by this chemical screening technique. The same techniques were also used in finding antifungal β-lactams (Roehl et al., 1987). By using Dragendorff's reagent for screening of microbial alkaloids, antifungal substances were discovered (see Chapter 7).

2.3 Early Identification of Known Compounds

No matter what methods are employed for the screening of antifungal compounds, early identification of known compounds or false-positive compounds is important. They should be identified as early as possible and be discarded. If false-positive but biologically interesting compounds are encountered, they may be studied further. When a compound, for example, a polyene, is expected to act as a false-positive, an additional screen is made to eliminate it. By a preliminary screening run, information may be obtained on the frequency at which particular compounds, whether false positives or known positives, are picked up. Antifungal screening is hampered especially when actinomycetes are used as microbial sources because they very often produce toxic antifungal compounds such as polyenes, antimycins, and cycloheximides. The characteristics of these compounds, and simple methods for their identification, are described next.

Polyenes

Potent antifungal activity with a broad spectrum   Polyenes are potently active against a variety of fungi, and often induce abnormal morphology and lysis of fungal cells, which can be observed microscopically.

UV absorption spectrum   A cultured supernatant solution is extracted with an equal volume of n-BuOH, and the BuOH layer is used to record a UV absorption spectrum. Oxopolyenes are less efficiently extracted into n-BuOH. To detect them, a small portion of cultured broth is evaporated into dryness and the residue is dissolved in MeOH. The MeOH layer is used to measure the UV absorption spectrum. Typical polyenes and oxopolyenes exhibit characteristic spectra with three to four absorption maxima.

Hemolytic activity   Samples contained in paper disks are placed on the surface of a blood agar plate and incubated at 37°C overnight. A hemolytic zone appears around the disks. The sensitivity of this method is not very great.

No activity against resistant fungal strains   When polyene macrolide-resistant fungal strains are available, they are very useful (Etienne et al., 1990).

Cycloheximides

Antimicrobial activity   Cycloheximides are active against *Saccharomyces* with large but hazy inhibition zones. They are inactive against *Candida albicans*.

Herbicidal activity   As evaluated by the methods described in Chapter 12.

Antimycins

Antimicrobial activity   Antimycins are active against *Saccharomyces, Candida*, and highly active against *Pyricularia*. They are also active against *Bacillus* and *Micrococcus*.

UV absorption spectrum   EtOAC extracts show strong absorption maxima at 224–226 nm and 320–322 nm.

TLC on SiO_2   After development with a solvent system with very low polarity and drying, a greenish-yellow color appears. This color may be intensified by exposure to iodine vapor.

Polyoxins and nikkomycins

Antimicrobial activity   The polyoxins are active against *Pyricularia* but inactive against *Candida* and *Aspergillus*. The nikkomycins are active against *Candida, Mucor*, and *Pyricularia* but are inactive against *Aspergillus*. They induce pseudospheroplasts in *Pyricularia*, which are observed microscopically.

Reversal of antifungal activity   Antifungal activity decreases markedly on nutrient-rich agar media.

Extraction   They are not extracted with organic solvents.

Polyethers

Antimicrobial activity   The polyethers show a characteristic antifungal and antibacterial activity. They are inactive against gram-negative bacteria, except *Xanthomonas oryzae*.

UV spectra and color reactions   They are extracted with organic solvents but show no characteristic UV absorption spectrum. On Ehrlich reaction, polyethers give various colors of green, blue, yellow, and orange. This reaction is neither very sensitive nor specific.

2.4 Concluding Remarks

The requirements for success in antifungal screening are the same in principle as those for antibacterial screening. They are (1) unique microorganisms, (2) specific fermentation methods, and (3) innovation in the methods for detection of activity. In view of the eucaryotic nature of fungal cells, more specificity is required in the assay methods for antifungal activity in vitro and in vivo. This chapter has focused attention on this point.

Recently, progress has been made in three areas of antifungal chemotherapy, which are expected to facilitate the discovery of newer and safer antifungal antibiotics. These are fungal physiology, chemical modification of naturally occuring antibiotics, and synthetic antifungal drugs.

During the course of study on the mode of action of ketoconazole, a synthetic antifungal agent of the triazole series, Shigematsu et al. (1981) found that ketoconazole inhibited the proliferation of filamentous mycelia of *Candida albicans* at sublethal doses. Screening for microbial metabolites with a similar mechanism of action is important.

Amino acid antagonists may attract attention. It has not been expected that they will find clinical usefulness because their antifungal activity is usually lost under the nutrient-rich conditions of the tissues and blood. However, this notion was revised recently by the discovery of RI-331, an inhibitor of homoserine dehydrogenase of *S. cerevisiae* (Yamaguchi et al., 1990). This antibiotic showed good in vivo efficacy against *Candida* infection in mice (Yamaguchi et al., 1988). It is suggested that amino acid antagonists provide a group of effective chemotherapeutics when they inhibit the common step from which biosynthetic pathways diverge leading to several amino acids.

Amphotericin B methyl ester was found to be as active against *C. albicans* as the mother compound, yet it showed far lower toxicity. Echinocandin B is a peptide with a fatty acid side chain connected by a ester bond. Gordee et al. (1988), of an Eli Lilly group, substituted the acyl side chain chemically to find LY 121019, named cilofungin, which is highly active against *Candida* infections. These examples strongly suggest it is quite likely that useful antifungal drugs can be developed by chemical modifications.

The azoles and arylamines are excellent synthetic drugs active against infectious fungi. Many azole derivatives inhibit ergosterol biosynthesis. It is challenging for the scientists engaged in screening to find antibiotics from microorganisms with similar modes of action. A result of such efforts has appeared recently (Aoki et al., 1992).

References

Aoki, Y., Yamazaki, T., Kondoh, M., Sudoh, Y., Nakayama, N., Sekine, Y., Shimada, H. and Arisawa, M. 1992. A new series of natural antifungals that inhibit p450 lanosterol C-14 demethylase. II. Mode of action. *Journal of Antibiotics* (Tokyo) 45:160–170.

Arima, K., Imanaka, H., Kousaka, M., Furuta, A. and Tamura, G. 1964. Pyrrolnitrin, a new antibiotic substance, produced by *Pseudomonas*. *Agricultural and Biological Chemistry* 28:575–576.

Buchel, K. H., Draber, W., Rogel, E. and Plempel, M. 1972. Syntheses and properties of clotrimazole and other antimycotic 1-triphenylmethyl imidazoles. *Drugs Made in Germany* 15:79–85.

Endo, A., Kakiki, K. and Misato, T. 1970. Mechanism of action of the antifungal agent polyoxin D. *Journal of Bacteriology* 104:189–196.

Etienne, G., Armau, E. and Tiraby, G. 1990. A screening method for antifungal substances using *Saccharomyces cerevisiae* strains resistant to polyene macrolides. *Journal of Antibiotics* (Tokyo) 43:199–206.

Farkas, V. 1985. Ultrastructural cytology of pathogenic fungi. In: Howard, D. H. (ed.), *Fungal Protoplasts. Applications in Biochemistry and Genetics*, pp. 3–30. Marcel Dekker, New York.

Fukuda, H., Kawakami, Y. and Nakamura, S. 1973. A method to screen anticholesterol substances produced by microbes and a new cholesterol oxidase produced by *Streptomyces violascens*. *Pharmaceutical and Chemical Bulletin* 21:2057–2060.

Gentile, G. C. 1958. Experimental ring worm in guinea pigs: Oral treatment with griseofulvin. *Nature* (London) 182:476.

Gordee, R. S., Zeckner, D. J., Howard, L. C., Alborn, W. E., Jr. and Debono, M. 1988. Anti-*Candida* activity and toxicology of LY 121019, a novel semisynthetic polypeptide antifungal antibiotic. *Annals of the New York Academy of Sciences* 544:295–309.

Gunji, S., Arima, K. and Beppu, T. 1983. Screening of antifungal antibiotics according to activities inducing morphological abnormalities. *Agricultural and Biological Chemistry* 47:2061–2069.

Hagenmaier, D. H., Hohne, H., Konig, W. A. and Zahner, H. 1976. Stoffwechselprodukte von Mikroorganismen. 154 Mitteilung. Nikkomycin, ein neuer Hemmstoff der Chitinsyntheses bei Pilzen. *Archiv fur Mikrobiologie* 107:143–160.

Hamamoto, T., Gunji, S., Tsuji, H. and Beppu, T. 1983. Leptomycins A and B, new antifungal antibiotics. I. Taxonomy of the producing strain and their fermentation, purification and characterization. *Journal of Antibiotics* (Tokyo) 36:639–645.

Ho, D. D. 1988. AIDS syndrome & therapy. *AIDS Journal* 1:157–165.

Hori, M., Kakiki, K. and Misato, T. 1974. Further study on the relation of polyoxin structure to chitin synthetase inhibition. *Agricultural and Biological Chemistry* 38:691–698.

Isono, K. and Suzuki, S. 1979. The polyoxins, pyrimidine nucleoside peptide antibiotics inhibiting fungal cell wall biosynthesis. *Heterocycles* 13:333–351.

King, H. D., Langharig, J. and Sanglier, J. 1986. Clavamycins, new clavam antibiotics from two variants of *Streptomyces hygroscopicus* I. Taxonomy of the producing organisms, fermentation, and biological activities. *Journal of Antibiotics* (Tokyo) 39:510–524.

Kobinata, K., Uramoto, M., Nishii, M., Kusakabe, H., Nakamura, G. and Isono, K. 1980. Neopolyoxins A, B and C, new chitin synthetase inhibitors. *Agricultural and Biological Chemistry* 44:1709–1711.

Mikami, Y. and Uno, J. 1988. *In vitro* and *in vivo* evaluation methods for antifungal agents. *Life Science & Biotechnology* (Biseibutsu) 4:44–49 (in Japanese).

Ogita, T., Gunji, S., Fukazawa, Y., Terahara, A. and Beppu, T. 1983. The structures of fosfazinomycins A and B. *Tetrahederon Letters* 24:2283–2286.

Roehl, F., Rabenhorst, J. and Zaehner, H 1987. Biological properties and mode of action of clavams. *Archives of Microbiology* 147:315–320.

Satomi, T., Kusakabe, H., Nakamura, G., Nishio, T., Uramoto, M. and Isono, K. 1982. Neopeptins A and B, new antifungal antibiotics. *Agricultural and Biological Chemistry* 46:2621–2623.

Shigematsu, M. L., Uno, J. and Arai, T. 1981. Correlative studies on *in vivo* and *in vitro* effectiveness of Ketoconazole (R 41400) against *Candida albicans* infection. *Japanese Journal of Mycology* 22:195–201.

Tanaka, Y., Hirata, K., Takahashi, Y., Iwai, Y. and Omura, S. 1987. Globopeptin, a new antifungal peptide antibiotic. *Journal of Antibiotics* (Tokyo) 40:242–244.

Tsuda, K., Kihara, T., Nishii, M., Nakamura, G., Isono, K. and Suzuki S. 1980. A new antibiotic, lipoleptin A. *Journal of Antibiotics* (Tokyo) 33:247–248.

Ubukata, M., Uramoto, M., Uzawa, J. and Isono, K. 1986. Structure and biological activity of neopeptins A, B and C, inhibitors of cell-wall glycan synthesis. *Agricultural and Biological Chemistry* 50:357–365.

Van Cutsem, J. and Thienpont, D. 1972. Miconazole, a broad spectrum antimycotic agent with antibacterial activity. *Chemotherapy* 17:392–399.

Yamaguchi, H., Uchida, K., Hiratani, T., Nagate, T., Watanabe, N. and Omura, S. 1988. RI-331, a new antifungal antibiotic. *Annals of the New York Academy of Sciences* 544:188–190.

Yamaguchi, M., Yamaki, H., Shinoda, T., Tago, Y., Suzuki, H., Nishimura, T. and Yamaguchi, H. 1990. The mode of antifungal action of (S)2-amino-4-oxo-5-hydroxypentanoic acid, RI-331. *Journal of Antibiotics* (Tokyo) 43:411–416.

Zocher, R., Keller, U. and Kleinkauf, H. 1983. Mechanism of depsipeptide formation catalyzed by enniatin synthetase. *Biochemical and Biophysical Research Communications* 110:292–299.

3

Antiviral Agents

Hideo Takeshima

3.1 Introduction

The pace of the discovery of clinically useful antiviral drugs has become markedly accelerated, in contrast to the slow progress made at the beginning of the search. It is not so long since the possibility of finding such agents seemed remote or even nonexistent. The arguments then mobilized were mainly that viruses were so closely associated with the host cell that there seemed no possibility of destroying one without the other. Today, however, it has become accepted that antiviral chemotherapy is essentially like antibacterial chemotherapy. Moreover, the progress was being made in a different field, the handling of viruses in the laboratory.

Today, the major disease targets for antiviral chemotherapy are influenza, rhinovirus colds, herpes, and human immunodeficiency virus (HIV), a causative agent of acquired immunodeficiency syndrome (AIDS), infections. Some corrective successes against these viral diseases have been achieved, for example, amantadine and its related compounds, rimantadine and adamantadine for influenza; 4',6 dichloroflavan for rhinovirus; vidarabine and acyclovir for herpes virus; and 2',3'-dideoxynucleotide analogs for HIV are clinically useful agents. There is also need for antiinfluenza and antiHIV compounds that are more effective in therapy than these agents. Thus, the problem of treating DNA virus infection is approaching solution, while much remains to be done in the field of RNA virus including retrovirus infections. Some important viral diseases of man and antiviral agents are listed in Table 3.1. Some chemical structures of the agents are also illustrated in Figure 3.1.

There are two fundamentally different strategies used in discovering antiviral agents. The first method is to test a large number of compounds or specimens prepared from the culture of microorganisms against a battery of viruses. However, this is not a logical method and has little intrinsic merit. In most cases, therefore, the results may be negative. The second method of screening is to select a known or assumed metabolic pathway of the virus, particularly in relationship to its mode of replication. With the discovery that viruses contain only one form of nucleic acid, either ribonucleic or deoxyribonucleic acid, the way was opened for specific attack on either RNA or DNA.

Table 3.1 Important viral diseases of man and antiviral agents

Virus	Disease spectrum	Antiviral agent
Influenza A	Acute respiratory	Amantadine
		Rimantadine
		Ribavirin
Influenza B	Mild respiratory	Interferon
Influenza C	Mild respiratory	None
Coronas	Mild respiratory	None
Arenaviruses	Acute generalize	Ribavirintriacetate
Rhinoviruses	Common cold	4',6-Dichloroflavan
Coxsackie		None
Papilloma viruses	Warts	None
Hepatitis A	Hepatitis	None
Herpes viruses:	Venereal	Acyclovir
HSV-2		Ara A
		BVDU
		IDU (keratitis only)
		Trifluorothymidine (keratitis only)
		Vidarabine
HSV-1	Superficial lesions (including venereal)	Asafone
Herpes zoster	Zoster, varicella	Acyclovir
Human immunodeficiency	Immunodeficiency	2',3'-Dideoxynucleotide analogs

In addition, enzymes present in viruses as one of their constituents or which are induced by added compounds are now regarded as fundamental to the logical use of antivirals. RNA and DNA polymerases, protein kinases, proteases, and thymidine kinase have been identified in most groups of viruses and are considered as good primary targets for discovering antiviral agents. Indeed, some specific inhibitors of these enzymes have been found. However, most of these were entirely negative in animal experiments. Thus, unfortunately, the mere identification of an enzyme is not proof that an inhibitor will prove effective. Under the circumstances, the activities of antiviral substances should be evaluated by a combination of several assay systems.

In this chapter we discuss various assay systems that are generally used for screening and evaluating antiviral compounds in the laboratory.

3.2 General Methods for the Laboratory Selection of Antiviral Agents

Primary testing methods are those based on detecting inhibition of virus replication in cell culture. Secondary screening is often done in organ culture, e.g., tracheal tissue explants. Finally, in vivo evaluation is done in animal models of the human disease, where available. Although animal models of

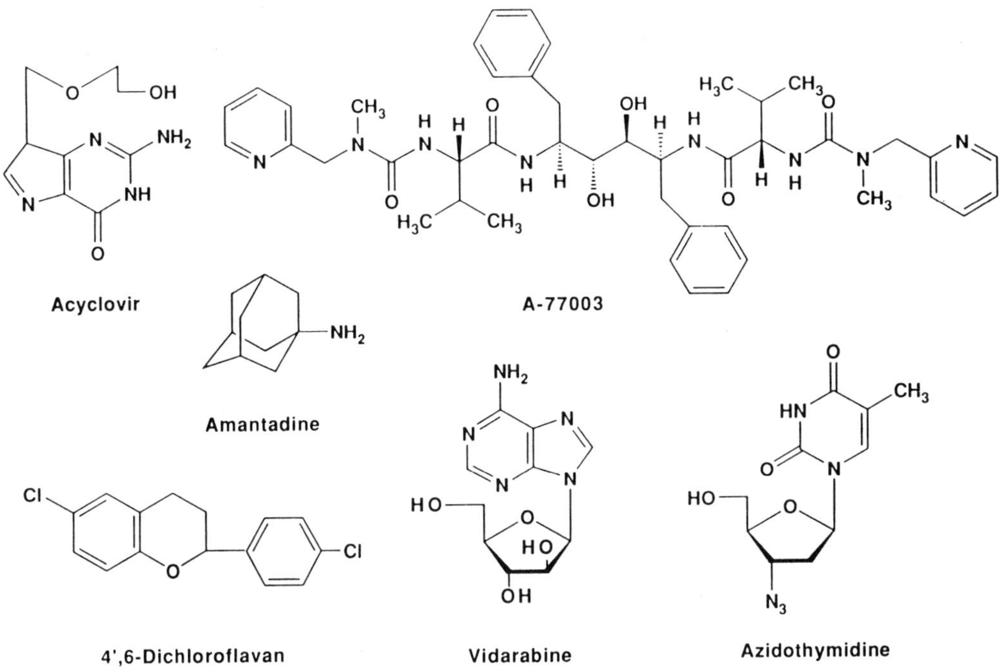

Figure 3.1 Structure of antiviral agents.

the virus disease may predict the therapeutic potential of a new compound and should reflect as closely as possible the main pathological feature of the human disease, many models of infection represent the human condition to only a limited extent.

Inhibition of cytopathic effect Many viruses, when affecting cells on infection, alter them in several ways accompanying the viral multiplication. One such change is called the cytopathic effect (CPE), which enables one to assess antiviral potency of the compounds as well as the degree of viral multiplication. The early refinements of these assays were described by Finter (1970) and Collier (1974). Practically, cell cultures are infected with viruses and thereafter exposed to therapeutic concentrations of the antiviral agents. The quantitative titration is defined as the concentration of antiviral agent that inhibits 50% of the CPE. This value is expressed in terms of 50% inhibitory (ID_{50}) or effective (ED_{50}) doses. This system is now widely used for the rapid estimation of infectivity titers and inhibition assays.

Plaque reduction assay Virus-infected cell cultures are overlaid with agar medium; after a few days, plaques form, resulting from each infective virus particle. These plaques become clearer when stained with neutral red. The antiviral effects of antiviral agents are measured by the number of plaques. However, use of the method has declined because this assay has the disadvantage of requiring a large number of cells; it is also time consuming and relatively insensitive.

Dye uptake reduction Dye uptake methods and those followiing are based principally on the method of Finter (1970), who used quantitative measurement of vital dye uptake and quantitative hemadsorption. Practically, virus-infected cell cultures, which give rise to oral cytopathic effects, are stained with neutral red or trypan blue, and the amount of dye incorporated is determined with a multichannel spectrophotometer. This assay can be automated and is rapid, compared with the plaque reduction assay.

Inhibition of hemadsorption Influenza and other myxovirus-infected cell cultures adsorb red blood cells from guinea pigs or chickens. This effect is observed as a cluster of red blood cells forming in cell culture. This assay is determined by the number of clusters, which can be estimated with the naked eye.

Focus reduction Human embryo fibroblast cell cultures infected with human cytomegalovirus or varicella zoster virus appear as a cell assembly called a "focus." By counting microscopically the number of foci formed, antiviral activity is determined. As described, the assay methods using cell cultures should be varied with respect to viral species and cell lines. Viruses, cell lines, and screening methods that are commonly used with in vitro cell culture assays are shown in Table 3.2.

Table 3.2 Assay methods using in vitro cell cultures for antiviral agents

Virus	Assay method	Cell culture
HSV	CPE	Primary rabbit kidney cells
	Plaque reduction	Vero, HeLa
	Dye uptake	Vero
VZV	Plaque reduction	MRC-5 (Human embryonic lung fibroblasts)
	Focus reduction	Human embryonic fibroblasts
HCMV	Plaque reduction	W1–38 (Human embryonic lung fibroblasts)
		Human embryonic fibroblasts
Influenza	Plaque reduction	MDCK (Madin-Derby canine kidney)
	Hemadsorption	Primary calf kidney cells
Rhino	Plaque reduction	Human embryonic lung fibroblasts
RSV	CPE	HeLa
	Plaque reduction	Human foreskin fibroblasts
HIV	CPE	ATH8, H9, MT4, etc.
	Syncytium formation	ATH8, Molt4, etc.

Enzyme-linked immunosorbent assays Several antiviral agents are now used to treat virus infections such as those caused by a herpes simplex virus. However, it has been found that viruses resistant to some of these antiviral agents occur in patients receiving antiviral chemotherapy. Therefore, it is necessary to determine the susceptibility of viruses to antiviral agents.

For this purpose, enzyme-linked immunosorbent assays (ELISA) have been developed. These methods can be used to assay viruses in cell culture systems that show little or no cytopathic effect. For antiviral assays, cell cultures are infected with the isolated virus and then incubated with varied concentrations of antiviral agents. After 1 or 2 days of incubation, the cells are fixed with glutaraldehyde and then treated sequentially with the antivirus antibody, the enzyme-conjugated antibody, and the substrate. Absorbance at 405 nm is then measured with a spectrophotometer. However, this method has little advantage over other methods already described. Therefore, because of its sensitivity and rapidity, it is suited only for routine viral diagnostic laboratories.

Nucleic acid hybridization assay The nucleic acid hybridization technique can be used to study the effect of antiviral agents on virus replication. This assay has the advantage of directly determining the amount of replication of virus DNA. The recombinant virus DNA clones are used as probes that are nick-translated with radioisotopes such as ^{32}P. Practically, cell cultures inoculated with the virus are exposed to antiviral agents; then the cells are destroyed with trypsin and filtered through a nitrocellulose filter. The DNAs on the filter are denatured in situ and fixed to the filter in a standard manner (Sambrook et al., 1989). The resulting DNA spots are hybridized with the radiolabeled and denatured recombinant DNA probes, and the amount of hybridized radioactivity is measured by autoradiography or scintillation counting. The hybridization technique can be applied to many DNA or RNA viruses if appropriate probes are available.

Radiochemical assays Radiolabeled nucleic acid precursors have been extensively applied to measure antiviral effects on the metabolism of the host cells with both DNA and RNA viruses (Bucknall, 1967). This approach can also be used to detect other drug effects on host cell metabolism.

Inhibition of virus-specific enzymes This type of assay system has seen an increase in use and being applied principally to the search for antiHIV, anti herpetic, and antiinfluenza agents. However, each assay is limited to only one part of the whole replicative process.

Cytotoxicity assay A measurement of the cytotoxicity of antiviral agents, judged as positive by any of the assay systems just described, is necessary to evaluate their intrinsic usefulness. This measurement can be determined quantitatively by a trypan blue exclusion test, the inhibitory effects on host cell DNA synthesis, or by other methods. The method of the trypan blue exclusion test is described briefly as follows. Antiviral agents are added to cell cultures without infection by a virus. After incubation for a few days, cells harvested by trypsinization are stained with trypan blue and dye excluded cells are counted. The ED_{50} is defined as the amount of antiviral agents causing a reduction of 50% in the viability of the cells as compared with the control cell cultures.

The potency of antiviral agents is estimated by the ratio of ED_{50} for cytotoxicity to ED_{50} for viral inhibition. This relationship is called the in vitro therapeutic index or selectivity index. Indeed, the selectivity index of antiviral agents must be more than 100 to 1000 to indicate a useful effect on viral inhibition in animal experiments.

Organ culture systems   In primary screening, almost all antiviral agents are commonly selected by in vitro screening systems, as described here because these screening assays are low in cost and easy to perform. However, most of the antiviral agents selected in vitro have not displayed the same antiviral effect in animal experiments as they do in vitro in cell culture. It seems that the antiviral and cytotoxic effects of antiviral agents observed in conventional cell cultures are different from those in susceptible animal cells. Therefore, the screening system using an antiviral organ culture, which reflects animal models, is of great value. The techniques of animal organ culture are now well established, and a whole range of cultures has been extensively used.

The following are some examples of organ cultures used to determine antiviral activity. The tracheal organ culture and the central nervous tissue culture are considered to be more similar to animal systems than are the usual cell culture systems.

Nasal and tracheal organ cultures   Organ cultures of nasal (DeLong and Reed, 1980) and tracheal (Burlington et al., 1982) tissues are available to evaluate antiviral agents against respiratory viruses, such as rhinovirus and influenza virus. These tissues are the targets of respiratory viral infection and have been well established to maintain morphology and function in vitro.

Two kinds of cultures are used in the study of antirhinovirus activity of antiviral compounds. One is a human embryonic nasal culture for antiviral tests and the other is a tracheal organ culture for toxicity tests. Tracheae are cut into rings and each ring is cultured in a screw-capped tube in a rotating drum. The antiviral agents are added to the medium of the tracheal cultures, and the ciliary activity is observed daily for 7 days and compared with that of a control. The ciliary activity is scored as the percentage of the activity of a complete normal ring.

In the other cultures, nasal epithelia are divided into several tissue fragments, each of which is placed with the ciliated surface uppermost in a petri dish. The nasal epithelium cultures are allowed to adsorb virus and are incubated with various concentrations of antiviral agents. The medium from infected or uninfected organ cultures is withdrawn and titrated in HeLa cells.

Central nervous tissue system Infection of herpes simplex virus (HSV) within central nervous system tissue induces neurological disease, resulting in a severe encephalitis.

To study neurological HSV infections, neural cell cultures, such as HSV-infected neuroblastoma cell lines, have been used. However, the amount of neuronal-specific enzymes in these neuroblastoma cell lines differs significantly from that of central nervous system tissues. Therefore, aggregated embryonic mouse brain cells have been used recently to study the effects of antiviral agents on replication of HSV in central nervous system tissue (Pulliam et al., 1986). The isolated neural cells from a fetal mouse brain are incubated with rotary shaking in a humidified atmosphere of 10% CO_2. After 30–35 days of cultivation, the aggregates of the brain cells are differentiated into neurons or glial cells on the basis of morphological characteristics.

For antiviral assays, the aggregates are infected with HSV and exposed to antiviral agents, following which they are pelleted by centrifugation and frozen by storage at $-70°C$. The infectious intracellular virus from the aggregates is measured using the plaque assay with Vero cells.

Use of animal models Animal models are useful for predicting the clinical efficacy of antiviral agents and may be more closely reflective of human disease. However, they are not always the best method for screening. In vivo screening systems for antiviral agents using animal organ cultures are shown in Table 3.3.

It is generally considered that infection and multiplication of viruses in target cells can proceed as follows: (1) adsorption to a target cell; (2) penetration into the cell; (3) uncoating; (4) virus nucleic acid synthesis; (5) viral protein synthesis and modification; and (6) virus assembly and release. Thus,

Table 3.3 Screening systems in vivo for antiviral agents

Virus	Infection (administration)	Animal	Determination of antiviral effect
Influenza	Respiratory (intranasal)	Mouse	Mortality or quantity of viruses in lungs
		Ferret	Quantity of viruses in nasal passages
Herpes	Ocular (subconjunctival)	Rabbit	Corneal infectivity titration
	Cutaneous (intradermal)	Guinea pig	Mortality or scores of cutaneous lesions
	Genital (intravaginal)	Mouse, Guinea pig	Quantity of viruses in vagina
	Encephalitis (intravenous or intraperitoneal)	Mouse	Mortality

antiviral agents could interact at any one of these steps, and it is desirable that each step be used in some screening systems. However, it is very difficult to reproduce the reactions at each of these steps in vitro. Even when it is possible, there is little application, except that this approach has been used to screen some systems for antiviral agents. Although there are several antiviral compounds whose mode of action and effect on these steps have been elucidated (Table 3.4), most of the modes of action were clarified after discovery. Thus, it is evident that the screening systems described in this section are not based on the specific steps of viral replication. For anti-HIV drugs, however, many efforts to develop assay systems based on these steps have been made (Mitsuya, et al., 1991a). In general, when searching for antiviral agents, the inhibitory activity against virus multiplication and the cytotoxicity to its host are first examined by methods such as CPE and plaque reduction. Second, for the selected compounds, an inhibition test for incorporation of a radiolabeled precursor, such as tritiated thymidine, or a nucleic acid hybridization method or an indirect immunofluorescence method is used for further confirmation of their activity.

Using these assay methods, several antibiotics derived from natural sources have shown inhibitory activity against virus multiplication in vitro (Table 3.5). The chemical structures of some antiviral antibiotics are shown

Table 3.4 Antiviral compounds and stages of inhibition in life cycle of viruses

Stage	Agent	Target virus
Binding to target cell	Pradimicin A	HIV
	CD_4 analogs	HIV
Entry and Uncoating	Amantadine, Rimantadine	Influenza A
	Arildone	Polio, Echo 12
	Dichloroflavan	Rhino
Transcription	Ribavirin	Measles, hepatitis B, herpes simplex
	AZT and other dideoxynucleotide analogs	HIV
	Acycloguanosine	Herpes virus group
	Adenine arabinoside	Herpes virus group
	Cytosine arabinoside	Herpes virus group
	Bromovinyldeoxyuridine	Varicella-zoster
	Enviroxime	Rhino
	Iododeoxyuridine	Herpes simplex
	Trifluorothymidine	Herpes simplex
	Methisazone	Smallpox, vaccinia
	Phosphonoformic acid	Herpes simplex, cytomegalo
	Rifampicin	Vaccinia
	Azidothymidine	HIV
Polyprotein cleavage	A-77003	HIV

Table 3.5 Antiviral antibiotics produced by various microorganisms

Virus		Antibiotic (Producing microorganism)	Assay method/cell type[1]
Orthomyxovirus	FPV	Kanamycin (*Streptomyces griseus*)	Plaque/CEC
	FPV	Hygromycin B (*Streptomyces sp.*)	Plaque/CEC
	Influenza	SF-2140 (*Actinomadura alboutea*)	HA/chorioallantoic membrane
Paramyxovirus	NDV	S-15-1 (*Streptomyces griseocarneus*)	Agar plaque diffusion/CEF
Herpesvirus	HSV-1	Aphidicolin (*Cephalosporium aphidicola*)	CPE/HEL
	HSV-1	Novobiocin (*Streptomyces niveus*)	Plaque/CV-1, C1300
	HSV-2	Oxetanocin (*Bacillus megatherium*)	CPE/Vero
	HSV-1,2	Daunomycin (*Streptomyces peucitius*)	Plaque/Vero
Poxvirus	Vaccinia	Augudin (*Fusarium* sp.)	Plaque/HEF
	Vaccinia, shope fibroma	Distamycin A (*Streptomyces distallicus*)	Plaque/BSC-1
Retrovirus	HIV	Pepstatin A (*Streptomyces testaceus*)	Protease activity
		Amphotericin B (*Streptomyces nodosus*)	IF/H9
		Sakyomicin A (*Nocardia* sp.)	Plaque, IF/MT-4
		Deoxorubicin (*Streptomyces peucitius*)	Dye uptake, IF/MT-4
		Oxetanocin A and its derivatives (*Bacillus megatherium*)	CPE/ATH8, *gag* protein expression
		Benanomicins A and B (*Actinomycete* sp. MH193-16F4)	Dye exclusion/MT-4, syncytium formation/Molt-4
		Pradimicin A (*Actinomadura hibisca*)	
RNA and DNA virus	Vaccinia, NDV, HSV-1,2	Virantomycin (*Streptomyces nitrosporeus*)	Plaque/CEC
	Vaccinia, NDV	Streptovirudins (*Streptomyces griseoflavus*)	Agar plaque diffusion/CEC

[1]Assay method: HA (hemagglutination), IF (immofluorescence); cell type: CEC (chick embryo cells), CEF (chick embryo fibroblasts), HEL (human embryo lung cells), HEF (human embryo fibroblasts), C1300 (neuronal cell line), H9, ATH8, Molt-4, and MT-4 (human T-cell line).

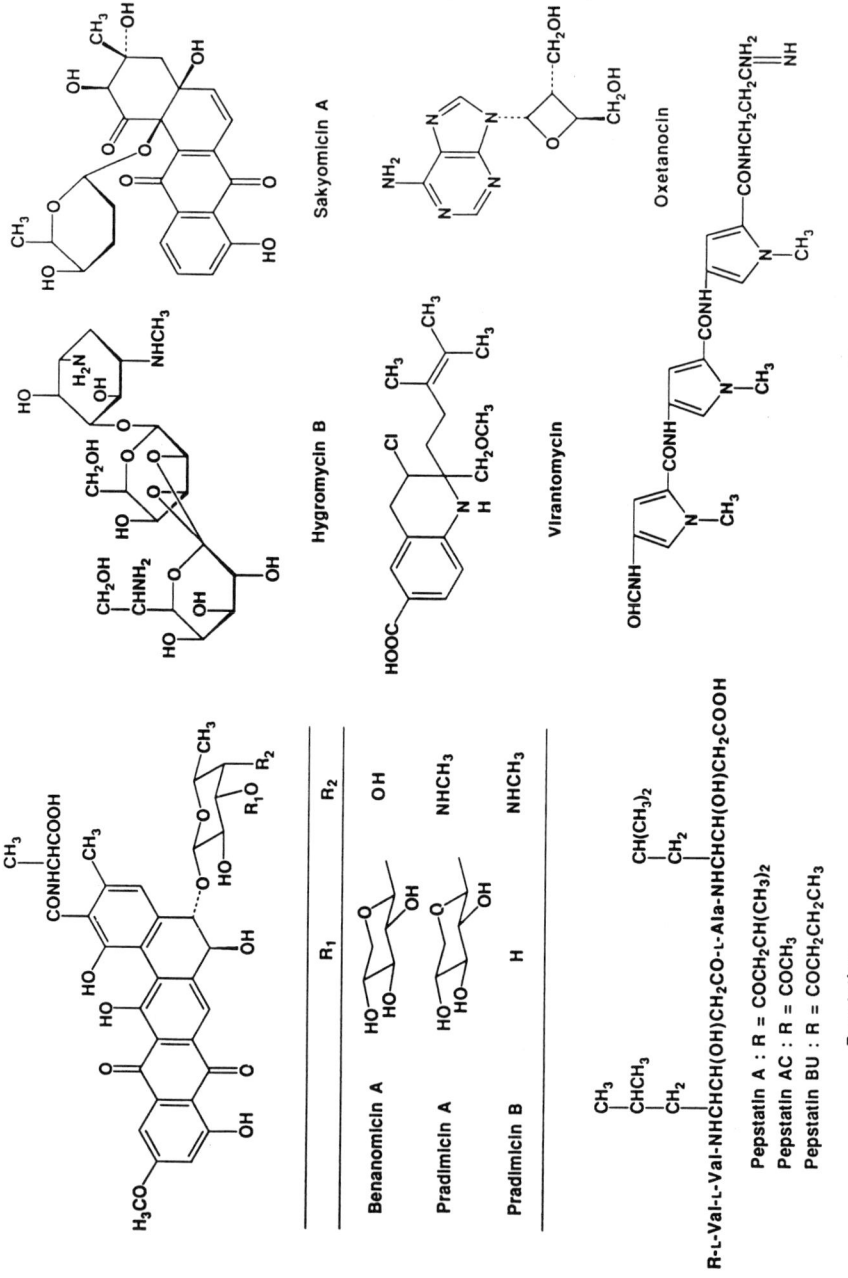

Figure 3.2 Structure of antiviral antibiotics produced by microorganisms.

in Figure 3.2. These antibiotics are previously known substances that have been found by other screening methods for antibacterial, antifungal, and other substances. For instance, as recently reported, pradimicin A (Tanabe-Toshikura et al., 1990) and benanomicin A and B (Hoshino et al., 1989) inhibited infection of HIV in human T cells as well as syncytium formation. These compounds were originally discovered as antifungal antibiotics in the culture filtrate of actinomytes related to the *Actinomadura* (Oki et al., 1988; Takeuchi et al., 1988). Therefore, in these cases, antiviral screening assays as shown in Table 3.5 have been applied to evaluate the potential of each substance for viral inhibitory effects rather than as research directed to the discovery of novel antiviral substances.

3.3 Detailed Methods for the Screening of Anti-HIV Substances

We describe in this section some practical methods used to evaluate antiviral compounds effective against HIV. This is as an example of a viral disease that is both highly topical and clinically severe. Most of the procedures described next can be applied to screening or evaluation for other antiviral agents by merely replacing the viruses and their target cells.

Since HIV was first recognized as a pathogenic retrovirus linked to AIDS, a number of potentially useful strategies for antiretroviral therapy of AIDS have been devised. In such development of antiretroviral drugs against AIDS, several stages of the replication of HIV have been considered targets for therapeutic interventions, as shown in Table 3.6. Among these, one of the most attractive targets has been the virally encoded reverse transcriptase. A number of drugs including 2',3'-dideoxynucleotide analogs, which target this enzyme, have been shown to be active against HIV (Mitsuya et al., 1991a). To design various potential antiviral agents, virtually every step in the life cycle of HIV has been used as a target for therapeutic intervention (Mitsuya et al., 1991a). Some principles for the development of drugs active against HIV have been reviewed (Mitsuya et al., 1991b).

Assay of HIV-1 binding to the target T cells (Mitsuya et al., 1988)
Radiolabeled viruses are prepared as follows: 2 week old, HX10 infected H9 cells are incubated for 48 h in medium containing 2 nmol [5-^{3}H]uridine ml^{-1} (26.7 Ci mmol^{-1}, New England Nuclear). The cell-free supernatant is filtered (HVLP, Millipore), centrifuged at 30,000 $\times$ g for 1 h, and dialyzed for 1 to 2 h at 4°C in NTE buffer (0.1 N NaCl, 0.1 M Tris, 0.001 M EDTA, pH 7.4). Preparations with high nonspecific binding are further clarified by centrifugation at 10,000 x g for 10–20 min. Preparations that show greater than 30% control binding after maximal OKT4A inhibition are not used. Virus preparations usually contain 0.83×10^4 to 2.5×10^4 Bq ml^{-1} or 0.5×10^{-6} to 2.5×10^{-5} Bq per virion.

Table 3.6 Stages in the HIV replicative cycle that may serve as targets for therapeutic intervention (Mitsuya et al., 1991a)

Stage	Possible intervention
Binding to target cell	Antibodies to HIV or cell receptors; CD4 analogs; calcium channel antagonists may rescue neuronal cells from possible gp120-induced cytocidal effect
Fusion to target cell	Antibodies or drugs that block the gp41 fusogenic domain function
Entry, uncoating of RNA, and functional release of HIV RNA	Drugs or antibodies that can block viral entry or uncoating; hypericin and pseudohypericin may block functional HIV RNA release
Transcription of RNA to DNA by reverse transcriptase	Reverse transcriptase inhibitors (e.g., AZT and other dideoxynucleoside analogs, TIBO compounds, and dipyridodiazepinone derivates, etc.)
Accumulation of unintegrated HIV in acute infection	Any intervention described here may suppress the cytopathic effect of HIV (if the DNA accumulation of unintegrated HIV DNA relates to premature cell death)
RNA degradation by RNase H activity	Specific inhibitors of HIV RNase H inhibitors
Migration of vial DNA to nucleus	Agents that block this step (as yet unidentified)
Integration of HIV DNA into host genome mediated by In protein	Agents that inhibit *pol*-encoded integrase (In protein) function (as yet unidentified)
Transcription and translation	Inhibitors of Tat or Rev activity; mutant Tat molecules; TAR inhibitors; TAR decoys; Rev protein inhibitors; HIV mRNA-specific destruction by ribozymes
Translation	Antisense constructs against regulatory HIV genes such as the new gene
Ribosomal frameshifting	Ribosomal frameshift inhibitors (as yet unidentified)
Gag-Pol polyprotein cleavage	HIV protease inhibitors such as transition-state mimetics and C_2 symmetric protease inhibitors
Myristoylation and glycosylation by cellular enzymes	Drugs (e.g., castanospermine and inhibitors of trimming glucosidase)
Dimerization, binding lysine tRNA	Inhibitors of these stages (as yet unidentified)
Packaging	Antisense constructs against the packaging sequence (as yet unidentified)
Viral budding	Interferons or interferon inducers; antibodies to viral antigens that block viral release
Extracellular processing of *Gag-Pol* polyproteins	HIV protease inhibitors

Binding inhibition assays are performed as follows: serial twofold dilutions of OKT4A and OKT8 (Ortho-Diagnostics) or drug solutions are made in binding assay buffer [(BAB) 2 mg ml^{-1} of bovine serum albumin, 2% heat-inactivated calf serum, and 1% sodium azide in Dulbecco's phosphate-buffered saline]. Labeled virus is diluted and 10 μl (4500–9000 cpm) is added to 2×10^5 H9 cells suspended in BAB. After incubation for 1 h at 37°C in 5% CO_2, cells are washed twice with BAB and lysed with distilled water and radioactivity is counted. Specific inhibition is calculated according to the formula:

$$\text{Specific inhibition} = 100 \times \frac{\text{cpm(control)} - \text{cpm(sample)}}{\text{cpm(control)} - \text{cpm(max)}} \quad (1)$$

where cpm(max) is the radioactivity observed with supramaximal concentration of OKT4A (50 μg ml^{-1}), cpm(control) is the radioactivity in the absence of drug or antibody, and cpm(sample) is the observed experimental radioactivity.

Inhibition assay for the HIV cytopathic effect (Pauwels et al., 1988)

Flat-bottom, 96-well plastic microtiter trays (Falcon, Becton Dickinson, Mountain View, CA) are filled with 100 μl of complete medium using a Titertek® Multidrop dispenser (Flow Laboratories). This eight-channel dispenser can fill a microtiter tray in less than 10 sec. Subsequently, stock solution (10× final test concentration) of compounds are added in 25-μl volumes to two series of triplicate wells so as to allow simultaneous evaluation of their effects on HIV- and mock-infected cells. Serial fivefold dilutions are made directly in the microtiter trays using an eight-channel Titertek® pipette (Flow Laboratories). Untreated control HIV- and mock-infected cell samples are included for each compound.

Exponentially growing MT-4 cells are centrifuged for 10 min at 140 × g and the supernatants are discarded. The pellet is either infected with 100 CCID$_{50}$ or is mock infected. The MT-4 cells are resuspended at 4×10^5 cells ml^{-1} in two flasks that are connected with an autoclavable dispensing cassette of a Titertek® Multidrop dispenser. Under slight magnetic stirring, 100-μl volumes are then transferred to the microtiter tray wells. The outer row wells are filled with 100 μl of medium. The cell cultures are incubated at 37°C in a humidified atmosphere of 5% CO_2 in air. The cells remain in contact with the test compounds during the entire incubation period. Five days after infection, the viability of mock- and HIV-infected cells is examined either microscopically in a hemacytometer by the trypan blue exclusion method or spectrophotometrically by the MTT method.

MTT assay for determination of HIV-protected cells (Pauwels et al., 1988)

The MTT assay is based on the reduction of the yellow 3-(4,5-dimethylthiazol-2-yl)-2,5-diphenyltetrazolium bromide (MTT) (Sigma Chemical Co., St. Louis, MO) by mitochondrial dehydrogenases of metabolically active cells to a blue formazan that can be measured spectrophotometrically. Therefore, to each well of the microtiter trays is added 20 μl of a solution of MTT (7.5 mg ml^{-1}) in phosphate-buffered saline using the Titertek® Multidrop. The trays are further incubated at 37°C in a CO_2 incubator for the indicated times. A fixed volume of medium (150 μl) is then removed from each cup using an eight-channel pipette without disturbing the MT-4 cell clusters containing the formazan crystals. If necessary, this step is preceded by centrifugation of the microtiter trays (450 $\times$ g, 5 min) in a plate holder (8 trays per run).

Solubilization of the formazan crystals is achieved by adding 100 μl of 10% (vol/vol) Triton X-100 in acidified isopropanol (2 ml concentrated HCl per 500 ml solvent) using the Titertek® Multidrop. Complete dissolution of the formazan crystals can be obtained after the trays have been placed on a plate shaker for 10 min (Flow Laboratories). Finally, the absorbances are read in an eight-channel computer-controlled photometer (Multiskan MCC, Flow Laboratories) at 540 and 690 nm. The absorbance measured at 690 nm is automatically subtracted from the absorbance at 540 nm to eliminate the effects of nonspecific absorption. Blanking is carried out directly on the microtiter trays with the first column wells that contain all reagents except the MT-4 cells. All data represent the average values for a minimum of three wells. The 50% cytotoxic dose (CD_{50}) is defined as the concentration of compound that reduced the absorbance (OD_{540}) of the mock-infected control sample by 50%. The percent protection achieved by the compounds in HIV infected cells is calculated by the following formula:

$$\frac{(OD_T)_{HIV} - (OD_C)_{HIV}}{(OD_C)_{mock} - (OD_C)_{HIV}} \text{ expressed as percentage} \quad (2)$$

where $(OD_T)_{HIV}$ is the optical density measured with a given concentration of the test compound in HIV-infected cells; $(OD_C)_{HIV}$ is the optical density measured for the control untreated HIV-infected cells; and $(OD_C)_{mock}$ is the optical density measured for the control untreated mock-infected cells. All OD values are determined at 540 nm. The dose achieving 50% protection according to these formula is defined as the 50% effective dose (ED_{50}).

Inhibition test for syncytium formation induced by HIV-1 (Mitsuya et al., 1988) After interaction of the viral gp120 and the cellular CD4 molecule, the hydrophobic fusogenic domain of gp41 interacts with the adjacent cell membrane and induces virion–cell or cell–cell fusion. Such fusion events could also serve as targets for evaluation of anti-HIV-1 agents. A practical example is as follows: ATH8 cells (5 $\times$ 10^5) are cultured with or without an equal number of chronically HIV-1-infected H9 cells in Costar 12 well culture plates after 2 h in the presence or absence of drugs. After 48 h of coculture, the number of giant cells is assessed on an inverted microscope.

Reverse transcriptase assay (Baba et al., 1988) The HIV-1 reverse transcriptase assay is performed at 37°C for 60 min with a 50-μl reaction mixture containing 50 mM TrisHCl at pH 8.4, 2 mM dithiothreitol, 100 mM KCl, 10 mM MgCl$_2$, 1 μCi of [methyl-^{3}H]dTTP (30 Ci mmol^{-1}), 0.01 A$_{260}$ unit of poly(rA)·oligo(dT), 0.1% Triton X-100, 10 μl of compound solution, and 10 μl of the enzyme [which had been partially purified by low centrifugation of the supernatant of HVT-78/HTLVIIIB cell cultures followed by filtration (0.45-μm pore) and ultracentrifugation (100,000 × g, 2 h)]. The reaction is stopped with 200 μl of trichloroacetic acid (5% vol/vol), and the precipitated material is analyzed for radioactivity.

Determination of HIV-1 *gag* protein expression in monocytes and macrophages (M/M) (Hayashi et al., 1990) The overall genomic structure of the provirus resembles that of all replication-competent retroviruses, as three essential genes have been identified. These include the *gag* gene, coding for the major nonglycosylated viral structural proteins, the *pol* gene, coding for the reverse transcriptase, and the *env* gene, coding for components of the viral envelope. Most of the antigens encoded by these open reading frames are specifically recognized by antibodies present in serum from many patients with AIDS and AIDS-related complex and from asymptomatic people infected with HIV. Of these antigens, p24 and p17 appear to be *gag* gene products; then, the determinative method of these proteins can be used also for screening of anti-AIDS substances. The target M/M (10^6) are preincubated with various concentrations of drugs for 16 h, exposed to 100 μl of an infectious M/M culture supernatant (where 1 μl of the supernatant represents the minimum infectious dose of HIV-1 under these conditions), and culture in 1 ml of complete medium. On day 6 in culture, the cells are extensively washed and further cultured in 1 ml of fresh complete medium. On day 10 in culture and beyond, the culture supernatants are collected and kept frozen until the amount of p24 *gag* protein in the supernatant is assessed by radioimmunoassay (Du Pont, NEN Research Products, Boston, MS).

HIV-1 protease assay (Meek et al., 1990) The late stages in the replication of HIV include crucial virus-specific secondary processing of certain viral proteins by an HIV-1 protease. This enzyme also represents a virus-specific target for new therapies. Peptide analogs have been synthesized (Meek et al., 1990; McQuade et al., 1990) based on the transition state mimetic concept (Szelke, et al., 1982). The following assay system is also one of the methods used for evaluation of those compounds on HIV-1 infectivity of T lymphocytes.

Cutures of uninfected Molt 4 (4 × 10^4 cells in 0.1 ml of RPMI 1640 medium containing 20% fetal calf serum) are treated with infections of HIV-1 (200 infectious units in 0.05 ml), which is recovered in the supernatant of a culture of H9III$_B$ cells by filtration through a 0.22-μm filter. Solutions of compounds (10 mM in DMSO) are diluted into medium and added in a single dose to the infected cell cultures in 0.05-ml aliquots to yield concentrations at day 1. On days 2 and 5 after infection, about 0.5 ml of fresh medium is added to each culture, and on day 7, the reverse transcriptase activity and amount of p24 antigen (Du Pont HIV-1 Antigen Capture Kit) in each culture are determined. Viral infectivity in the absence of inhibitors is determined in infected cell cultures treated with respective concentrations of DMSO, and a control sample contains no virus.

3.4 Future Prospects

The purpose of this chapter is to discuss some principles and assay methods for the screening and development of antiviral agents and to highlight some recent advances in this area.

As described, the principles of most methods for the evaluation of antiviral substances including ant-HIV drugs are basic and traditional. However, as seen in the development of AIDS therapy, a number of in vitro assay systems have been newly established and should further accelerate finding new antiviral agents. These are based on new knowledge on the replicative cycle of HIV, for example, HIV-1 protease inhibitors, are among the new category of rationally designed antiretroviral agents against HIV infection.

Further, it was recently made clear that the inhibitory activity of pradimicin A on the proliferation of HIV results from the binding of the antibiotic to mannose residue of HIV glycoprotein caused inhibition of the early stage of viral infection in T cells. Based on the result, a new in vitro assay system for finding new antiviral agents might be devised, for instance, to screen the compound that binds specifically to mannose residue of certain glycoprotein.

The T-cell surface glycoprotein, CD4, acts as the cellular receptor for HIV-1 (Maddon et al., 1986). In addition, a membrane binding protein found recently on human T-lymphocyte membrane was proposed as to be another receptor for HIV (Katunuma, 1990). The purified membrane protein, designated as tryptase TL2, recognized and bound specifically to V3 region in the outer membrane protein gp120 of HIV, and has serine protease activity. The antibody against the V3 region neutralized HIV-1 strongly and suppressed both membrane fusion and entry of the virus to target cell. Although much more research is required, such a finding may also prove useful in designing better viral binding inhibitors.

In past years, thus, molecular aspects of the replication of HIV have been made clear with unprecedented speed and a number of scientists have engaged in this work because it seems certain that more effective agents affecting individual or multiple steps in viral replication will soon become available.

References

Baba, M., Pauwels, R., Balzarini, J., Arnout, J., Desmyter, J., and DeClercq, E. 1988. Mechanism of inhibitory effect of dextran sulfate and heparin on replication of human immunodeficiency virus in vitro. *Proceedings of the National Academy of Sciences of United States of America* 85:6132–6136.

Bucknall, R. A. 1967. The effects of substituted benzimidazoles on the growth of viruses and the nucleic acid metabolism of the host cells. *Journal of General Virology* 10:89–99.

Burlington, D. B., Meiklejohn, B. and Mostow, S. R. 1982. Antiinfluenza A virus activity of amantadine hydrochloride and remantadine hydrochloride in ferret tracheal ciliated epithelium. *Antimicrobial Agents and Chemotherapy* 21:794–799.

Collier, H. B. 1974. Chlorpromazine as a sustitute for orthodianisdine and ortho-tolidine in the determination of chlorine, hemoglobin and peroxide activity. *Clinical Biochemistry* 7:331–338.

DeLong, D. C. and Reed, S. E. 1980. Inhibition of rhinovirus replication in organ culture by a potential antiviral drug. *Journal of Infectious Diseases* 141:87–91.

Finter, N. B. 1970. Methods for screening in vitro and in vivo for agents active against myxoviruses. *Annales of the New York Academy of Sciences* 173 (1):131–138.

Hayashi, S., Norbeck, D. W., Rosenbrook, W., Fine, R. L., Matsukura, M., Plattner, J. J., Broder, S. and Mitsuya, H. 1990. Cyclobut-A and cyclobut-G, carbocyclic oxetanocin analogs that inhibit the replication of human immunodeficiency virus in T cells and monocytes and macrophages in vitro. *Antimicrobial Agents and Chemotherapy* 34:287–294.

Hoshino, H., Seki, J. and Takeuchi, T. 1989. New antifungal antibiotics, benanomicins A and B inhibit infection of T-cell with human immunodeficiency virus (HIV) and syncytium formation by HIV. *Journal of Antibiotics* (Tokyo) 42:344–346.

Katunuma, N. 1990. New biological functions of intracellular proteases and their endogeneous inhibitors as bioreactant. *Advances in Enzyme Regulation* 30:377–392.

Maddon, P. J., Dalgleish, A. G., McDougal, J. S., Clapham, P. R., Weiss, R. A. and Axel, R. 1986. The T4 gene encodes the AIDS virus receptor and is expressed in the immune system and the brain. *Cell* 47:333–348.

McQuade, T. J., Tomasselli, A. G., Lin, L., Karacostas, V., Moss, B., Sayer, T. K., Heinrikson, R. L. and Tarpley, W. G. 1990. A synthetic HIV-1 protease inhibitor with antiviral activity arrests HIV-like particle maturation. *Science* 247:454–456.

Meek, T. D., Lambert, D. M., Dreyer, G. B., Carr, T. J., Tomaszek, T. A., Moore, M. L., Strickler, J. E., Debouck, C., Hyland, L. J., Mattews, T. J., Metcalf, B. W. and Petteway, S. R. 1990. Inhibition of HIV-1 protease in infected T-lymphocytes by synthetic peptide analogues. *Nature* (London) 343:90–92.

Mitsuya, H., Yarchoan, R., Kageyama, S. and Broder, S. 1991a. Targeted therapy of human immunodeficiency virus-related disease. *FASEB Journal* 5:2369–2381.

Mitsuya, H., Yarchoan, R. and Broder, S. 1991b. Molecular targets for AIDS therapy. *Science* 249:1533–1544.

Mitsuya, H., Looney, D. J., Kuno, S., Ueno, R., Wong-Staal, F. and Broder, S. 1988. Dextran sulfate suppression of viruses in the HIV family: Inhibition of virion binding to $CD4^+$ cells. *Science* 240:646–649.

Oki, T., Konishi, M., Tomatsu, K., Tomita, K., Saitoh, K., Tsunakawa, M., Nishio, M., Miyaki, T. and Kawaguchi, H. 1988. Pradimicin, a novel class of potent antifungal antibiotics. *Journal of Antibiotics* (Tokyo) 41:1701–1704.

Pauwels, R., Balzarini, J., Baba, M., Snoeck, R., Schols, D., Herdewijn, P., Desmyter, J. and DeClercq, E. 1988. Rapid and automated tetrazolium-based colorimetric assay for the detection of anti-HIV compounds. *Journal of Virological Methods* 20:309–321.

Pulliam, L., Panitch, H. S., Baringer, J. R. and Dix, R. D. 1986. Effect of antiviral agents on replication of herpes simplex virus type 1 in brain cultures. *Antimicrobial Agents and Chemotherapy* 30:840–846.

Sambrook, J., Fritsch, E. and Maniatis, T. 1989. *Molecular Cloning: A Laboratory Manual.* Cold Spring Harbor Laboratory, New York.

Szelke, M., Leckie, B., Hallett, A., Jones, D. M., Sueiras, J., Atrash, B. and Lever, A. F. 1982. Potent new inhibitors of human renin. *Nature* (London) 299:555–557.

Takeuchi, T., Hara, T., Naganawa, H., Okada, M., Hamada, M. and Umezawa, H. 1988. New antifungal antibiotics, benanomicins A and B from an *Actinomycete*. *Journal Antibiotics* (Tokyo) 41:807–811

Tanabe-Tochikura, A., Tochikura, T., Yoshida, O., Oki, T. and Yamamoto, N. 1990. Pradimicin A inhibition of human immunodeficiency virus: Attenuation by mannan. *Virology* 176:467–473.

4

Antiparasitic Agents

Kazuhiko Otoguro and Haruo Tanaka

4.1. Introduction

The large variety of parasitic diseases caused by invasion of the human body and domestic animals by protozoa and helminths undoubtedly constitute a major medical and public health problem, especially in the tropical and subtropical areas of the world. Millions of peoples living in these areas are affected by parasitic diseases that rarely occur in nontropical areas. Malaria (population currently infected, 150 million), schistosomiasis (more than 200 million), filariasis (130 million), leishmaniasis (1.2 million), and African and American trypanosomiasis (more than 24 million) are the main parasitic diseases in these areas (Edward et al., 1986).

In recent years, new types of parasitic diseases have appeared in advanced nations. These include opportunistic infections and imported parasitic diseases. The acquired immunodeficiency syndrome (AIDS), a type of immunocompromise, is a typical example of conditions that lead to opportunistic infectious diseases. AIDS patients are susceptible not only to a number of viral and bacterial diseases and malignancies such as Kaposi's sarcoma, but also to many parasitic diseases such as toxoplasmosis, pneumocytosis, cryptosporidiosis, and isosporiasis (Edward et al., 1986). Imported parasitic diseases include malaria, filariasis, amebiasis, and schistosomiasis. The patients with these parasitic infections in nontropical areas often are those who have returned from trips to tropical and subtropical areas. These patients are increasing in number every year (Bruce-Chwatt, 1978).

Several antiprotozoal and anthelmintic agents are available to reduce the incidence of many of the parasitic diseases. These drugs can be divided into four classes: herbal preparations, inorganic and metallorganic compounds, synthetic organic compounds, and antibiotics.

Antibiotics of practical use include the macrocyclic lactones avermectins (broad-spectrum nematocidal and insecticidal activities), tetracyclines (antiprotozoal activities against *Plasmodium*, *Entamoeba*, and *Anaplasma*), the macrolide spiramycin (antiprotozoal activities against *Cryptosporidia* and *Toxoplasma*), the polyene macrolides amphotericin B (antiprotozoal activities against *Naegleria* and *Leishmania*) and trichomycin (anti-*Trichomonas*), the aminoglycosides paromomycin (antiprotozoal and cestocidal activities against *Entamoeba, Giardia, Balantidia, Leishmania,* and *Taenia*), hygromycin B and destomycin (antinematodal activity against *Ascaris*), clindamycin (antiprotozoal activities against *Plasmodium, Babesia,* and *Toxoplasma*), and the polyether ionophores monensin and salinomycin (anticoccidial activity) (Bossche et al., 1986; Campbell and Rew, 1986). These antibiotics, except for the avermectins, were, however, first discovered as antibiotics active against microorganisms and then found to exhibit antiprotozoal and anthelmintic activities. Of the several thousand antibiotics that have been discovered, only a few have been reported to have antiprotozoal and anthelmintic activities. Examples of effective screening for new antibiotics active against protozoa and helminths are relatively rare in comparison with those against bacteria, viruses, and fungi because in vitro cultivation of protozoa and helminths is difficult or even impossible. Table 4.1 summarizes the antiparasitic antibiotics discovered by screening methods using parasites as test organisms.

This chapter focuses on useful screening methods for anthelmintic substances including new ones reported recently by us. Although parasitic or-

Table 4.1 Antiparasitic antibiotics discovered by screening methods using parasites as test organisms

Test organism	Method[1]	Antibiotics	References
Protozoa			
Tetrahymena pyriformis	A	Ikarugamycin	Jomon et al., 1972
Trichomonas foetus	B	Setamycin	Otoguro et al., 1988a
Trypanosoma cruzi	B	Trypacidin	Balan et al., 1963
Trypanosoma cruzi	B	Vermiculine	Fuska et al., 1972
Leishmania brasiliensis	B	Dectylarin	Kettner et al., 1973
Entamoeba histolytica	C	Protomycin	Hirabayashi 1959
Eimeria tenella	E	WS-5995A and -B	Ikushima et al., 1980
Toxoplasma gondii	E	Xanthomycin-like	Okami et al., 1955
Nematodes			
Nematospiroides dubius	E	Avermectins	Burg et al., 1979
Caenorhabditis elegnas	D	LL-F28249	Greestein et al., 1987
Bursaphelenchus lignicolus	D	Jietacin A and B	Otoguro et al., 1988b
Rhabditis macrocerca(?)	E	Thaimycins A, B, and C	Cassinelli et al., 1979
Haemonchus contortus	D	VM44857, VM44864, and VM44866	Hood et al., 1989
Cestodes			
Hymenolepis nana(?)	D,E	Axenomycins	Bruna et al., 1973
H. diminuta(?)	D,E	L-155,175	Goetz et al., 1985

[1] A: Single culture (agar plate); B: Single culture (liquid medium); C: Monoxenic (synxenic) culture; D: Immobilization assay; E: In vivo (using infected animals).

ganisms fall into the following groups: protozoa (single-celled organisms), nematodes (roundworms), cestodes (tapeworms), trematodes (flukes), acarines (ticks and mites), and insects, this chapter considers only screening methods for antiprotozoal and anthelmintic (nematocidal, cestocidal, and trematocidal) agents from microorganisms. Acaricidal and insecticidal agents are described in the chapter on agrochemicals (Chapter 13).

4.2 Screening for Antiprotozoal Antibiotics

General aspects

In vitro screening Zähner et al. (1953) first reported a screening method for antiprotozoal antibiotics employing protozoa as test organisms. Before this they used an indirect approach in which antibacterial agents had been isolated and then tested for antiprotozoal activity. This approach was tried because of the difficulty in cultivating protozoa and the paucity of such species that can be grown in pure culture. Using the direct assay they used *Herpetomonas culicidarum, Euglena gracilis,* and *Tetrahymena geleii,* which can grow in pure culture, as test organisms. Using the direct assay they showed that about 30% of actinomycetes obtained from soil samples and animal dung were active. In the direct screening method a protozoan culture was added to the semisolid liquid cultures of the actinomycetes in test tubes.

Saburi (1954) established a screening method using *Trichomonas vaginalis* as test organism. In this method, a liquid medium inoculated with a seed culture of the organism was added to an agar slant of an actinomycete and then was examined for the growth of the organism under the microscope.

Nemec et al. (1963) reported a screening method using *Trypanosoma cruzi* as test organism. Paper disks containing butanol extracts from cultured broths of soil-isolated fungi were put into a small volume of *T. cruzi* suspension, and the antiprotozoal activities were examined microscopically after 18 h. They selected antiprotozoal antibiotics without antibacterial or antifungal activity because antibiotics having both antimicrobial and antiprotozoal activities were considered to be more toxic. Nemec et al. (1969a) also established additional screening methods using *Leishmania brasiliensis, Strigomonas culicidarum, Euglena gracilis, Astesia chattoni* and *Tetrahymena pyriformis.* Trypacidin (Figure 4.1) (Balan et al., 1963) and vermiculine (Fuska et al., 1972) were found as a result of screening work using *T. cruzi,* and dactylarin (Kettner et al., 1973) using *L. brasiliensis.* The structures are shown in Figure 4.1.

Thiemann and Beretta (1967) used a paper disk method with a semisolid agar plate containing the cells of *Ochromonas malhamensis.* Antiprotozoal activities of *Streptomyces* cultures against *O. malhamensis* and *T. vaginalis* were almost parallel to each other although the activities against the former were a little lower than those against the latter.

Jomon et al. (1972) chose *Tetrahymena pyriformis* as the test organism for

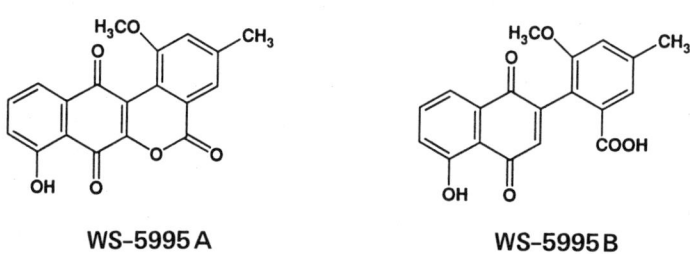

Figure 4.1 Structures of antiprotozoal antibiotics discovered with antiprotozoal screening.

the following reasons in addition to its lack of pathogenicity: The organism is suitable for microscope observation because it is relatively large (20–30 µm). Also, it is not sensitive to most of the toxic antibiotics frequently found in the culture broths of streptomycetes. Jomon et al. (1972) discovered ikarugamycin (see Figure 4.1), a new antibiotic with specific antiprotozoal activity, by this method.

For the screening of antiamoebic antibiotics, a synxenic culture of *Entamoeba histolytica* with several bacteria was used by Hirabayashi (1959). This was necessary because the parasite could not grow as a single culture. The antiamoebic activity of those antibiotics that have an antibacterial activity cannot be assayed by this method. Hirabayashi (1959) discovered a new antiamoebic antibiotic, protomycin (Figure 4.1), using the method.

A systemic screening method including in vitro and in vivo assays for new antitrichomonal substances was reported by Ōmura and coworkers (Otoguro et al., 1988a). The assays employed *Trichomonas foetus*, a pathogen of cattle trichomoniasis, which is nonpathogenic for humans. Primary screening is based on in vitro antitrichomonal activities of culture broths of actinomycetes. With secondary screening, after crude materials obtained from culture broths were administered orally to mice, excretion of antitrichomonal activity into urine was examined. Tertiary screening was for therapeutic activity against experimental trichomoniasis in mice. These methods are described next.

Applying the screening method to 6000 soil isolates, Otoguro et al. (1988a) discovered a new antibiotic, setamycin (Figure 4.1), which is produced by the new genus *Kitasatosporia*, and found that a known antifungal antibiotic, trichostatin A, is also active against *T. foetus* both in vitro and in vivo.

In vivo screening   Among protozoa, there are obligate parasites such as *Toxoplasma* that cannot grow in a free-living state. For the screening of antiprotozoal substances against such protozoa in vitro methods are impossible; in vivo methods are required.

Okami et al. (1955) established an in vivo screening system for antitoxoplasmic substances using *Toxoplasma gondii* as the test organism assayed against soil actinomycetes. They showed that 15 strains among 255 soil actinomycetes screened completely inhibited the growth of *T. gondii* when the latter cultures were mixed with the test organism and injected intraperitoneally.

For many years no further reports appeared concerning direct application of in vivo assays to screening for antiprotozoal antibiotics from soil microorganisms. However, Ikushima et al. (1980) have reported on the results of their screening program for anticoccidial substances. They discovered the new anticoccidial compounds WS-5995 A and B (Figure 4.1) produced by a soil isolate, *Streptomyces auranticolor*, in screening work using an in vivo method.

Tejmar-Kolar and Zähner (1984) have reported a new economical in vivo screening model for *Eucoccidium dinophili* using an archiannelidium, *Dino-*

philus gyrociliatus, as a host. The parasite-host system is useful for the recognition of substances that are inactive against free-living protozoa but active against the protozoa only in the host.

The specific procedures reported by these authors as typical examples of systemic screening methods, and also the new model of Tejmar-Kolar and Zähner (1984), are described next.

Typical examples of screening methods for antiprotozoal antibiotics

A screening method for antitrichomonal antibiotics (Otoguro et al., 1988a)
Trichomonas foetus, a pathogen of cattle trichomoniasis, is used as test organism. It is maintained at 37°C in Diamond's medium (Diamond, 1957) supplemented with 10% heat-inactivated calf serum, 100 U ml^{-1} of benzylpenicillin and 1.0 mg ml^{-1} of streptomycin sulfate, using 48-h transfers.

In the primary screening (in vitro assay), the inhibitory activities of cultured broths of soil actinomycetes against the organism are assayed. An aliquot (200 μl) of a culture broth of an actinomycete, 100 μl of heat-inactivated calf serum containing 2000 U ml^{-1} of benzylpenicillin and 20 mg ml^{-1} of streptomycin sulfate are added to 1.6 ml of Trichosel broth (BBL); 100 μl of a diluted cell suspension (2 × 10^5 cells ml^{-1}) prepared from an actively growing *T. foetus* culture is then delivered into this medium. After incubation for 48 h at 37°C, the number of *T. foetus* cells is counted under a microscope with a hemocytometer. The cultured broths showing reduction of the number more than 50% are selected for secondary screening.

In the secondary screening, urine excretion of antitrichomonal activity after administration of antitrichomonal substances is examined. A water extract of a dried culture filtrate combined with a methanol extract of the mycelium is used for the test. A sample solution (0.5 ml) possessing in vitro activity equivalent to that of 200–1000 g ml^{-1} of metronidazole, which is used as a chemotherapeutic agent for the treatment of trichomoniasis, is administered orally to mice, and urine is then collected for 4 h. The antitrichomonal activity of the urine against *T. foetus* is determined by the foregoing in vitro assay method.

In the tertiary screening, in vivo trichomonal activity is assayed. A mouse is infected intraperitoneally with 1 ml of a 48-h *T. foetus* culture containing about 2 million cells immediately after the oral administration of the test sample. The samples suspended in 0.5% methylcellulose or distilled water are administered orally once daily for 4 successive days. The experiments are terminated at 7 days after the mice are inoculated with trichomonads. The criteria of antitrichomonal efficacy are based on the presence of living trichomonads in peritoneal fluid. The living trichomonads were detected microscopically in a 48-h culture of a washing fluid with sterilized physiological saline solution from infected abdominal cavity in the treated mice.

An in vivo screening method for anticoccidial substances (Ikushima et al., 1980) Two-week-old White Leghorn chickens are used as a host. Samples are administered orally three times a day from day 0 to day 2. Chickens are infected orally with 30,000 sporulated oocysts of *Eimeria tenella* after the first medication at day 0. Mortality rate and weight gain of the treated, infected chickens are compared with those of nontreated, infected, and nontreated, uninfected controls at day 5. For the criteria of efficacy of the samples on coccidiosis, the gross cecal lesions produced by *E. tenella* are scored as follows: 0, no lesion; 1, a few discrete pinpoint lesions; 2, moderate cecal involvement; 3, marked cecal involvement, moderate hemorrhage; 4, maximal cecal involvement, massive hemorrhage, or death due to the infection.

An in vivo screening model (Tejmar-Kolar and Zähner, 1984) The entire development of the parasite *Eucoccidium dinophili*, which is a regular change of gametogony to sporogony, occurs extracellularly in the body of the host worm *Dinophilus gyrociliatus*. The various stages of the development can be observed under a microscope through the transparent cuticle of the worm. The cycle of development of the parasite is shown in Figure 4.2.

The host *D. gyrociliatus* is grown in seawater and fed with heat-killed green alga *Dunaliella*. The spores of *E. dinophili* are kept in seawater at 20°–24°C. The mixture of 50 μl of the worm suspension, 50 μl of the spore suspension, and 50 μl of a solution per hole in microtiter dishes is incubated at 27°C. The whole development of the parasite caused by massive infection was completed within 8–10 days after addition of the spores. The development of the parasite is observed under a microscope every second day and the progress of the infection is compared with the control (without antibiotic). The test is carried out in two stages. In the first stage, toxicity against the host is determined. In the second stage, the inhibitory effect on the parasite at one-fifth of the concentration of the toxic dose is determined.

4.3 Screening for Anthelmintic Antibiotics

General aspects

In vitro screening In the 1950s, several assay methods for nematocidal activity were reported (Santmyer, 1956; Tarjan, 1956; Taylor et al., 1957; Bishop, 1958), but they were not applied to the large-scale screening of nematocidal antibiotics. However, Tarjan (1956) recommended that a free-living nematode, *Panagrellus* sp., is suitable as an assay organism for nematocides because it is small in size (to about 1.4 mm long) and can be cultured easily.

In the 1960s, in vitro tests employing free-living plant parasitic nematodes were applied to the screening of nematocidal antibiotics. Mori (1961) reported an in vitro method using *Rhabditis* sp. and *Panagrellus redivivus*, which are grown in an Erlenmeyer flask for the screening of actinomycete

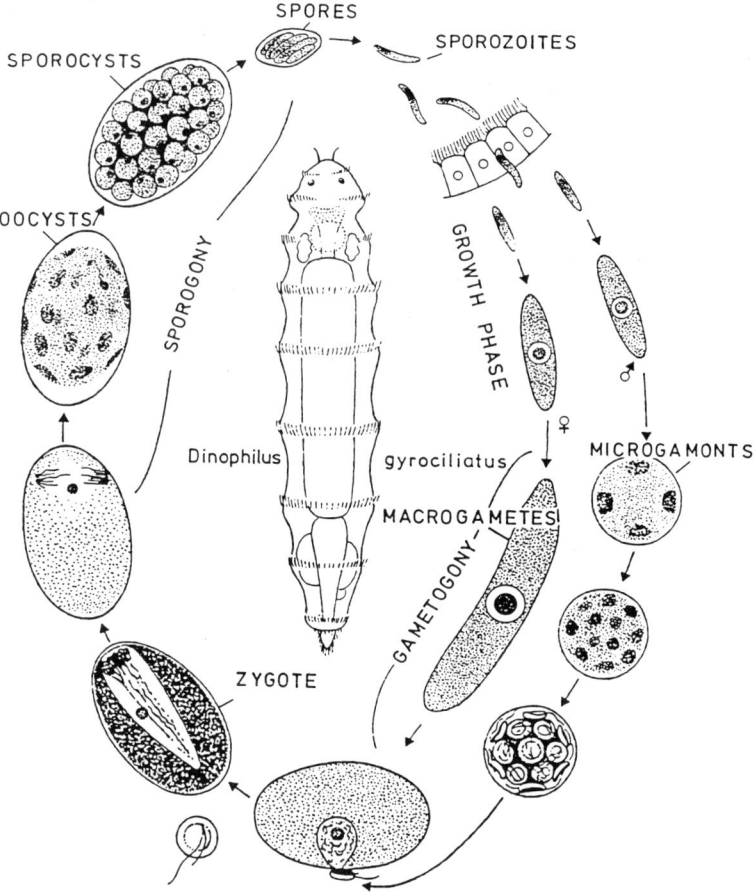

Figure 4.2 Developmental cycle of *Eucoccidium dinophili*. Reprinted from Tejmar-Kolar and Zähner, FEMS Microbiol Lett 24:21–24, 1984, by permission of Elsevier Science Publishers.

cultures. Through his screening work for nematocidal substances from about 2000 strains of actinomycetes, 4 strains were selected. One strain among them was found to be an aureothin producer. The substances produced by the other strains were not well characterized. Bacikova et al. (1965a, 1965b) reported an in vitro screening method using *Anguillula aceti* collected from beech woodshavings in a vinegar generator of a vinegar factory, and applied it to the evaluation of the nematocidal activity of many derivatives of isothiocyanate and other antibiotics. Nemec et al. (1969a, 1969b) and Balan et al. (1969) attempted screening for antivermal antibiotics from various microorganisms by Bacikova's method with some modifications, and found that fungi imperfecti, Oomycetes, and the genus *Penicillium* are rich sources of antivermal substances.

Cassienelli (1970) developed a screening program for the detection of anthelmintic substances produced by microorganisms from various natural sources and found that *Streptomyces michiganensis* produced the new compounds thaimycins A, B, and C, active against helminths (*Rhabditis macrocerca*) and protozoa (*Entamoeba histolytica* and *Trichomonas foetus*) in vitro. These compounds exhibited good therapeutic activity on infection by *Hymenolepis nana* in mice. Although the screening method has not been published, it is believed to be an in vitro method using *Rhabditis macrocerca* as test organism, because thaimycin production was followed by testing the in vitro anthelmintic activity against this organism.

Greenstein et al. (1987) discovered the LL-F28249 antibiotic complex (Figure 4.3) using the free-living nematode *Caenorhabditis elegans* as an antinematode screen. The compounds are novel macrocyclic lactones related to the avermectins described next.

Hood et al. (1989) discovered the additional related new compounds VM 44857, VM 44864, and VM 44866 (Figure 4.3) using the infective larval stage (L_3) of *Haemonchus contortus* as the test organism, which was prepared by culturing the feces from experimentally infected sheep. VM 44857 and VM 44866 were shown to be potent anthelmintics against mixed nematode infections in sheep.

In the screening work using *C. elegans*, Ondeyka et al. (1990) and Schaeffer et al. (1990) found that known fungal metabolites paraherquamides and cochlioquinone A, respectively, have nematocidal activities.

Recently, Ōmura and coworkers (Otoguro et al., 1988b) established a new rapid and simple method for nematocidal antibiotics, using the pinewood nematode *Bursaphelenchus lignicolus* described next, and used it for practical screening work. Broth filtrates of about 10,000 strains of actinomycete soil isolates were screened for nematocidal activity using this method. Consequently, two new antibiotics, jietacins A and B (see Figure 4.3), which are produced by *Streptomyces* sp. (Ōmura et al., 1987; Imamura et al., 1989) were discovered.

Several other new anthelmintic antibiotics, anthelmycin (Hamill and Hoehn, 1964), anthelvencins A and B (Probst et al., 1966), axenomycins A, B, and D (Bruna et al., 1973), aspiculamycin (Haneishi et al., 1974), and L-155,175 (Goetz et al., 1985) have been reported, but the screening methods used were not published.

In vivo screening In vivo screening systems are not usually used to screen a large number of microorganisms because they are troublesome and expensive. However, when a fermentation broth is directly administered to an animal, it can be tested simultaneously for efficacy and for toxicity. A fermentation broth which passes these criteria receives high priority for assignment to chemists to recover the active product as a pure material.

Considering the need for useful antiparasitic substances, Ōmura's group at The Kitasato Institutes in Japan and the Merck Sharp & Dohme Research Laboratories in the United States cooperated in screening for antiparasitic

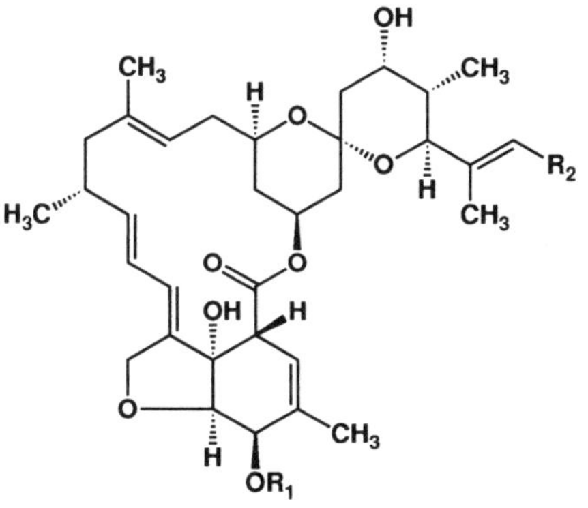

	R_1	R_2
LL-F28249 α	H	$CH(CH_3)_2$
LL-F28249 β	H	CH_3
LL-F28249 γ	CH_3	CH_3
LL-F28249 λ	CH_3	$CH(CH_3)_2$

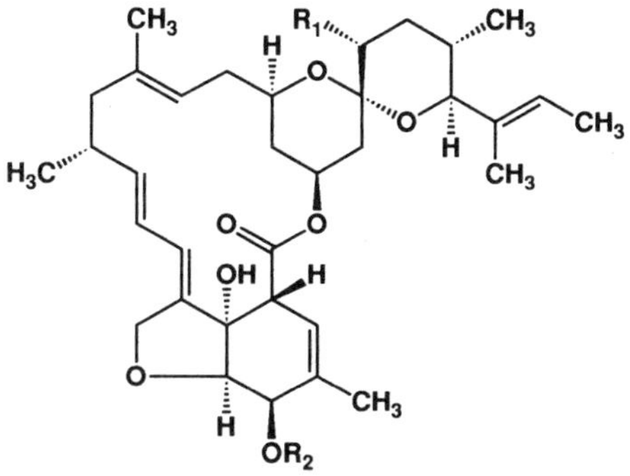

	R_1	R_2
VM 44857	H	H
VM 44864	OH	CH_3
VM 44866	OH	H

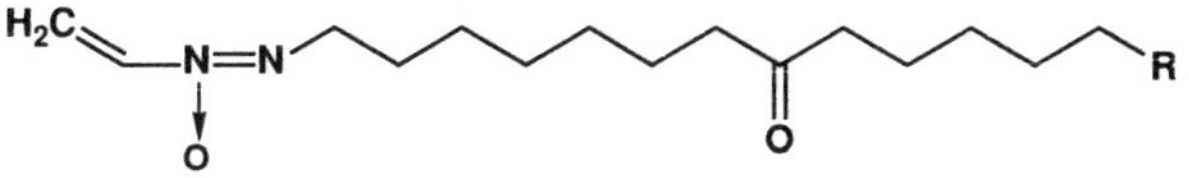

	R
Jietacin A	$CH(CH_3)_2$
Jietacin B	$CH_2CH(CH_3)_2$

Figure 4.3 Structures of anthelmintic antibiotics discovered with anthelmintic screening.

antibiotics, using not only some in vitro systems but also an in vivo system, and thereby discovered the new antibiotics named avermectins. Concerning the cooperative screening work, Stapley and Woodruff (1982) gave the following special mention. There were three important points. (1) Unusual soil microorganisms were selectively isolated from soil samples. In this way, the chance was increased that new substances could be discovered. (2) Cultures were simultaneously screened for several kinds of biological activities. (3) Because of the confidence that the choice of unusual soil microorganisms would yield a high frequency of detection of interesting substances, fermentation broths were tested directly in animals.

During the screening work, a fermentation broth was found to possess a potent nematocidal activity against *Nematospiroides dubius* grown in mice. The active principles were isolated and named avermectins (see Figure 4.3) (Burg et al., 1979; Fisher and Mrozik, 1984). The most active avermectin component, avermectin B_{1a}, has been modified chemically, yielding ivermectin

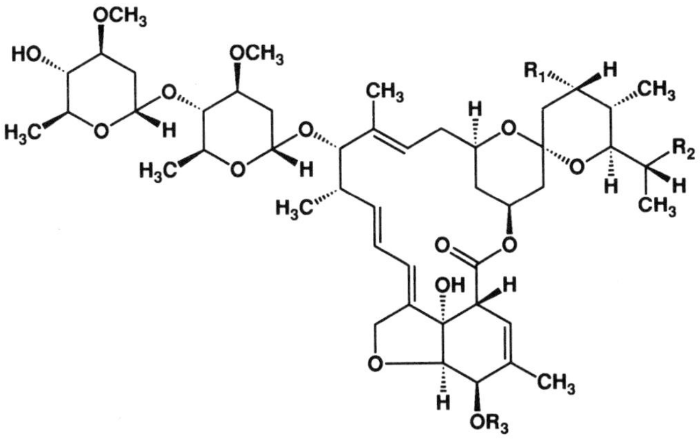

	R₁	R₂	R₃
Avermectin A₁ₐ		C₂H₅	CH₃
Avermectin A₁ᵦ		CH₃	CH₃
Avermectin A₂ₐ	OH	C₂H₅	CH₃
Avermectin A₂ᵦ	OH	CH₃	CH₃
Avermectin B₁ₐ		C₂H₅	H
Avermectin B₁ᵦ		CH₃	H
Avermectin B₂ₐ	OH	C₂H₅	H
Avermectin B₂ᵦ	OH	CH₃	H

Figure 4.3 *(cont.)*

(22,23-dihydroavermectin B_{1a}), which was developed and is used as a broad-spectrum antiparasitic agent for animals and also for humans. Ivermectin can eliminate multiple infection by nematodes and arthropodes when it is given to the host in single doses. It can relieve animals from such infection and humans from onchocerciasis (river blindness) and bancroftosis, which are major parasitic diseases in tropical and subtropical areas.

Typical examples of screening methods for anthelmintic antibiotics

A screening method using Bursaphelenchus lignicolus *(Otoguro et al., 1988b)* Kimura et al. (1981) devised a method to distinguish live nematodes from dead ones, which is based on the nature of living nematodes in that they penetrate through a certain type of Japanese paper. Otoguro et al. (1988b) successfully developed a more rapid and simple screening method for nematocidal antibiotics using the pinewood nematode *Bursaphelenchus lignicolus* and Kimura's method with modifications.

Figure 4.4 summarizes the method using *B. lignicolus* for in vitro nematocidal activity. The test organism *B. lignicolus* is grown for about 10 days on slants of *Botrytis cinerea* grown on potato-glucose agar and harvested from the slants by the Baermann funnel technique. After incubation of a nematode

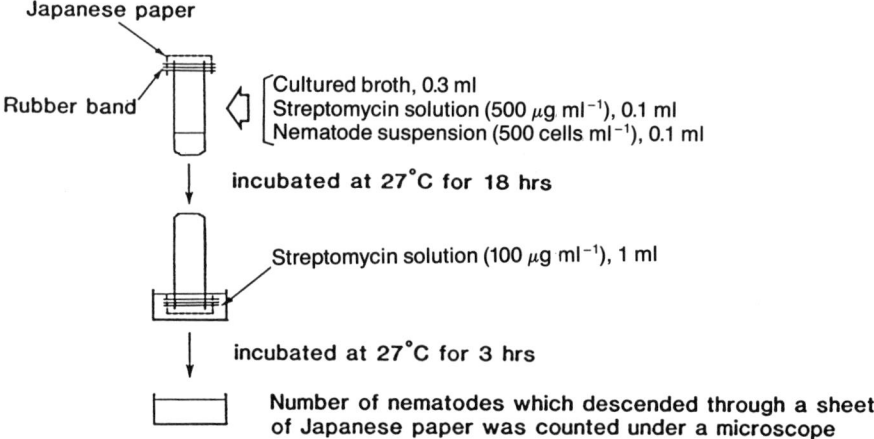

Figure 4.4 New screening method for nematocidal substances using *Bursaphelenchus lignicolus*. Reprinted from Otoguro et al., 1988b, by permission of Japan Antibiotic Research Association.

suspension with a cultured broth of an actinomycete, the number of nematodes in the well that penetrate through Japanese paper (Kokuyo typewriting paper, Tai-19) is counted under the microscope as living nematodes. The method is more simple than Kimura's assay because it uses a small test tube and a 24-well microplate and thus saves assay time; it takes about 24 h while Kimura's method needs 4 days.

An in vitro method using *Caenorhabditis elegans* (Pong et al., 1980)
Caenorhabditis elegans N_2 is grown on agar plates covered with a lawn of *Escherichia coli*. Nematodes are harvested from plates with *C. elegans*. Ringer's solution and 0.5 ml aliquots of the suspension (200–400 adult nematodes) are distributed on a Falcon 24-well plate. Sample solutions are added to each well. After 30 min of incubation at room temperature, mobility of the nematodes in each well is examined under a microscope. The percentages of the population paralyzed by the assay samples are determined.

An in vivo method using *Nematospiroides dubius* (Burg et al., 1979)
Male CD-1 white mice (15–20 g, Charles River) are infected by gavage with larvae of *N. dubius*. After infection, they are fed for 6 days with milled Purina Lab Chow in which the test material has been mixed. The mice are returned to a normal diet and, at 14 days post infection, examined for the production of eggs and the presence or absence of nematodes. The results of the assay are scored subjectively as fully active, moderately active, slightly active, and inactive.

4.4 Future Prospects

We have listed antiparasitic antibiotics discovered by screening methods using parasites in Table 4.1. Among them, only the avermectin complex is practically useful. The derivative ivermectin (22,23-dihydroavermectin B_{1a}) was developed and used as an anthelmintic chemotherapeutic against nematodes and

Table 4.2 Identified targets in different parasites for chemotherapy

Biochemical targets	Parasites	Effective drugs
Neuromuscular system		
Cholinergic agonists	Helminth	Bephenium, thenium
Choline antagonists	Helminth	Levamisole, pyrantel, morantel, methyridine
Acetylcholinesterase inhibitors	Helminth	Metrifonate, dichlorovos, haloxon
Muscle hyperpolarizers	Helminth	Piperazine
GABA agonist	Helminth and ectoparasite	Avermectins
Energy metabolism		
Inhibitor of glucose uptake	Helminth	Mebendazole, praziquantel, pyrvinium, styrylprridinium
Inhibitor of glycolysis	Helminth	Organoarsenics (stibophen, tartar emetic, etc.)
Inhibitor of glycogen metabolism	Helminth	Levamisole
Inhibitor of fumarate reductase	Helminth	Thiabendazole, mebendazole, levamisole
Uncoupler of phosphorylation	Helminth	Niclosamide, bithionol, disophenol
Inhibitor of mitochondrial electron transport	Coccidia	Metictorpindol, decoquinate
Inhibitor of pyruvate: ferredoxin oxidoreductase	Anaerobic protozoa	Metronidazole
Other mechanisms		
Inhibitor of protein kinase	Helminth	Suramin, levamisole
Inhibitor of folate metabolism	Helminth	Suramin
Inhibitor of tubulin polymerization	Helminth	Benzimidazole anthelmintics
Opsonizer of immune system	Helminth	Diethylcarbamazine
DNA binders	Helminth	Hycanthone
Inhibitor of dihydro-pteroate metabolism	Sporozoa	Sulfathiazole
Inhibitor of nuceloside phosphotransferase	Flagellated protozoa	Allopurinol, formycin B, thiopurinol riboside
Inhibitor purine phosphoribosyltransferase	Protozoa and trematode	Allopurinol
Inhibitor of ornithine decarboxylase	Protozoa	α-Difluoromethyl ornithine
Inhibitor of thiamine transport	Coccidia	Amprolium

arthropodes in animals and humans, as described. However, because it is inactive against protozoa and cestodes, antiparasitic drugs active against such parasites are still required.

Many chemically synthesized drugs (Table 4.2) have become available for the treatment of parasitic infections in animals and humans (Campbell and Rew, 1986; Sharma, 1987). However, as described in the Introduction, the rate of incidence of parasitic diseases in humans has not declined, particularly in the developing countries, although oncocerciasis and bancroftosis are expected to be conquered in the near future by the use of ivermectin. At present, because effective vaccines are not available for any of the parasitic infections, their control has to rely on chemotherapy. Thus, there is an urgent need for new antiparasitic antibiotics, and the development of new useful methods is very important in future screening work.

Fortunately, much information is now available about the metabolism of the parasites and the mode of action of various classes of antiparasites, as shown in Table 4.2 (Wang, 1984; Sharma, 1987). For example, Wang (1984) suggested that unique enzymes found only in the parasites, indispensable enzymes found in the parasites, and transporter, receptor, or cellular structural components with different pharmacological properties found in the parasites, all become potential targets for antiparasitic chemotherapy. Thus, the development of new screening methods using such enzyme systems and their use for practical screening work is expected

References

Bacikova, D., Nemec, P., Drobnica, L., Antos, K., Kristian, P. and Hulka, A. 1965a. Antiworm activity of some natural and synthetic compounds. I. Effect of aliphatic and mononuclear aromatic isothiocyanates on *Turbatrix aceti*. *Journal of Antibiotics* (Tokyo) Series A 18:162-170.

Bacikova, D., Betina, V. and Nemec, P. 1965b. Anthelminthic activity of antibiotics. *Nature* (London) 206:1371-1372.

Balan, J., Balanova, J., Nemec, P., Barathova, H. and Ulicna, V. 1969. Incidence of antiprotozoal and antivermal antibiotics in fungi. III. Genus *Penicillium*. *Journal of Antibiotics* (Tokyo) 22:355-357.

Balan, J., Ebringer, L., Nemec, P., Kovac, S. and Dobias, J. 1963. Antiprotozoal antibiotics. II. Isolation and characterization of trypacidin, a new antibiotic, active against *Trypanosoma cruzi* and *Toxoplasma gondii*. *Journal of Antibiotics* (Tokyo) Series A 16:157-160.

Bishop, D. 1958. A technique for screening antibiotics against eelworm. *Nematologica* 3:143-148.

Bossche, V. H., Thienpont, D. and Janssens, G. P. (editors). 1986. *Chemotherapy of Gastrointestinal Helminths*. Handbook of Experiment Pharmacology, p.77. Springer-Verlag, New York.

Bruce-Chwatt, L. J. 1978. Mass travel and imported diseases. *Annales de la Societe Belges de Medecine Tropicale* 58:77-88.

Bruna, C. D., Ricciardi, M. L. and Sanfilippo, A. 1973. Axenomycins, new cestocidal antibiotics. *Antimicrobial Agents and Chemotherapy* 3:708-710.

Burg, R. W., Miller, B. M., Baker, E. E., Birnbaum, J., Currie, S. A., Hartman, R., Kong, Y-L., Monaghan, R. L. Olson, G., Putter, I., Tunac, J. B., Wallick, H., Stapley, E. O., Ōiwa, R. and Ōmura, S. 1979. Avermectins, new family of potent anthelmintic

Chapter 4 Antiparasitic Agents 77

agents: Producing organism and fermentation. *Antimicrobial Agents and Chemotherapy* 15:361–367.

Campbell, W. C. and Rew, R. S. (editors). 1986. *Chemotherapy of Parasitic Diseases.* Plenum Press, New York.

Cassinelli, G., Cotta, E.D, Amico, G., Bruna, C. D., Grein, A., Mazzoleni, R., Ricciardi, M. L. and Tintinelli, R. 1970. Thaimycins, new anthelmintic and antiprotozoal antibiotics produced by *Streptomyces michiganensis* var. *amylolyticus* var. *nova. Archives of Mikrobiology* 70:197–210.

Diamond, L. S. 1957. The establishment of various trichomonads of animal and man in axenic cultures. *Journal of Parasitology* 43:488–490.

Edward, M. K., Marietta, V. and David, J. T. 1986. *Medical Parasitology*, 6th Ed., pp. 1–19. W. B. Saunders, Philadelphia.

Fisher, M. H. and Mrozik, H. 1984. The avermectin family of macrolide-like antibiotics. In: Ōmura, S. (editor), *Macrolide Antibiotics*, pp. 553–606. Academic Press, San Diego.

Fuska, J., Nemec, P. and Kuhr, I. 1972. Vermiculine, a new antiprotozoal antibiotic from *Penicillium vermiculatum. Journal of Antibiotics* (Tokyo) 25:208–211.

Goetz, M. A., McCormick, P. A., Monaghan, R. L., Ostlind, D. A., Hensens, O. D., Liesch, J. M. and Albers-Schonberg, G. 1985. L- 155,175: a new antiparasitic macrolide. Fermentation, isolation and structure. *Journal of Antibiotics* (Tokyo) 38:161–168.

Greenstein, M., Johnson, L., Lechevalier, M., Lechevalier, H. and Maiese, W. M. 1987. LL-F28249 antibiotic complex: A new family of antiparasitic macrocyclic lactones. I. Taxonomy and bioactivity. In: *Program and Abstracts of the 27th Interscience Conference on Antimicrobial Agents and Chemotherapy*, No. 996, pp. 270, Oct. 4–7, 1987, New York.

Hamill, R. L. and Hoehn, M. M. 1964. Anthelmycin, a new antibiotic with anthelmintic properties. *Journal of Antibiotics* (Tokyo) Series A 17:100–103.

Haneishi, T., Arai, M., Kitano, N. and Yamamoto, S. 1974. Aspiculamycin, a new cytosine nucleoside antibiotic. III. Biological activities, *in vitro* and *in vivo. Journal of Antibiotics* (Tokyo) 27:339–342.

Hirabayashi, A. 1959. Studies on the antiamoebic effect of protomycin, a new antibiotic isolated from the culture filtrate of a species of *Streptomycetes. Journal of Antibiotics* (Tokyo) Series A 12:298–309.

Hood, J. D., Banks, R. M., Brewer, M. D., Fish, J. P., Manger, B. R. and Poulton, M. E. 1989. A novel series of milbemycin antibiotics from *Streptomyces* strain E225. I. Discovery, fermentation and anthelmintic activity. *Journal of Antibiotics* (Tokyo) 42:1593–1598.

Ikushima, H., Okamoto, M., Tanaka, H., Ohe, O., Kohsaka, N., Aoki, H. and Imanaka, H. 1980. New anticoccidial antibiotics, WS-5995A and B. I. Isolation and characterization. *Journal of Antibiotics* (Tokyo) 33:1107–1113.

Imamura, N., Kuga, H., Otoguro, K., Tanaka, H. and Ōmura, S. 1989. Structures of jietacins. Unique α, β-unsaturated azoxy antibiotics. *Journal of Antibiotics* (Tokyo) 42:156–158.

Jomon, K., Kuroda, Y., Ajisaka, M. and Sakai, H. 1972. A new antibiotic, ikarugamycin. *Journal of Antibiotics* (Tokyo) 25:271–280.

Kettner, M., Nemec, P., Kovac, S. and Balanova, J. 1973. Dactylarin, a new antiprotozoal antibiotic from *Dactylaria lutea. Journal of Antibiotics* (Tokyo) 26:692–696.

Kimura, Y., Mori, M., Hyeon, S., Suzuki, A. and Mitsui, M. 1981. A rapid and simple method for assay of nematocidal activity and its application to measuring the activities of dicarboxylic esters. *Agricultural and Biological Chemistry* 45:249–251.

Mori, R. 1961. Studies on nematocidal antibiotics. I. Screening and isolation of nematocidal substances produced by actinomycetes. *Journal of Antibiotics* (Tokyo) Series A 18:162–170.

Nemec, P., Balan, J. and Ebringer, L. 1963. Antiprotozoal antibiotics. I. Method of

specific screening. *Journal of Antibiotics* (Tokyo) Series A 16:155–156.
Nemec, P., Krizkova, L., Balan, J., Balanova, J. and Kutkova, M. 1969a. Incidence of antiprotozoal and antivermal antibiotics in fungi. I. Class fungi imperfecti. *Journal of Antibiotics* (Tokyo) 22:345–350.
Nemec, P., Krizkova, L., Balan, J., Balanova, J. and Kutkova, M. 1969b. Incidence of antiprotozoal and antivermal antibiotics in fungi. II. Class Oomycetes. *Journal of Antibiotics* (Tokyo) 22:351–354.
Okami, Y., Utahara, R., Oyagi, H., Nakamura, S. and Umezawa, H. 1955. The screening of anti-toxoplasmic substance produced by *Streptomycete* and anti-toxoplasmic substance No. 534. *Journal of Antibiotics* (Tokyo) Series A 8:126–131.
Ōmura, S., Otoguro, K., Imamura, N., Kuga, H., Takahashi, Y., Masuma, R., Tanaka, Y., Tanaka, H., Su, X-H. and You, E-T. 1987. Jietacins A and B, new nematocidal antibiotics from a *Streptomyces* sp. Taxonomy, isolation, and physico-chemical and biological properties. *Journal of Antibiotics* (Tokyo) 40:623–629.
Ondeyka, J. G., Goegelman, R. T., Schaeffer, J. M., Kelemen, L. and Zitano, L. 1990. Novel antinematodal and antiparasitic agents from *Penicillium charlesii*. I. Fermentation, isolation and biological activity. *Journal of Antibiotics* (Tokyo) 43:1375–1379.
Otoguro, K., Ōiwa, R., Iwai, Y., Tanaka, H. and Ōmura, S. 1988a. Screening for new antitrichomonal substances of microbial origin and antitrichomonal activity of trichostatin A. *Journal of Antibiotics* (Tokyo) 41:461–468.
Otoguro, K., Liu, Z-H., Fukuda, K., Li, Y., Iwai, Y., Tanaka, H. and Ōmura, S. 1988b. Screening for new nematocidal substances of microbial origin by a new method using the pine wood nematode. *Journal of Antibiotics* (Tokyo) 41:573–575.
Pong, S. S., Wang, C. C. and Fritz, L. C. 1980. Studies on the mechanism of action of avermectin B: stimulation of release of γ-aminobutyric acid from brain synaptosomes. *Journal of Neurochemistry* 34:351–358.
Probst, G. W., Hoehn, M. M. and Woods, B. L. 1966. Anthelvencins, new antibiotics with anthelmintic properties. *Antimicrobial Agents and Chemotherapy* 1965:789–795.
Saburi, Y. 1954. A new method of antibiotic screening against protozoa (*Trichomonas vaginalis*). *Journal of Antibiotics* (Tokyo) Series A 7:127–131.
Santmyer, P. H. 1956. Studies on the metabolism of *Panagrellus redivivus*. *Proceedings of the Helminthological Society of Washington* 23:30–36.
Schaeffer, J. M., Frazier, E. G., Bergstrom, A. R., Williamson, J. M., Liesch, J. M. and Goetz, M. A. 1990. Cochlioquinone A, a nematocidal agent which competes for specific [^{3}H]ivermectin binding sites. *Journal of Antibiotics* (Tokyo) 43:1179–1182.
Sharma, S. 1987. Treatment of helminth diseases challenges and achievements. *Progress in Drug Research* 31:9–100.
Stapley, E. O. and Woodruff, H. B. 1982. Avermectins, antiparasitic lactones produced by *Streptomyces avermitilis* isolated from a soil in Japan. In: Umezawa, H., Demain, A. L., Hata, T. and Hutchinson, C. R. (editors), *Trends in Antibiotic Research*, pp. 154–170, Japan Antibiotic Research Association, Tokyo.
Tarjan, A. C. 1956. Evaluation of various nematodes for use in contact nematocide tests. *Proceedings of the Helminthological Society of Washington* 22:33–37.
Taylor, A. L., Fejdmesser, J. and Feber, W. A. 1957. A new technique for preliminary screening of nematocides. *Plant Disease Reporter* 41:527.
Tejmar-Kolar, L. and Zähner, H. 1984. Search for effective substances against parasitic protozoa: An attempt to develop a new screening model. *FEMS Microbiology Letter* 24:21–24.
Thiemann, J. E. and Beretta, G. 1967. Antiprotozoal antibiotics. *Ochromonas malhamensis* as test organism for the broth screening of anti-trichomonas agents. *Journal of Antibiotics* (Tokyo) 20:191–193.
Wang, C. C. 1984. Parasite enzymes as potential targets for antiparasitic chemotherapy. *Journal of Medicinal Chemistry* 27:1–7.
Zähner, F., Isenberg, H. D., Rosenfeld, M. H. and Schatz, A. 1953. The distribution of soil actinomycetes antagonistic to protozoa. *Journal of Parasitology* 39:33–37.

Part 2
Antitumor Substances

5

Antitumor Agents

Kanki Komiyama and Shinji Funayama

5.1 Introduction

Until now many antitumor substances have been discovered, including synthetic substances, natural products of higher plant origin, and antitumor antibiotics. These antitumor antibiotics have played an important role in the development of antitumor agents.

Although actinomycin D (**1**), mitomycin C (**2**), doxorubicin (**3**), bleomycin (**4**), etc., have been discovered (Figure 5.1) and are used clinically, these antibiotics possess undesirable side effects. Thus, novel antitumor antibiotics possessing a more specific activity or a new type of structure are always being looked for. Antimicrobial agents can discriminate between mammalian cells and bacteria while antitumor agents can only identify and differentiate tumor cells from normal cells on the basis of the rapidly dividing nature of tumor cells.

Antitumor antibiotics can be classified roughly into three groups: direct cytocidal substances, antimetabolites, and immunopotentiators. This chapter mainly describes screening methods used to discover antitumor antibiotics of the first two groups. To evaluate the antitumor effects of substances or to screen for antitumor agents, two screening systems are known, i.e., one using tumor cells and the other using other biomaterials. These screening systems also can be classified into three approaches: assaying antibacterial activity, cytocidal activity, and inhibitory effects using intact experimental animals.

Various methods using mammalian cells and bacteria to evaluate antitumor activity in vitro have been reported, and some of these are listed in Tables 5.1 and 5.2. However, each of these screening methods possesses deficiencies when applied to the screening and purification procedures for antitumor antibiotics.

To search for novel antitumor antibiotics, it is desirable with the initial screening method to obtain in vivo antitumor activity data as well as cytocidal activity results against tumor cells in vitro. This helps in distinguishing simple toxins from antitumor substances. In the in vivo evaluation of a sample, tumor-bearing animals are generally given test materials (cultured broth or

Mitomycin C (2)

Doxorubicin (3)

Actinomycin D (1)

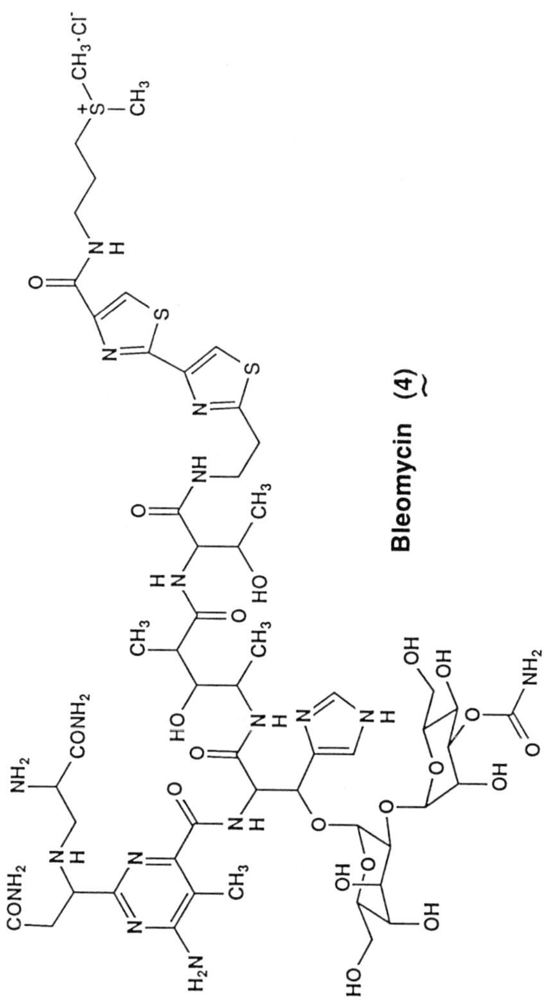

Bleomycin (4)

Figure 5.1 Clinically effective antitumor antibiotics (see text for discussion to bolded numbers).

Table 5.1. Screening systems used to search for antitumor compounds of microbial origin using tumor cells

Screening System	Materials	Evaluation	Reference
Morphological change	Giemsa, HeLa cells	Microscopic observation	Hata, 1971
Dye staining	Neutral red, L1210 cells	Spectrophotometry	Cantino et al., 1986
	KB cell		Oku et al., 1982
	MTT, Human tumor cell line	Microculture plate reader	Alley et al., 1988
	Giemsa, B16	Microculture plate reader	Mirabelli et al., 1985
	Mouse melanoma cells and HT-29 human colon carcinoma cells		
	Methylene blue human colon carcinoma and melanoma	Microculture plate reader	Finlay et al., 1984
	Crystal violet, human and mouse tumor	Visual examination of stain pattern	Cantino et al., 1985
HTCA	Human tumor cells and agar medium	Number of colonies formed	Salmon et al., 1978 Shoemaker, 1986
Uptake of radioactive substance	[^{14}C]-adenine into EAC/P388	Radioactivity	Fuska et al., 1975
Colony formation	Agar medium and paper disk	Diameter of inhibitory zone	Sato et al., 1967
Inhibition of focus formation of chicken fibroblasts by Rous sarcoma virus	Chicken fibroblast and Rous sarcoma virus	Number of foci	Takeuchi et al., 1984
Drug-resistant cell	P388/Adriamycin	Count cell number	Uchida et al., 1985
Conversion of transformed morphology to normal morphology	Rous sarcoma virus infected rat kidney cells	Morphological observation	Uehara et al., 1985
Cell and bacteria	KB, L-1210, *E. coli*, *Bacellus subtilis*	Cytotoxicity and antimicrobial activity	Hanka et al., 1978

Table 5.2 Screening systems used to search for antitumor compounds of microbial origin using bacteria

Screening System	Materials	Evaluation	Reference
Antimicrobial activity	Macromolecule permeable mutant of *E. coli*	OD at 660 nm paper disk	Tsunakawa et al., 1985
Biochemical induction assay	Prophage	Colorimetric assay	Elespuru and Yarmolinsky, 1979
Bacterial mutant	Yeast mutant with distorted cell membranes	Growth inhibition in agar plate	Gause et al., 1976
	Saccharomyces cerevisiae FL599–1B	Growth inhibition	Gause et al., 1976
	E. coli mutant with altered cytoplasmic membrane and increased permeability	Growth inhibition	Gause, 1975
Stimulation of bioluminescence	*Photobacterium leiognathi*	Paper disk growth inhibition bioluminescent response	Steinberg et al., 1985
Prescreening using genetically modified *E. coli*	*E. coli* BR513	Growth inhibition	Bartus et al., 1984
Antiphage	RNA phage f2; *E. coli* K-12	Growth zone by phage disk	Koenuma et al., 1974

crude extract) intraperitoneally (i.p.). Thereafter the efficacy of the sample is evaluated by the number of survival days of the experimental animals or the growth rates of the tumors.

After obtaining these results, purification of the antitumor component(s) can proceed by checking the cytocidal activity alone. To check the cytotoxicity of the test material(s), tumor cells or bacteria, etc., growing in vitro are exposed to the samples and incubated for a certain period; then, inhibition of their growth rate is evaluated.

To discover an enzyme inhibitor, a cell-free system is usually used. In many cases, a system relating to nucleic acid synthesis has been used. Even when specific enzymes of tumor cells have not been discovered, toxic substance(s) can sometimes be purified. Therefore, before the purification of active substances using this method, it is desirable to check the antitumor activity and toxicity of the fermentation components in vivo before the isolation of an active principle is started.

After the initial screening, the stability of the active component in the fermentation broth to high and low pH, high temperature, light, etc. must be checked. Following this the usual techniques applied to the isolation of natural products are used for the purification of the active component(s).

The test methods for the initial screening must be simple, rapid, reproducible, quantitative, and inexpensive. A test system using laboratory animals take a relatively long time to obtain results and requires a large amount of the test material. In contrast, time and money will be saved dramatically by using test systems employing cultured cells (in vitro) or cell-free systems (using enzymatic reactions) for the purification of active substance(s). So, if it is demonstrated that antitumor activity observed in vivo parallels that seen in vitro, one should apply in vitro activity tests for subsequent purification of the active component.

It is essential to check the novelty of the active substance to be isolated, even during the purification process, by observing its specific activity against tumor cells, color reactions, mobility on TLC, etc. If new substances are discovered in the fermentation broth, these compounds can be further evaluated using the experimental tumor systems listed in Table 5.1. After obtaining these results, the agent may be evaluated toxicologically for further evaluation leading to clinical trials.

5.2 Screening Methods

Animal tumor systems The screening methods using experimental animals (in vivo methods) are classified as shown in Figure 5.2. Among these, a transplantable tumor is commonly used as a tool for initial evaluation (Goldin, 1967; Geran et al., 1972) because evaluating the antitumor activity of

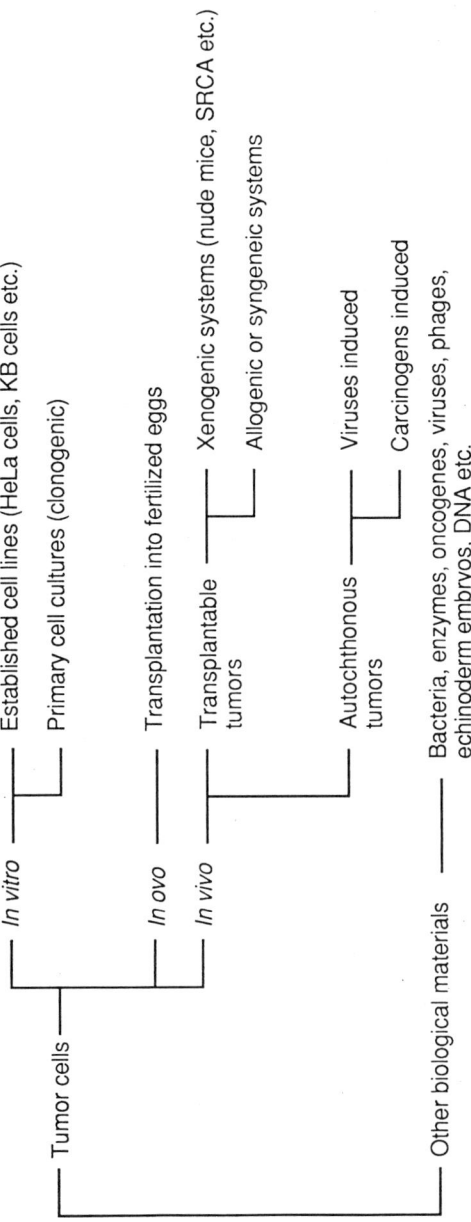

Figure 5.2 Screening systems used to search for antitumor compounds of microbial origin.

test sample(s) on an autochthonous tumor under the same experimental condition is quite difficult.

Transplantable tumors Because of the variety of available tumor strains and their ease of usage, mouse tumors are used by many researchers for the initial in vivo evaluation of antitumor substances. There are three principal kinds of transplantable tumor systems: xenogeneic, allogeneic, and syngeneic tumors. The selection of the tumor strain is important for screening in vivo. For the initial test, allogeneic or syngeneic tumors are usually used. In the initial screening, it is also essential to use a sensitive tumor because then it becomes possible to detect even a small amount of active substance(s) in the fermentation broth. Because it is impossible to recognize all the useful antitumor drugs by using a single tumor system, it is desirable to use a large battery of tumor systems. However, we prefer to increase the number of test samples rather than to increase the number of screening systems employed. So, in general, one to three kinds of sensitive tumor are used in the first screening system.

The sensitivity of a tumor to an antitumor agent is varied, according to the tumor line, site of transplanted tumor, tumor growth, number of tumor cells implanted, treatment schedule (number of treatments, period of treatment), treatment route [intravenous (i.v.), i.p., orally (p.o.)], etc. Among various experimental conditions, i.p. inoculation of the tumor followed by i.p. injection of the toxic agent is usually the most sensitive. Sarcoma 180, carcinoma 755, Ehrlich carcinoma and P388 leukemia are examples of transplantable tumors with relatively high sensitivity, and these tumor cell lines are generally used for the screening of antitumor antibiotics.

Tumor transplantation For ascites tumor transplantation, donor tumor cells are transplanted on the day most suitable for the tumor transplantation (when the tumor-bearing host is relatively healthful, and the accumulation of ascitic fluid is sufficient). The fluid is withdrawn from the tumor-bearing host using a syringe that has been kept in a sterile tube containing heparin and 10-to 100-fold (vol/vol) Hank's balanced salt solution (HBSS). A quantity of the ascitic fluid is placed in a small test tube containing 0.5% trypan blue solution, and the cell number is counted. The ascites fluid kept in an ice bath is diluted with HBSS to give a desired inoculum concentration of tumor cells (1×10^7 cells ml^{-1}). The diluted ascites fluid is injected i.p. (0.1 ml per mouse).

Before a solid tumor transplantation, the donor tumor is excised and necrotic material is removed. The tumor tissue is minced with scissors and filtered using 80 mesh stainless steel net and a spatula. The tumor cells are washed with HBSS and finally diluted to a 10% (vol/vol) cell suspension. The inoculum volume of tumor suspension is usually 0.1 ml per animal, and tumors are inoculated subcutaneously (s.c.) in the axillary region.

Treatment   Test materials are usually given on day 1, on days 1, 5, and 9, or 1–9 days after tumor inoculation. To determine the morphological change of tumor cells in an ascitic tumor, test samples are given when ascitic tumor cells accumulate in the abdominal cavity (usually on day 4).

Test evaluation   To evaluate the antitumor activities of the test samples on ascitic tumor when using the in vivo screening method, the following three methods are applicable.

   1. The evaluation of survival effect is calculated from the equation as follows:

$$\text{increase in life span (\%)} = (T/C - 1) \times 100$$

where T is the mean or median survival days (MSD) of the treated group and C is the MSD of the control group.

   2. In the case of ascitic tumors, at certain days after treatment the animal is sacrificed, all ascites cells are collected into a centrifugation tube, and the growth inhibitory ratio is calculated by measuring the total packed cell volume.

   3. To observe morphological changes in ascitic tumor cells, a mouse is given an i.p. injection with the test sample on the day when the ascitic fluid is inoculated. One or 2 days later, a small amount of ascitic fluid is withdrawn with a syringe and a small drop of the fluid is placed on a microscope slide to make a smear. After the smear is fixed with methanol and stained with Giemsa solution, morphological changes of the tumor cells are examined using a light microscope. Compared with the evaluation method by ILS, this assay system is more rapid and more reproducible in determining the activity of a test sample. In addition, the number of experimental animals can be reduced because usually two animals (in one dose) are sufficient for the evaluation, and also the toxicity of the sample can be roughly elucidated.

   By way of comparison, to evaluate test samples using solid tumors, the following three methods are commonly used.

   1. Observing the survival days and mean or median survival days as compared with that of the control. The ILS is calculated as in the case with the ascitic tumor.

   2. At appropriate days after the last injection of the test sample, the tumor is excised and weighed. The mean tumor weight is calculated and compared with that of the control group.

   3. The longest diameter (length) and shortest diameter (width) of the tumor are measured with vernier calipers, and the tumor weight can be estimated according to the following formula:

$$\text{Tumor weight (mg)} = \text{length (mm)} \times [\text{width (mm)}]^2 / 2 \text{ (mg)}$$

Autochthonous tumors   There are many methods to induce autochthonous tumors using chemical carcinogens and oncoviruses. However, these methods are not suitable for the primary screening of antitumor antibiotics.

An exception is the Friend virus, which has sometimes been employed in chemotherapeutic studies. The Friend virus rapidly produces malignant prolongation in reticulum cells involving the spleen and liver and a marked polycythemia is also induced (Sugiura, 1959: Mirand, 1966: Sidwell et al., 1969).

Animals and viruses Mice (ICR, DBA) are inoculated i.v. with a cell-free spleen suspension containing Friend leukemia virus. Two weeks after the virus inoculation, the mice are sacrificed, their spleens are minced with scissors and suspended in ice-cold HBSS to make a 10% homogenate. The homogenate is centrifuged at $3000 \times g$ for 15 min. The supernatant (virus) is stored at $-80°C$ until used.

Virus inoculation and treatment Mice are inoculated i.v. with 0.1 ml of the virus solution. Several days after the inoculation, the mice are treated with a test sample, singly, intermittently, or successively.

Test evaluation For the evaluation of drug effectiveness, mice are killed several weeks after virus inoculation, and the spleen and plasma are removed. The spleens are weighed and the reduction in splenomegaly is calculated. Also, blood hematocrit values are observed. Though effective agents against tumor viruses in vivo might be selected by this method, compounds that possess selective toxicity on spleen cells are also selected.

Cell culture systems

Cell lines The variety of established cell lines available have increased rapidly in the last few years. It would seem most desirable to use human tumor cells rather than animal cells for the screening of antitumor antibiotics.

Cell culture media A large variety of cell culture media are commercially available, and usually minimal essential medium (MEM) or RPMI 1640 supplemented with 10% calf serum, penicillin (100 U ml^{-1}), and streptomycin (100 µg ml^{-1}) are used.

Evaluation

Morphological changes A sample is added to tumor cells attached to the surface of a culture plate, which is then incubated. Thereafter, cells are fixed and stained with dye, and morphological changes are observed. For example, after trypsinization of HeLa cells to make a single suspension in MEM supplemented with 10% calf serum, 5×10^3 of the cells in 0.2 ml are plated into a 96-well microtiter plate. One day after cultivation at 37°C under a 5% CO_2:95% air atmosphere, 5 µl of each sample solution is added to each well and the plates are reincubated for 2–3 days. The medium is removed by decantation, and the microtiter plate is dipped into 100% MeOH for more than 5 min to fix the cells. After the HeLa cells are washed with water, they are stained with Giemsa solution, washed with water, and examined for morphological changes under the light microscope before being dried. Because some antitumor antibiotics cause specific morphological change(s), it is sometimes useful to compare the active substance(s) with known antitumor antibiotic(s).

Several staining methods to detect cytotoxicity in vitro have been devised, including neutral red, Giemsa, crystal violet, and MTT (Tetrazolium).

Determination of dye uptake (neutral red) (Oku et al., 1982, Cantino et al., 1986) After the incubation of cells and addition of samples, as just described, the culture medium is removed by decantation and the cells are washed once with HBSS. Each well is treated with 100 μl of 0.04% neutral red stain in HBSS and incubated for 20 min at 37°C in 5% CO_2 (balance, air). Each well is washed once with 0.2 ml cold 0.9% saline, and the cells are lysed with a mixture of 0.2 ml of 0.1 M acetic acid:ethanol (1:1,vol/vol) and mixed well. Optical densities are then determined at 540 nm on a microtiter plate reader.

Determination of dye staining (Giemsa) (Mirabelli et al., 1985) Following the appropriate incubation period, culture supernatants are aspirated and the cells washed once with 0.2 ml of HBSS which is aspirated. About 0.2 ml of absolute methanol is added for fixation and allowed to stand for more than 5 min. Following staining with Giemsa solution, each well is washed three times with 0.2 ml of water to remove unbound stain. Stained cells are solubilized with 0.2 ml of 0.1 N HCl. Plates are placed in a microtiter plate shaker apparatus for 20 min to allow the stain to dissolve into the HCl. Absorbances are read by a microtiter plate reader at a wavelength of 600 nm (reference wavelength at 405 nm).

Determination of dye uptake (MTT method) (Alley et al., 1988) The original MTT colorimetric assay was described by Mosmann (1983). In principle, the viable cell number per well is directly proportional to the production of formazan from MTT.

After cells are exposed to the test material for an appropriate incubation period as described above, 20 μl of MTT stock solution (5 mg MTT/PBS, 1 ml) is added to each well and the plate is incubated at 37°C for 4 h. After aspiration of the medium, 100 μl of 0.04 N HCl–isopropanol is added to each well and mixed. After 30–50 min at room temperature, the plate is read on a microplate reader, using a test wavelengh of 570 nm (reference wavelength at 630 nm).

Bacterial systems

Antibacterial activity in completely synthetic media Several microorganisms are used for the screening of antitumor agents. Among them, the antimicrobial activity test using *Bacillus subtilis* is a very sensitive system, in particular when the bacteria are cultivated in completely synthetic media. This system is suitable for detecting antimetabolites of purines-pyrimidines and of amino acids.

Preparation of spores and media   Bacillus subtilis is plated on a heart infusion agar medium (150 ml 500 ml^{-1} Roux flask), incubated at 37°C for 48 h, and allowed to stand at room temperature. After 10–30 days, 100 ml of sterilized water is added to the flask to make a spore suspension. The suspension is heated at 55–60°C for 30 min and stored in a refrigerator. This spore suspension can be used for 1 year.

Test method and evaluation   One milliliter of a spore suspension of B. subtilis is mixed with 100 ml of Davis synthetic medium at 42–43°C, and the mixture is immediately plated on a petri dish. After this agar medium is prepared, a paper disk containing the test sample is put on the plate. One day after incubation at 37°C, the inhibitory zone of bacterial growth is measured.

In vitro screening methods for antitumor or antitumorigenic substances involving the transformation of chick embryo fibroblasts infected with Rous sarcoma virus

Screening method using Rous sarcoma virus   Virus-induced Rous sarcoma in chickens has been proposed as a model for experimental therapy of tumors (Pienta et al., 1963). This method can readily discriminate specific inhibitors from nonspecific and toxic substances by determining the inhibitory effect of a test compound on (a) focus formation on a plate containing a layer of chick embryo fibroblasts infected with RSV and on (b) total cell proteins using the same plate. The result (a) represents the effect on tumorigenic processes as well as on tumor cell growth and also (b) nonspecific toxic effects on normal cells because the population of virus-induced tumor cells is relatively small.

Test method and evaluation

Materials   Rous sarcoma virus (Schmidt-Ruppen strain type A) and eggs of a White Leghorn line S (leukemia, Mareck disease-free) are used.

Media   Eagles MEM supplemented with 10% tryptose phosphate broth, 5% calf serum, 0.5% chick serum, and 60 mg liter^{-1} kanamycin.

Culture conditions   A secondary culture of chick embryo fibroblasts (1 × 10^6 cells 4 ml^{-1} medium) is infected with 200 focus-forming units of SR-ASV-A and plated into a 60-mm petri dish. After 20 h incubation (37°C, 5% CO$_2$), the culture medium is removed and 5 ml of medium containing 0.75% Bacto agar is added to form an overlayer of gelled medium. A test sample is added on the top of the gelled layer within 24 h after the cell seeding.

Evaluation   The number of foci and the amount of protein in a dish are determined by a colony counter and Lowry's and Oyama's method after day 8 of cultivation.

Screening methods for antitumor substances using cell-free systems
An effective antitumor antibiotic is one of the most exciting and elusive targets for researchers in the microbial chemistry field. There are roughly three methods used in screening for antitumor antibiotics. One is in vivo screening method against experimental animal tumors such as Yoshida sarcoma, Ehrlich carcinoma, etc. The second is a cell cultivation method with observation of morphological changes or cytotoxic action against tumor cells such as HeLa S3, P388 leukemia, L1210, or KB in culture or in disk plate agar. The third method is the screening type of method using cell-free systems.

Until now, the several antitumor antibiotics used for cancer chemotherapy, such as actinomycin D (**1**), mitomycin C (**2**), doxorubicin (**3**), and bleomycin (**4**) (see Figure 5.1) have been discovered through the screening of microbial culture filtrates for inhibitory activity against experimental animal tumors in vitro or in vivo, as described. However, one frequently encounters very important, serious problems. One is that there is no direct relationship between a particular type of human tumor and such experimental animal tumors, and another is that almost all antitumor antibiotics found through these screening methods exhibit cytotoxic action.

If we assume that a specific enzyme plays an important role in the cell division of cancer cells, it might be worthwhile to screen microbial culture filtrates for inhibitory activity against such a enzyme.

Using this point of view, inhibitors against enzymes that affect cell division or cell growth have been investigated. However, as few effective antibiotics have been discovered with cell-free systems, only the name of the antibiotics and the screening system were listed in Table 5.3. Among effective antitumor agents, some show inhibitory activity agaisnt DNA topoisomerases I and II, which alter DNA conformation through a concerted breaking and rejoining of DNA strand. From this point of view, DNA topoisomerase II has been considered as the primary cellular target for a screening of antitumor agent. Recently, saintopin was discovered in the fermentation broth of *Paecilomyces* (Yamashita et al., 1990).

In addition to the screening systems described here, some of the antitumor antibiotics obtained through the study of cell-free systems are shown in Chapter 3.

5.3   Recently Discovered Antitumor Antibiotics

Novel antitumor antibiotics from microbial sources have increasingly been found by means of the various screening methods described in this chapter. Table 5.4 summarizes some of those isolated between 1984 and 1990.

Table 5.3. Screening system used to search for antitumor compounds of microbial origin using cell-free system

Target enzyme	Inhibitor	Reference
Aminopeptidase B	Bestatin	Ishizuka et al., 1980
Alkaline phosphatase	Forphenicinol	Ishizuka et al., 1982
Tyrosine-specific protein kinase	Erbstatin	Umezawa et al., 1986
	Lavendustin A	Onoda et al., 1989
Adenosine deaminase	2'-Deoxycoformycin	LePage et al., 1976
	Adecypenol	Tanaka et al., 1989
Protein synthesis	Phenomycin	Nakamura et al., 1967
Glyoxalase	MS-3	Kurasawa et al., 1975a,b
	Glyo-II	Takeuchi et al., 1975
DNA toposomerase II	Saintopin	Yamashita et al., 1990
Phosphatidyl inositol turnover	Inostamycin	Imoto et al., 1990

5.4 Concluding Remarks

The primary screening system for antitumor substances of microbial origin must be sensitive, reproducible, inexpensive, and simple, so that (a) one can detect the existence of even a small amount of a valuable active component(s); (b) one can trust the data; and (c) one can manage large numbers of samples within a relatively short time. For example, at present the use of a human tumor transplanted into a nude mouse is considered one of the best screening systems to use predicting clinically useful antitumor activity. However, we can only use this system in case of a few samples that have been selected from a great number of candidates.

There are mainly three different kinds of screening methods: in vivo screening against experimental animal tumors, cell cultivation methods in culture or on disc agar plate (in vitro), and assays for inhibitors against enzymes that affect cell division or cell growth.

By useing experimental animals in screening for antitumor antibiotics, we can select the microbes that produce a compound(s) with promising activity in increasing the life span of tumor-bearing experimental animals. However, adoption of this screening system is not always suitable for primary screening because it requires large amounts of sample and longer times to obtain results. Screening systems using tumor cell lines (in vitro) are superior to the in vivo screening system in that they require much smaller amounts of test material and much shorter times for their evaluation. However, these systems possess several weak points. One is that, in general, a cytotoxic substance and an antitumor substance cannot be distinguished. The third screening method was established on the basis that a specific enzyme may play an important role in the cell division of cancer cells. Although some of them do not seem to be suitable for primary screening, one of the merits of screening for inhibitors of enzymes is that a compound possessing a specific biological activity will be chosen.

Table 5.4 Recently isolated novel antitumor antibiotics

Name	Producing organism	Activity	References
Akrobomycin	*Actinomadura roseoviolacea*	P388	Imamura et al., 1984
Chicamycin	*Streptomyces albus*	P388, S180	Konishi et al., 1984
CI920	*S. pulveraceus* sp. *fostreus*	L1210 and P388 leukemia	Fry et al., 1984
FR-900405	*Actinomadula pulveracea*	P388, L1210, B16, P815 mastocytoma	Kiyoto et al., 1984
FR-900406			
Awamycin	*Streptomyces* sp. No. 80-217	S180, IMC	Funayama et al., 1985a
Capoamycin	*S. capoamus*	Ehrlich	Hayakawa et al., 1985a
Kazusamycins	*Streptomyces* sp. No. 81-484	P388	Komiyama et al., 1985
			Funaishi et al., 1987
			Yoshida et al., 1987
Kerriamycin	*S. violaceolatus*	Ehrlich	Hayakawa et al., 1985b
Lactoquinomycin	*S. tanashiensis*	Ehrlich	Okabe et al., 1985
Sohbumycin	*Streptomyces* sp. No. 82-85	HeLa S3[1]	Umezawa et al., 1985a
Trienomycins	*Streptomyces* sp. No. 83-16	P388, S180 Ehlich	Umezawa et al., 1985b
			Funayama et al., 1985b
			Komiyama et al., 1987
			Funayama et al., 1988
			Nomoto et al., 1989
			Smith et al., 1990
Guanine-7-oxide	*Streptomyces* sp. No. 3780	P388	Nishii et al., 1985
7-Hydroxyguanine	*S. purpurascens*	L1210	Kitahara et al., 1985
Chromoxymycin	*S. libani*	P388, B16	Hori et al., 1986
DC-86-M	*S. luteogriseus*	Sarcoma 180	Takahashi et al., 1986
Elsamicins	Unidentified actinomycete	P388, L1210, B16	Konishi et al., 1986
PD124,895	*Streptomyces* sp.	B16, P388, L1210	Hurley et al., 1986
PD116,152	*Streptomyces* sp.	P388	Smitka et al., 1986
Rhizoxin	*Rhizopus*	Human and murine tumor cells[1]	Tsuruo et al., 1986
Valanimycin	*S. viridifaciens*	L1210, Ehrlich	Yamato et al., 1986
Arugomycin	*S. violaceochromogenes*	S180, Ehrlich, P388	Shimosaka et al., 1987

Continued next page

Table 5.4 (cont.)

Name	Producing organism	Activity	References
Azinomycins	S. griseofuscus	L5178Y, P388, LLC, P815, B16, Meth A, Ehrlich	Ishizeki et al., 1987
FR-900,482	S. sandaensis	P388, L1210, B16, MM46, Ehrlich, EL-4	Shimomura et al., 1987
Hydroxychlorothricin	Streptomyces sp. K818	Ehrlich	Yamamoto et al., 1987
Rebeccamycin	Nocardia (Saccharothrix)	P388, L1210, B16	Bush et al., 1987
WF3405	Amauroascus aureus F3405	P388, LLC, L1210, M5076 reticulo sarcoma	Kiyoto et al., 1987
BMY-28438 (3,7-dihydroxytroporone)	S. tropolofaciens No. K611-97	B16, P388	Sugawara et al., 1988
C-1027	S. globisporus C-1027	L1210, P388	Hu et al., 1988
			Otani et al., 1988
			Zhen et al., 1989
DC89-A1	Streptomyces sp.	P388, S180 solid tumor	Ichimura et al., 1988
DC92-B	Actinomadura sp.	P388, S180 solid tumor	Takahashi et al., 1988b
Duocarmycin A	Streptomyces sp. DO-88	S180 solid tumor	Takahashi et al., 1988c
Glidobactins	Polyangium brachysporum sp. nov. No. K481-B101	P388, L1210	Ogawa et al., 1989
			Oka et al., 1988
Pyrindamycins	Streptomyces sp. SF2582	P388, P388/ADM[1]	Ohba et al., 1988
Sibanomycin	Micromonospora sp. SF2364	P388	Itoh et al., 1988
Thrazarine	S. coerulescens MH 802-fF5	Human and mouse tumor cell line[1]	Takahashi et al., 1988a
Ankinomycin	Streptomyces sp. SF2587	P388, P388/ADM, L1210, EL-4, B16, M-5076	Sato et al., 989
			Ishii et al., 1989
AT-2433A1, -A2, -B1, -B2	Actinomadura melliaura	P388	Matson et al., 1989b
Calicheamicins	Micromonospora echinospora sp. calichensis	P388, B16	Maiese et al., 1989
			Lee et al., 1989
Chrysomycin M, V	S. albaduncus C38291	P388, L1210, B16	Matson et al., 1989a
FR-900462	S. tokashikiensis No. 7124	P388, B16	Iwami et al., 1989a
			Iwami et al., 1989b

Table 5.4 (cont.)

Name	Producing organism	Activity	References
FR-900840	*Streptomyces* No. 8727	P388, L-1210, B16	Nishimura et al., 1989a Nishimura et al., 1989b Nishimura et al., 1989c
FR-66979	*S. sandaensis* No. 6897	P388, B16, BHK-21	Terano et al., 1989
Lavanducyanin	*Streptomyces* sp. CL190	P388, L1210	Imai et al., 1989
Leinamycin (DC-107)	*Streptomyces* sp. S-140	P388, S180	Hara et al., 1989
Myrocin C	*Myrothecium verrucaria* No.55	Ehrlich	Nakagawa et al., 1989 Hsu et al., 1989
OH-1049R	*Streptomyces* sp. No. 1049	S180	Komiyama et al., 1989 Funayama et al., 1989b
Phenazinomycin	*Streptomyces* sp. WK-2057	S180, IMC carcinoma	Ōmura et al., 1989 Funayama et al., 1989a
Phospholine	*S. hygroscopicus*	P388, L1210, EL-4[1]	Ozaka et al., 1989a Ozaka et al., 1989b
SAGP	*S. pyrogenes*	Ehrlich, Meth A, S180	Yoshida et al., 1989
Sandramycin	*Nocardioides* sp. (ATCC39419)	P388	Matson and Bush, 1989
DC-92B, DC-92D	*Actinomadura* sp. DO92	P388, S180	Yasuzawa et al., 1990
Duocarmycin SA	*Streptomyces* sp. DO113	P388	Ichimura et al., 1990
Himastatin	*S. hygroscopicus* C39108-P210-51	P388, B16	Lam et al., 1990 Leet et al., 1990
Kapurimycins	*Streptomyces* sp. DO-115	P388	Hara et al., 1990 Yoshida et al., 1990
LL-D49194	*S. vinaceus-drappus*	P388, B16	Maiese et al., 1990 Suzuki et al., 1990
Resorthiomycin	*S. collinus*	L5178[1]	Tahara et al., 1990a Tahara et al., 1990b

[1]In vitro activity.

Because each of the screening systems discussed possesses merit(s) and demerit(s), it is difficult to select the screening system most suitable for each screening goal. A combination of screening systems may be essential in the search for antitumor antibiotics of microbial origin.

References

Alley, M. C., Scudiero, D. A. Monks, A., Hursey, M. L., Czerwinski, M. J., Fine, D. L., Abbott, B. J., Mayo, J. G., Shoemaker, R. H. and Boyd, M. R. 1988. Feasibility of drug screening with panels of human tumor cell lines using a microculture tetrazolium assay. *Cancer Research* 48:589–601.

Bartus, H. R., Mirabelli, C. K., Auerbach, J. I., Shatzman, A. R., Taylor, D. P., Johnson, R. K., Rosenberg, M. and Crooke, S. T. 1984. Improved genetically modified *Escherichia coli* strain for prescreening antineoplastic agents. *Antimicrobial Agents and Chemotherapy* 25:622–625.

Bush, J. A., Long, B. H., Catino, J. J. and Bradner, W. T. 1987. Production and biological activity of rebeccamycin, a novel antitumor agent. *Journal of Antibiotics* (Tokyo) 40:668–678.

Cantino, J. J., Francher, D. M. and Schuring, J. E. 1986. Evaluation of *cis*-diamminedichloroplatinum (II) combined with metoclopramide or sodium thiosulfate on L1210 leukemia *in vitro* and *in vivo*. *Cancer Chemotherapy and Pharmacology* 18:1–4.

Cantino, J. J., Francher, D. M., Edinger, K. J. and Stringfellow, D. A. 1985. A microtiter cytotoxicity assay useful for the discovery of fermentation-derived antitumor agents. *Cancer Chemotherapy and Pharmacology* 15:240–243.

Driscoll, J. S. 1984. The preclinical new drug research programof the national cancer institute. *Cancer Treatment Reports* 68:63–76.

Elespuru, R. K. and Yarmolinsky, M. B. 1979. A colorimetric assay of lysogenic induction designed for screening potential carcinogenic and carcinostatic agents. *Enviromental Mutagenesis* 1:55–78.

Finlay, G. J., Baguley, B. C. and Wilson, W. R. 1984. A semiautomated microculture method for investigating growth inhibitory effects of cytotoxic compounds on exponentially growing carcinoma cells. *Analytical Biochemistry* 139:272–277.

Fry, D. W., Theodore, J. B. and Jackson, R. C. 1984. Studies on the biochemical mechanism of the novel antitumor agent, CI-920. *Cancer Chemotherapy and Pharmacology* 13:171–175.

Funaishi, K., Kawamura, K., Sugiura, Y., Nakahori, N., Yoshida, E., Okanishi, M., Umezawa, I., Funayama, S. and Komiyama, K. 1987. Kazusamycin B, a novel antitumor antibiotic. *Journal of Antibiotics* (Tokyo) 40:778–785.

Funayama, S., Okada, K., Oka, H., Tomisaka, S., Miyano, T., Komiyama, K. and Umezawa, I. 1985a. Structure of awamycin, a novel antitumor ansamycin antibiotic. *Journal of Antibiotics* (Tokyo) 38:1284–1286.

Funayama, S., Okada, K., Komiyama, K. and Umezawa, I. 1985b. Structures of trienomycins A, B and C. Novel cytocidal ansamycinantibiotics. *Journal of Antibiotics* (Tokyo) 38:1677–1683.

Funayama, S., Anraku, Y., Mita, A., Yang, Z.-B., Shibata, K., Komiyama, K., Umezawa, I. and Ōmura, S. 1988. Structure- activity relationship of a novel antitumor ansamycin antibiotic trienomycin A and related compounds. *Journal of Antibiotics* (Tokyo) 41:1223–1230.

Funayama, S., Eda, S., Komiyama, K., Ōmura, S. and Tokunaga, T. 1989a. Structure of phenazinomycin, a novel antitumor antibiotic. *Tetrahedron Letters* 30:3151–3154.

Funayama, S., Anraku, Y., Mita, A., Komiyama, K. and Ōmura, S. 1989b. Structural study of isoflavonoids possessing antioxidant activity isolated from the fermentation broth of *Streptomyces* sp. *Journal of Antibiotics* (Tokyo) 42:1350–1355.

Fuska, J., Vesely, P., Ivanikaja, L. and Fuska, A. 1975. *In vitro* screening of cytotoxic substances using different tumor cells. In: Hellmann, K. and Connors, T. A. (editors), *Chemotherapy* 7, pp. 327–333. Plenum Press, New York.

Gause, G. 1975. Recent experience with microbial systems and cancer cellls *in vitro* in the screening for antitumor antibiotics. In: Hellmann, K. and Connors, T. A. (editors), *Chemotherapy* 7, pp. 299–302. Plenum Press, New York.

Gause, G. F., Laiko, A. V. and Selesneva, T. I. 1976. Yeast mutants with distorted cell membranes as test in the screening for antitumor antibiotics. *Cancer Treatment Reports* 60:637–638.

Geran, R. I., Greenberg, N. H., Macdonald., M. M., Schumacher, A. M. and Abbott, B. J. 1972. Protocols for screening chemical agents and natural products against animal tumors and other biological systems (3rd Edition). *Cancer Chemotherapy Reports* 3:1-88.

Goldin, A. 1967. Preclinical methodology for the selection of Anticancer agents. In: Busch, H. (editor), *Methods in Cancer Research*, Vol. IV, pp. 193–255. Academic Press, New York.

Hanka, L. J., Bhuyan, B. K., Martin, D. G., Neil, G. L. and Douros, J. D. 1978. A multi-end point in vitro system for detection of new antitumor drugs. *Antibiotics and Chemotherapy*(Basel) 23:26–32.

Hara, M., Takahashi, I., Yoshida, M., Asano, K., Kawamoto, I., Morimoto, M. and Nakano, H. 1989. Leinamycin, a new antitumor antibiotic from *Streptomyces*; producing organism, fermentation and isolation. *Journal of Antibiotics*(Tokyo) 42:1768–1774

Hara, M., Mokudai, T., Kobayashi, E., Gomi, K. and Nakano, H. 1990. The kapurimycins, new antitumor antibiotics produced by *Streptomeces*. Producing organism, fermentation, isolation and biological activities. *Journal of Antibiotics* (Tokyo) 43:1513–1518.

Hata, T. 1971. Cancer chemotherapy: Studies on anticancer agents particularly anticancer antibiotics. *Kitasato Archives of Experimental Medicine* 44:135–154.

Hayakawa, Y., Iwakiri, T., Imamura, K., Seto, H. and Otake, N. 1985a. Studies on the isotetracenone antibiotics I. Capoamycin, a new antitumor antibiotic. *Journal of Antibiotics* (Tokyo) 38:957–959.

Hayakawa, Y., Iwakiri, T., Imamura, K., Seto, H. and Otake, N. 1985b. Studies on the isotetracenone antibiotics II. Kerriamycins, A, B and C, new antitumor antibiotics. *Journal of Antibiotics* (Tokyo) 38:960–963.

Hori, Y., Hino, M., Kawai, Y., Kiyoto, S., Terano, H., Kohsaka, M., Aoki, H., Hashimoto, M. and Imanaka, H. 1986. A new antitumor antibiotic, chromoxymycin. II. Production, isolation, characterization and antitumor activity. *Journal of Antibiotics* (Tokyo) 39:12–16.

Hsu, Y., Hirota, A., Shima, S., Nakagawa, M., Adachi, T., Nozaki, H. and Nakayama, M. 1989. Myrocin C, a new diterpene antitumor antibiotic from Myrothecium verrucaria. II. Physicochemical properties and structure determination. *Journal of Antibiotics* (Tokyo) 42:223–229.

Hu, J., Xue, Y., Xie, M., Zhang, R., Otani, T., Minami, Y., Yamada, Y. and Marunaka, T. 1988. A new macromolecular antitumor antibiotic, C-1027. I. Discovery, taxonomy of producing organism, fermentation and biological activity. *Journal of Antibiotics* (Tokyo) 41:1575–1579.

Hurley, T. R., Bunge, R. H., Willmer, N. E., Hokanson, G. C. and French, J. C. 1986. PD124,895 and PD124,966, two new antitumor antibiotics. *Journal of Antibiotics (Tokyo)* 39:1651–1656.

Ichimura, M., Muroi, K., Asano, K., Kawamoto, I., Tomita, F., Morimoto, M. and Nak-

ano, H. 1988. DC89-A1, A new antitumor antibiotic from *Streptomyces*. *Journal of Antibiotics* (Tokyo) 41:1285-1288.
Ichimura, M., Ogawa, T., Takahashi, K., Kobayashi, E., Kawamoto, I., Yasuzawa, T., Takahashi, I. and Nakano, H. 1990. Duocarmycin SA, a new antitumor antibiotic from *Streptomyces* sp. *Journal of Antibiotics* (Tokyo) 43:1037-1038.
Imai, S., Furihata, K., Hayakawa, Y., Noguchi, T. and Seto, H., 1989. Lavanducyanin, a new antitumor substance produced by *Streptomyces* sp. *Journal of Antibiotics* (Tokyo) 42:1196-1197.
Imamura, K., Odagawa, A., Tanabe, K., Hayakawa, Y. and Otake, N. 1984. Akrobomycin, a new anthracycline antibiotic. *Journal of Antibiotics* (Tokyo) 37:83-84.
Ishii, S., Nagasawa, M., Kariya, Y., Itoh, O., Yamamoto, H., Inouye, S. and Kondo, S. 1989. Antitumor activity of ankinomycin. *Journal of Antibiotics* (Tokyo) 42:1518-1519.
Ishizeki, S., Ohtushika, M., Irinoda, K., Kukita, K., Nagaoka, K. and Nakashima, T. 1987. Azinomycins A and B, new antitumor antibiotics. III. Antitumor activity. *Journal of Antibiotics* (Tokyo) 40:60-65.
Ishizuka, M., Masuda, T., Kanbayashi, N., Fukasawa, S., Takeuchi, T., Aoyagi, T. and Umezawa, H. 1980. Effect of bestatin on mouse immune system and experimental murine tumors. *Journal of Antibiotics* (Tokyo) 33:642-652.
Ishizuka, M., Masuda, T., Nakabayashi, N., Watanabe, Y., Matsuzaki, M., Sawazaki, Y., Ohkura, A., Takeuchi, T. and Umezawa, H. 1982. Antitumor effect of forphenicinol, a low molecular weight immunomodifier, on murine transplantable tumors and microbial infections. *Journal of Antibiotics* (Tokyo) 35:1049-1054.
Itoh, J., Watabe, H., Ishii, S., Gomi, S., Nagasawa, M., Yamamoto, H., Shomura, T., Sezaki, M. and Kondo, S. 1988. Sibanomicin, a new pyrrolo[1,4]-benzodiazepine antitumor antibiotic produced by a *Micromonospora* sp. *Journal of Antibiotics* (Tokyo) 41:1281-1284.
Iwami, M., Nakayama, O., Okuhara, M., Terano, H. and Kohsaka, M. 1989a. New antitumor antibiotic, FR-900462. I. Taxonomy of the producing strain. *Journal of Antibiotics* (Tokyo) 42:680-685.
Iwami, M., Nakayama, O., Okuhara, M., Terano, H. and Kohsaka, M. 1989b. New antitumor antibiotic, FR-900462. II. Production, isolation, characterization and biological activity. *Journal of Antibiotics* (Tokyo) 42:686-690.
Kitahara, M., Ishii, K., Kawaharada, H., Watanabe, K., Suga, T., Hirata, T. and Nakamura, S. 1985. 7-Hydroxyguanine, a novel antimetabolite from a strain of *Streptomyces purpurascens*. II. Physicochemical properties and structure determination. *Journal of Antibiotics* (Tokyo) 38:977-980.
Kiyoto, S., Shibata, T., Kawai, Y., Hori, Y., Nakayama, O, Terano, H., Kohsaka, M., Aoki, H. and Imanaka, H. 1984. New antitumor antibiotic, FR-900405. III. Mechanism of action of FR-900405. *Journal of Antibiotics* (Tokyo) 38:955-956.
Kiyoto, S., Murai, H., Tsurumi, Y., Terano, H., Kohsaka, M., Takase, S., Uchida, I., Hashimoto, M., Aoki, H. and Imanaka, H. 1987. WF-3405, A novel antitumor antibiotic. Taxonomy, isolation, structure elucidation and biological properties. *Journal of Antibiotics* (Tokyo) 40:290-295.
Koenuma, M., Kinashi, H. and Otake, N. 1974. An improved screening method for antiphage antibiotics and isolation of sarkomycin and its relatives. *Journal of Antibiotics* (Tokyo) 27:801-804.
Komiyama, K., Funayama, S., Anraku, Y., Mita, A., Takahashi, Y. and Ōmura, S. 1989. Isolation of isoflabonoids possessing antioxidant activity from the fermentation broth of *Streptomyces* sp. *Journal of Antibiotics* (Tokyo) 42:1344-1349.
Komiyama, K., Okada, K., Oka, H., Tomisaka, S., Miyano, T., Funayama, S. and Umezawa, I. 1985. Structural study of a new antitumor antibiotic, kazusamycin. *Journal Antibiotics* (Tokyo) 38:220-223.
Komiyama, K., Hirokawa, Y., Yamaguchi, H., Funayama, S., Masuda, K., Anraku, Y.,

Umezawa, I. and Ōmura, S. 1987. Antitumor activity of trienomycin A on murine tumors. *Journal Antibiotics* (Tokyo) 40:1768–1772.
Konishi, M., Ohkuma, H., Naruse, N. and Kawaguchi, H. 1984. Chicamycin, a new antitumor antibiotic. II. Structure determination of chicamycins A and B. *Journal of Antibiotics* (Tokyo) 37:200–206.
Konishi, M., Sugawara, K., Kofu, F., Nishiyama, Y., Tomita, K., Miyaki, T. and Kawaguchi, H. 1986. Elsamicins, new antitumor antibiotics related to chartreusin. I. Production, isolation, characterization and antitumor activity. *Journal of Antibiotics* (Tokyo) 39:784–791.
Kurasawa, S., Takeuchi, T. and Umezawa, H. 1975a. Studies on glyoxalase inhibitor. Isolation of a new active agent, MS-3, from a mushroom culture. *Agricultural and Biological Chemistry* 39:2003–2008.
Kurasawa, S., Naganawa, H., Takeuchi, T. and Umezawa, H. 1975b. The structure of MS-3: A glyoxalase I inhibitor produced by a mushroom. *Agricultural and Biological Chemistry* 39:2009–2014.
Lam, K. S., Hesler, G. A., Mattei, J. M., Mamber, S. W., Forenza, S. and Tomita, K. 1990. Himastatin, a new antitumor antibiotic from *Streptomyces hygroscopicus*. I. Taxonomy of procucing organism, fermentation and biological activity. *Journal of Antibiotics* (Tokyo) 43:956–960.
Lee, M. D., Manning, J. K., Williams, D. R., Kuck, N. A., Testa, R. T. and Borders, D. B. 1989. Calicheamicins, a novel family of antitumor antibiotics. 3. Isolation, purification and characterization of calichemicins 1Br, 1Br, 2I, 3I, I, 1I and 1I. *Journal of Antibiotics* (Tokyo) 42:1070–1087.
Leet, J. E., Schroeder, D. R., Krishnan, B. S. and Matson, J. A. 1990. Himastatin, a new antitumor antibiotic from *Streptomyces hygroscopicus*. II. Isolation and characterization. *Journal of Antibiotics* (Tokyo) 43:961–966.
LePage, G. A., Worth, L. S. and Kimball, A. P. 1976. Enhancement of the antitumor activity of arabinofuranosyladenine by 2′-deoxy- conformycin. *Cancer Research* 36:1481–1485.
Maiese, W. M., Lechevalier, M. P., Lechevalier, H. A., Korshalla, J., Kuck, N., Fantini, A., Wildey, M. J., Thomas, J. and Greenstein, M. 1989. Calicheamicins, a novel family of antitumor antibiotics: Taxonomy, fermentation and biological properties. *Journal of Antibiotics* (Tokyo) 42:558–563.
Maiese, W. M., Laveda, D. P., Korshalla, J., Kuck, N., Fantini, A., Wildey, M. J., Thomas, J. and Greenstein, M. 1990. LL-D49194 antibiotics, a novel family of antitumor angents: Taxonomy, fermentation and biological properties. *Journal of Antibiotics* (Tokyo) 43:253–258.
Matson, J. A. and Bush, J. A. 1989. Sandramycin, a novel antitumor antibiotic produced by *Nocardioides* sp. Production, isolation, characterization and biological propertiers. *Journal of Antibiotics* (Tokyo) 42:1763–1767.
Matson, J. A., Rose, W. C., Bush, J. A., Myllymaki, R., Bradner, W. T. and Doyle, T. W. 1989a. Antitumor activity of chrysomycins M and V. *Journal of Antibiotics* (Tokyo) 42:1446–1448.
Matson, J. A., Claridge, C., Bush, J. A., Titus, J., Bradner, W. T., Doyle, T. W., Horan, A. C. and Patel, M. 1989b. AT2433-A1, AT2433-A2, AT2433-B1, and AT2433-B2. Novel antitumor antibiotic compounds produced by *Actinomadura melliaura*. Taxonomy, fermentation, isolation and biological propertiers. *Journal of Antibiotics* (Tokyo) 42:1547–1555.
Mirabelli, C. K., Bartus, H., Bartus, J. O. L., Johnson, R., Mong, S. M., Sung, C. P. and Crooke, S. T. 1985. Application of a tissue culture microtiter test for the detection of cytotoxic agents from natural products. *Journal of Antibiotics* (Tokyo) 38:758–766.
Mirand, A. E. 1966. Erythropoietic response of animals infected with various strains of Friend virus. *National Cancer Institute Monograph* 22:483–503.
Mosmann, T. 1983. Rapid colorimetric assay for cellular growth and survival: Appli-

cation to proliferation and cytotoxicity assays. *Journal of Immunological Methods* 65:55–63
Nakagawa, M., Hsu, Y., Hirota, A., Shima, S. and Nakayama, M. 1989. Myrocin C, a new diterpene antitumor antibiotic from *Myrothecium verrucaria*. I. Taxonomy of the producing strain, fermentation, isolation and biological properties. *Journal of Antibiotics* (Tokyo) 42:218–222.
Nakamura, S., Yajima, T., Hamada, M., Nishimura, T., Ishizuka, M., Takeuchi, T., Tanaka, N. and Umezawa, H. 1967. A new antitumor antibiotic, phenomycin. *Journal of Antibiotics* (Tokyo) Series A 20:210–216.
Nishii, M., Inagaki, J., Nohara, F., Isono, K., Kusakabe, H., Kobayashi, K., Sakurai, T., Koshimura, S., Sethi, S. K. and McCloskey, J. A. 1985. A new antitumor antibiotic, guanine 7-N- oxide produced by *Streptomyces* sp. *Journal of Antibiotics* (Tokyo) 38:1440–1443.
Nishimura, M., Nakada, H., Nakajima, H., Hori, Y., Ezaki, M., Goto, T. and Okuhara, M. 1989a. A new antitumor antibiotic, FR900840. I. Discovery, identification, isolation and characterization. *Journal of Antibiotics* (Tokyo) 42:542–548.
Nishimura, M., Nakada, H., Takasee, S., Katayama, A., Goto, T., Tanaka, H. and Hashimoto, M. 1989b. A new antitumor antibiotic, FR900840. II. Structural elucidation of FR900840. *Journal of Antibiotics* (Tokyo) 42:549–552.
Nishimura, M., Nakada, H., Kawamura, I., Mizota, T., Shimomura, K., Nakahara, K., Goto, T., Yamaguchi, I. and Okuhara, M. 1989c. A new antitumor antibiotic, FR900840. II. Antitumor activity against experimental tumors. *Journal of Antibiotics* (Tokyo) 42:553–557.
Nishioka, H., Sawa, T., Hamada, M., Shimura, N., Imoto, M. and Umezawa, K. 1990. Inhibition of phosphatidylinositol kinase by toyokamaycin. *Journal of Antibiotics* (Tokyo) 43:1586–1589.
Nomoto, H., Katsumata, S., Takahashi, K., Funayama, S., Komiyama K., Umezawa, I. and Omura, S. 1989. Structural studies on minor components of trienomycin group antibiotics trienomycins D and E. *Journal of Antibiotics* (Tokyo) 42:79–481.
Ogawa, T., Ichimura, M., Katsumata, S., Morimoto, M. and Takahashi, K. 1989. New antitumor antibiotics, duocarmycins B1 and B2. *Journal of Antibiotics* (Tokyo) 42:1299–1301.
Ohba, K., Watabe, H., Sasaki, T., Takeuchi, Y., Kodama, Y., Nakazawa, T., Yamamoto, H., Shomura, T., Sezaki, M. and Kondo, S. 1988. Pyrindamycins A and B, new antitumor antibiotics. *Journal of Antibiotics* (Tokyo) 41:1515–1519.
Oka, M., Ohkuma, H., Kamei, H., Konishi, M., Oki, T. and Kawaguchi, H. 1988. Glidobactins D, E, F, G and H; Minor Components of the antitumor antibiotic glidobactin. *Journal of Antibiotics* (Tokyo) 41:1906–1909.
Okabe, T., Nomoto, K., Funabashi, H., Okuda, S., Suzuki, H. and Tanaka, N. 1985. Lactoquinomycin, a novel anticancer antibiotic II. Physicochemical properties and structure assignment. *Journal of Antibiotics* (Tokyo) 38:1333–1336.
Oku, T., Imanishi, J. and Kishida, T. 1982. Assessment of antitumor cell effect of human leucocyte interferon in combination with anticancer agents by a convenient assay system in monolayer cell culture. *Japanese Journal of Cancer Research* 73:667–674.
Ōmura, S., Eda, S., Funayama, S., Komiyama, K., Takahashi, Y. and Woodruff, H. B. 1989. Studies on a novel antitumor antibiotic, phenazinomycin: Taxonomy, fermentation, isolation, and physicochemical and biological activity. *Journal of Antibiotics* (Tokyo) 42:1037–1042.
Onoda, T., Iimura, H., Sasaki, Y., Hanada, M., Isshiki, K., Naganawa, H., Takeuchi, T., Tatsuta, K., and Umezawa, K. 1989. Isolation of a novel tyrosine kinase inhibitor, lavendustin A, from *Streptomyces griseolavendus*. *Journal of Natural Products* 52:1252–1257.
Otani, T., Munami, Y., Marunaka, T., Zhang, R. and Xie, M. 1988. A new macromo-

lecular antitumor antibiotic, C-1027. II. Isolation and physico-chemical properties. *Journal of Antibiotics* (Tokyo) 41:1580-1585.

Ozaka, T., Suzuki, K., Sasamata, M., Tanaka, K., Kobori, M., Kadota, S., Nagai, K., Saito, T., Watanabe, S. and Iwanami, M. 1989a. Novel antitumor antibiotic phospholine. 1. Production, isolation and characterization. *Journal of Antibiotics* (Tokyo) 43:1331-1338.

Ozaka, T., Tanaka, K., Sasamata, M., Kaniwa, H., Shimizu, M., Matsumoto, H. and Iwanami, M. 1989b. Novel antitumor antibiotic phospholine. 2. Structure determination. *Journal of Antibiotics* (Tokyo) 42:1339-1343.

Pienta, R. J., Bernstein, E. H. and Groupe, V. 1963. Experiences with virus-induced Rous sarcoma as a model in experimental therapy. *Cancer Chemotherapy Report* 31:25-39.

Salmon, S. E., Hamburger, A. W., Soehnlen, B., Durie, B. G. M., Alberts, D. S. and Moon, T. E. 1978. Quantitation of differential sensitivity of human-tumor stem cells to anticancer drugs. *New England Journal of Medicine* 298:1321-1327.

Sato, H., Goto, M. and Kuroki, T. 1967. Culture of rat ascites hepatoma cells in agar medium and screening for anticancer substances. *Japanese Cancer Association GANN Monograph* 2:127-140.

Sato, Y., Watabe, H., Nakazawa, T., Shomura, T., Yamamoto, H., Sezaki, M. and Kondo, S. 1989. Ankinomycin, a potent antitumor antibiotic. *Journal of Antibiotics* (Tokyo) 42:149-152.

Shimomura, K., Hirai, O., Mizota, T., Matsumoto, S., Mori, J., Shibayama, F. and Kikuchi, H. 1987. A new antitumor antibiotic, FR-900482. III. Antitumor activity in transplantable experimental tumors. *Journal of Antibiotics* (Tokyo) 40:600-606.

Shimosaka, A., Kawai, H., Hayakawa, Y., Komeshima, N., Nakagawa, M., Seto, H. and Otake, N. 1987. Arugomycin, a new anthracycline antibiotic. III. Biological antivities of arugomycin and its analogues obtained by chemical degradation and modification. *Journal of Antibiotics* (Tokyo) 40:1283-1291.

Shoemaker, R. H. 1986. New approaches to antitumor drug screening: The human tumor colony forming assay. *Cancer Treatment Reports* 70:9-12.

Sidwell, R. W., Dixon, G. J., Compton, P. and Schabel, F. M., Jr. 1969. The effect of treatment with a combination of 6-mercapto-purine and porfiromycin on an established Friend leukemia virus infection. *Cancer Research* 29:497-502.

Smith, A. B., III, Wood, J. L., Wong, W., Gould, A. E., Rizzo, C. J., Funayama, S. and Omura, S. 1990. (+)-Trienomycins A, B and C: Relative and absolute stereochemistry. *Journal of the American Chemical Society* 112:7425-7426.

Smitka, T. A., Bunge, R. H., Wilton, J. H., Hokanson, G. C. and French, J. C. 1986. PD116,152, A new phenazine antitumor antibiotic. Structure and antitumor anctivity. *Journal of Antibiotics* (Tokyo) 39:800-803.

Steinberg, D. A., Peterson, G. A., White, R. J. and Maiese, W. M. 1985. The stimulation of bioluminescence in *Photobacterium leiognathi* as a potential prescreen for antitumor agents. *Journal of Antibiotic* (Tokyo) 38:1401-1407.

Sugawara, K., Ohbayashi, M., Shimizu, K., Hatori, M., Kamei, H., Konishi, M., Oki, T. and Kawaguchi, H. 1988. BMY-28438 (3,7- dihydroxytropolone), a new antitumor antibiotic active against B16 melanoma. I. Production, isolation, structure and biological activity. *Journal of Antibiotics* (Tokyo) 41:862-868.

Sugiura, K. 1959. Effects of compounds on the Friend mouse virus leukemia. *Gann* 50:251-264.

Suzuki, H., Tahara, M., Takahashi, M., Matsumura, F., Okabe, T., Shimazu, A., Hirata, A., Yamaki, H., Yamaguchi, H., Tanaka, N. and Nishimura, T. 1990. Resorthiomycin, a novel antitumor antibiotic. I. Taxonomy, isolation and biological activity. *Journal of Antibiotics* (Tokyo) 43:129-134.

Tahara, M., Okabe, T., Furihata, K., Tanaka, N., Yamaguchi, H., Nishimura, T. and

Suzuki, H. 1990a. Resorthiomycin, a novel antitumor antibiotic. II. Physicochemical properties and structure elucidation. *Journal of Antibiotics* (Tokyo) 43:135–137.
Tahara, M., Tomida, A., Nishimura, T., Yamaguchi, H. and Suzuki, H. 1990b. Resorthiomycin, a novel antitumor antibiotic. III. Potentiation of antitumor drugs and its mechanism of action. *Journal of Antibiotics* (Tokyo) 43:138–142.
Takahashi, K., Takahashi, I., Morimoto, M. and Tomita, F. 1986. DC-86-M, A novel antitumor antibiotic. II. Structure determination and biological activities. *Journal of Antibiotics* (Tokyo) 39:624–628.
Takahashi, A., Nakamura, H., Ikeda, D., Naganawa, H., Kameyama, T., Kurasawa, S., Okami, Y. and Takeuchi, T. 1988a. Thrazarine, a new antitumor antibiotic. II. Physicochemical properties and structure determination. *Journal of Antibiotics* (Tokyo) 41:1568–1574.
Takahashi, I., Takahashi, K., Asano, K., Kawamoto, I., Yasuzawa, T., Ashizawa, T., Tomita, F. and Nakano, H. 1988b. DC92-B, A new antitumor antibiotic from *Actinomadura*. *Journal of Antibiotics* (Tokyo) 41:1151–11531.
Takahashi, I., Takahashi, K., Ichimura, M., Morimoto, M., Asano, K., Kawamoto, I., Tomita, F. and Nakano, H. 1988c. Duocarmycin A, A new antitumor antibiotic from *Streptomyces*. *Journal of Antibiotics* (Tokyo) 41:1915–1917.
Takeuchi, M., Sato, Y. and Nitta, K. 1984. An *in vitro* screening method for antitumor and/or antitumorigenic substances involving the transformation of chick embryo fibroblasts infected with rous sarcoma virus. *Journal of Antibiotics* (Tokyo) 37:235–238.
Takeuchi, T., Chimura, H., Hamada, M., Umezawa, H., Yoshioka, O., Oguchi, N., Takahashi, Y. and Matsuda, A. 1975. A glyoxalase I inhibitor of a new structural type produced by *Streptomyces*. *Journal of Antibiotics* (Tokyo) 28:737–742.
Tanaka, H., Kawakami, T., Yang, Z., Komiyama, K. and Omura, S. 1989. Potentiation of cytotoxicity and antitumor activity of adenosin analogs by the adenosine deaminase inhibitor adecypenol. *Journal of Antibiotics* (Tokyo) 42:1722–1724.
Terano, H., Takase, S., Hosoda, J. and Kohsaka, M. 1989. A new antitumor antibiotic, FR-66979. *Journal of Antibiotics* (Tokyo) 42:145–148.
Tsuruo, T., Oh-hara, T., Iida, H., Tsukagoshi, S., Sato, Z., Matsuda, I., Iwasaki, S., Shimizu, F., Sasagawa, K., Fukami, M., Fukuda, K. and Arakawa, M. 1986. Rhizoxin, a macrocyclic lactoneantibiotic, as a new antitumor agent against human and murine tumor cells and their vincristine-resistant sublines. *Cancer Research* 46:381–385.
Uchida, T., Imoto, M., Watanabe, Y., Miura, K., Dobashi, T., Matsuda, N., Sawa, T., Naganawa, H., Hamada, M., Takeuhi, T. and Umezawa, H. 1985. Saquayamycins, new aquayamycin-group antibiotics. *Journal of Antibiotics* (Tokyo) 38:1171–1181.
Uehara, Y., Hori, M., Takeuchi, T. and Umezawa, H. 1985. Screening of agents which convert transformed morphology of Rous sarcoma virus-infected rat kidney cells to 'normal morphology': Identification of an active agent as herbimycin and its inhibition of intracellular src kinase. *Japanese Journal of Cancer Research* 76:672–675
Umezawa, H., Imoto, M., Sawa, T., Isshiki, K., Matsuda, N., Uchida, T., Iimura, H., Hamada, M., and Takeuchi, T. 1986. Studies on a new epidermal growth factor-receptor kinase inhibitor, Erbstatin, produced by MH435-hF3. *Journal of Antibiotics* (Tokyo) 39:170–173.
Umezawa, I., Funayama, S., Okada, K., Iwasaki, K., Satoh, J., Masuda, K. and Komiyama, K. 1985a. Studies on a novel cytocidal antibiotic, trienomycin A. Taxonomy, fermentation, isolation, physicochemical and biological characteristics. *Journal of Antibiotics* (Tokyo) 38:699–705.
Umezawa, I., Tronquet, C., Funayama, S., Okada, K. and Komiyama, K. 1985b. A novel antibiotic, sohbumycin. Taxonomy, fermentation, isolation, physicochemical and biological characteristics. *Journal of Antibiotics* (Tokyo) 38:967–971.
Woo P. W. K., Dion H. W., Lange M. S., Dahl L. F., Durham L. J. 1974. A novel adenosine and Ara-A deaminase inhibitor, (R)-3-(2-deoxy- &B-D-erythro-pentotura-

nosyl)-3,6,7,8-tetrahydroimidazo [4,5-d] [1,3] diazepin-8-01. *Journal of Heterocyclic Chemistry* 11:641–643.

Yamashita, Y., Saitoh, Y., Ando, K., Takahashi, K., Ohno, H. and Nakano, H. 1990. Saintopin, a new antitumor antibiotic with topoisomerase II dependent DNA cleavage activity, from *Paecilomyces. Journal Antibiotics* (Tokyo) 43:1344–1346.

Yamamoto, I., Nakagawa, M., Hayakawa, Y., Adachi, K. and Kobayashi, E. 1987. Hydroxychlorothricin, a new antitumor antibiotic. *Journal of Antibiotics* (Tokyo) 40:1452–1454.

Yamato, M., Iinuma, H., Naganawa, H., Yamagishi, Y., Hamada, M., Masuda, T., Umezawa, H., Abe, Y. and Hori M. 1986. Isolation andproperties of valanimycin, a new azoxy antibiotic. *Journal of Antibiotics* (Tokyo) 39:184–191.

Yasuzawa, T., Saitoh, Y. and Sano, H. 1990. Structures of the novel anthraquinone antitumor antibiotics, DC92-B and DC92-D. *Journal of Antibiotics* (Tokyo) 43:485–491.

Yoshida, E., Komiyama, K., Naito, K., Watanabe, Y., Takamiya, K., Okura, A., Funaishi, K., Kawamura, K., Funayama, S. and Umezawa, I. 1987. Antitumor effect of kazusamycin B on experimental tumors. *Journal of Antibiotics* (Tokyo) 40:1596–1604.

Yoshida, M., Hara, M., Saitoh, Y. and Sano, H. 1990. The kapurimycins, new antitumor antibiotics produced by *Streptomyces*. Physicochemical properties and structure determination. *Journal of Antibiotics* (Tokyo) 43:1519–1523.

Zhen, Y., Ming, X., Yu, B., Otani, T., Saito, H. and Yamada, Y. 1989. A new macromolecular antitumor antibiotic, C-1027. III. Antitumor activity. *Journal of Antibiotics* (Tokyo) 42:1294–1298.

6

Cell Differentiation Inducers

Haruki Yamada

6.1 Introduction

The discovery of a new antitumor compound of microbial origin depends on an in vitro assay system, because antitumor activity in extremely low amounts of test samples of culture fluids is not always detectable by in vivo assays. Cancer has been considered to be a disease of altered maturation in which the rate of proliferation of the cell population is increased relative to that of its maturation, resulting in an increase in tissue mass. Therefore, differentiation-inducing agents of tumor cells are suggested to be new candidates for cancer chemotherapy by reason of their specificity against tumor cells and low cytotoxicity.

Neoplastic cells such as teratocarcinoma, neuroblastoma, and leukemia can be induced to differentiate in vitro and in vivo by a wide variety of structurally diverse chemical agents including solvents [dimethyl sulfoxide (DMSO) and dimethyl formamide (DMF)], hormones, vitamins (A and D_3), tumor promoters [phorbol 12-myristate 13-acetate (TPA), teleocidine, and lyngbyatoxin], immunopotentiators *Mycobaterium bovis* (BCG), cell wall skelton (CWS), lipopolysaccharide (LPS), and interferon), and a number of chemotherapeutic agents (aclacinomycin A, marcellomycin, bleomycin, mitomycin C, methotrexate, cytosine arabinoside, 6-mercaptopurine, 6-thioguanine, and actinomycin D).

This differentiation-inducing activity of a wide range of cancer chemotherapeutic agents suggests that their antineoplastic effects may be the result of a combination of both cytotoxic and maturative actions. These differentiated cells are similar in their nature to normal mature cells and exhibit reduced ability for growth and malignant formation. It was expected that new antitumor agents will be obtained from the culture of microorganisms by screening for cell differentiation-inducing substances. There are excellent reviews covering this topic, and the reader is referred to Hozumi (1982) and Sartorelli et al. (1987).

6.2 Leukemic Cells and General Screening Methods

Screening by the use of M1 cells   The mouse myeloid leukemia cell line M1 was originally established in vitro from spontaneous leukemia SL strain mice (Ichikawa, 1969). It has been shown that M1 cells can be induced to differentiate into macrophages and granulocytes when treated with proteinous factors (D factor) in conditioned media from various cells and in various body fluids, and with chemicals such as glucocorticoid hormones, $1\alpha,25$-dihydroxyvitamin D, and lipopolysaccharides (Ichikawa, 1969; Sachs, 1978; Abe and Miyamura, 1981). Differentiation of M1 cells is accompanied by the induction of phagocytic activity, locomotive activity, morphological changes, the appearance of Fc and C3 receptors on the cell surface, and the synthesis of lysosomal enzymes (Ichikawa, 1969; Sachs, 1978). These are typical characteristics of mature macrophages and granulocytes. M1 cells are relatively unstable among the cloned leukemic cells, and the quality of serum for the culture is very important for the growth of the cells. The cell differentiation has been assayed by phagocytosis, lysozyme activity, morphological changes, and locomotive activity.

Cell culture (Ichikawa, 1969; Hayakawa et al., 1985a)   The cells have been cultured in Eagle's minimum essential medium with twice the normal concentration of amino acids and vitamins, and supplemented with 10% heat-inactivated (56°C, 30 min) calf or horse serum at 37°C in a humidified atmosphere containing 5% CO_2.

Phagocytic activity (Ichikawa, 1969; Hayakawa et al., 1985b)   M1 cells were incubated at 37°C with the test sample for the indicated periods, and the percentages of phagocytic activities were measured by using carbon or polystyrene latex particles. The numbers of phagocytic cells were determined under a microscope, the phagocytic cells being defined as those containing carbon particles or more than five latex particles. Cell growth was determined from the cell number after trypan blue-stained cells had been excluded.

Lysozyme activity (Osserman and Lawlor, 1966)   Lysozyme activity in the cells was measured by the following lysoplate method. Heat-killed *Micrococcus lysodeikticus* organisms, suspended uniformly in a small volume of phosphate buffer, are added to melted (60–70°C) 1% agar in M/15 phosphate buffer, pH 6.3, to a final concentration of 50 mg of organisms in 100 ml of buffered agar, which is poured into petri dishes to a depth of 4 mm. After the agar solidifies, sample wells, 2 mm in diameter, are cut with a thin-walled brass tube machined to a beveled cutting edge. (Each sample well is filled with approximately 25 μl of test sample.) After being filled, the plates are left at room temperature (24–26°C) for 12 to 18 h, during which period zones of clearing develop in the initially translucent gel as the result of bacterial lysis.

The diameters of the cleared zones are proportional to the log of concentration of lysozyme.

Morphological change (Hayakawa et al., 1985a) Morphological changes were determined by examining cells stained with a May–Grunwald–Giemsa solution.

Locomotive activity (Ichikawa, 1969) Locomotive activity was observed of dispersed colonies 10 days after adding phosphate-buffered saline containing test samples to 10 day-old colonies in soft agar.

Lactosillan During the course of screening for differentiation inducers of M1 cells, Hayakawa et al. (1985a) isolated a new polysaccharide, lactosillan (1) (Figure 6.1) from the culture filtrate of a bacterium, *Alcaligenes latus* G66A. Lactosillan is a heteroglycan composed of repeating units of a pentasaccharide as shown in Figure 6.1 (Hayakawa et al., 1982). Lactosillan strongly induced differentiation of M1 cells but not HL-60 and Friend leukemia cells. On treatment with 5–20 µg ml^{-1} of lactosillan for 72 h, 70–80% of the cells were induced to phagocytize latex beads; no cytotoxicity was observed with the cells treated with 100 µg ml^{-1} of lactosillan. M1 cells treated with lactosillan were induced to form dispersed colonies in soft agar and to synthesize lysosomal enzymes. The morphology of 65.5% of the cells changed when treated with 10 µg ml^{-1} of lactosillan for 72 h; further, 14.4% of cells developed into macrophage-like cells. Lactosillan has antitumor activity against sarcoma-180 solid tumor but was ineffective on mouse leukemia L-1210 and P-388 cells.

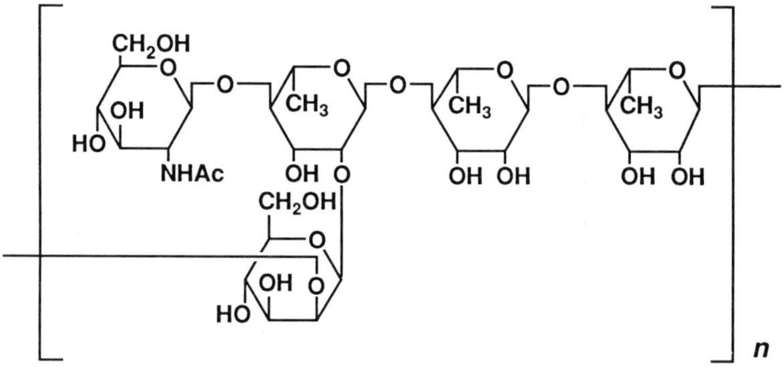

Lactosillan (1)

Figure 6.1 Structure of lactosillan (1).

Screening by the use of HL-60 cells A human promyelocytic leukemia cell line (HL-60) isolated from peripheral blood leukocytes of a patient with acute promyelocytic leukemia can be induced to differentiate into mature granulocytes and macrophages by various inducers such as dimethyl sulfoxide (DMSO), dimethylformamide (DMF), vitamin A, anthracyclines, actinomycin D, etc. This cell differentiation has been assayed by the same methods as with M1 cells. HL-60 cells also are relatively unstable among the known leukemia cells, and they have long growth times.

Cell culture (Hayakawa et al., 1985b)   HL-60 cells were cultivated in RPMI 1640 medium supplemented with 10% heat-inactivated fetal calf serum at 37°C in a humidified atmosphere containing 5% CO_2.

Spicamycin   Spicamycin (2) (Figure 6.2), which was isolated from the culture broth of *Streptomyces alanosinicus* 879-MT3 by Hayakawa et al. (1985b), at 40–320 ng ml^{-1} induced marked differentiation of M-1 cells. Spicamycin at 2.5–640 ng ml^{-1} also induced marked differentiation of HL-60 cells. It also showed antitumor activity against p388 leukemia in mice when administered intraperitoneally in 0.25–2.0 mg kg^{-1}day^{-1} doses. In addition, spicamycin has antiyeast activity but no antibacterial activity.

The pyrromycinone group of anthracyclines   The oligosaccharide-containing anthracycline antibiotics are effective inducers of the differentiation of HL-60 promyelocytic leukemia cells, both in vitro (Schwartz and Sartorelli, 1982; Morin and Sartorelli, 1984) and in vivo (Schwartz et al., 1983). The most active drugs of this class as initiators of the maturation of these leukemic cells were in the pyrromycinone group of anthracyclines, which are shown in Figure 6.3 (Sartorelli et al., 1987). The trisaccharide-containing antibiotics, marcellomycin (3) and aclacinomycin A (4), were the most efficacious members of this class, while the disaccharide musettamycin (5) and 10-descar-

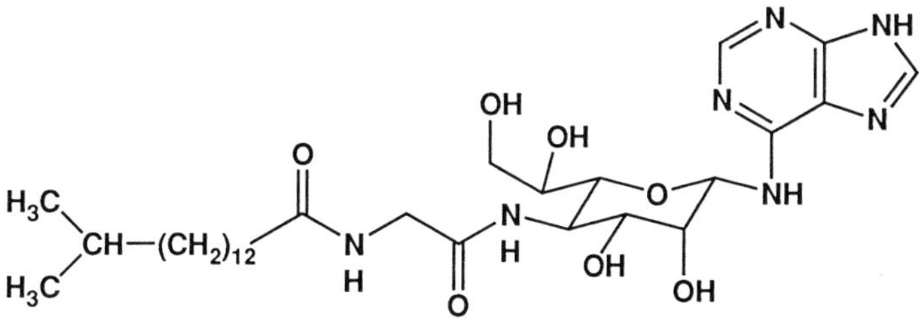

Figure 6.2 Structure of spicamycin (2).

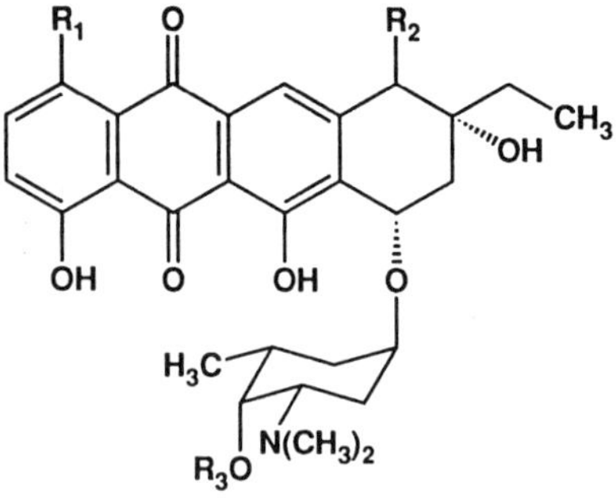

	R₁	R₂	R₃
Marcellomycin (3)	OH	COOCH₃	2-deoxyfucose-2-deoxyfucose
Aclacinomycin (4)	H	COOCH₃	2-deoxyfucose-2-cinerulose
Musettamycin (5)	OH	COOCH₃	2-deoxyfucose
10-Descarbomethoxymarcellomycin (6)	OH	H	2-deoxyfucose-2-deoxyfucose
Pyrromycin (7)	OH	COOCH₃	H

Figure 6.3 Structures of pyrromycinone group of anthracyclines (3-7).

bomethoxymarcellomycin (6) were slightly less active. The monosaccharide pyrromycin (7) had minimal activity, and other monosaccharide-containing anthracyclines, such as adriamycin and carcinomycin, were inactive as inducers of the differentiation of HL-60 leukemia cells (Schwartz and Sartorelli, 1982).

These studies of the relationship between structure and activity suggested that the oligosaccharide side chain of the pyrromycinone class of anthracyclines was the predominant determinant of differentiation-inducing activity (Sartorelli et al., 1987). A biochemical action unique to the trisaccharide-containing anthracyclines is the inhibition of the synthesis of glycoproteins containing asparagine-linked oligosaccharides (Morin and Sartorelli, 1984). The preferential interference with the biosynthesis of asparagine-linked oligosaccharides by aclacinomycin A and marcellomycin corresponded to the inhibitory effect of these substances on the incorporation of tritiated mannose into lipid-linked oligosaccharides (Morin and Sartorelli, 1984). Adriamycin and pyrromycin, at equivalent growth-inhibiting concentrations, did not interfere with the formation of dolichol-linked oligosaccharides. As a consequence of the interference with glycoprotein biosynthesis by the active anthracycline inducers of maturation, the levels of cell-surface transferrin receptor, an asparagine-linked glycoprotein, decreased significantly (Morin and Sartorelli,1984).

Screening by the use of Friend murine leukemia cells   Murine erythroleukemia cells established by Friend and Patuleia (1966) are transformed cells that appear to be blocked in differentiation at a relatively late stage in erythropoiesis (Markes and Rifkind, 1978). Erythroleukemia cells can differentiate into hemoglobin-positive, normal erythroblast-like cells on chemical stimulation. The erythroid differentiation of Friend cells has been assayed by benzidine staining of hemoglobin accumulated in the cells (Orkin et al., 1975). The differentiated Friend leukemia cells lost their colony forming ability in the soft agar medium and simultaneously failed to cause leukemia in the syngenic hosts, as reported by Sugano et al. (1975). F5-5, a kind of Friend leukemia cell, was the most suitable for the assay system because its response was stable on repeated culture passages. Colony formation in the soft agar medium was markedly inhibited by 1% DMSO, but cell growth in the liquid medium was not. Antibiotics such as mitomycin C, adriamycin, and actinomycin C, but not cycloheximide, did not induce detectable benzidine-positive cells among the F5-5 cells in the concentration range tested (Morioka et al., 1985d).

Cells and culture conditions (Morioka et al., 1985d) The Friend leukemia cells, F5-5, differentiated to synthesize hemoglobin with the stimulation of DMSO, but cell line C9-6 did not. Ham's F-12 culture medium supplemented with 10% fetal calf serum has been used for the growth of the F5-5 and C9-6 cell lines, and Dulbecco's minimal essential medium (MEM) supplemented with 10% fetal calf serum for line C1745A which originated from the mouse DBA12 line.

The cells were cultured at 37°C for 5 days at initial cell densities of 2×10^4 cells ml^{-1} for F5-5 and C9-6, and at 1×10^5 cell ml^{-1} for C1745A in a humidified atmosphere of 5% CO_2. Cell viability was measured by the trypan blue dye exclusion method and counted on a Burker-Turk counter.

Benzidine staining method (Orkin et al., 1975) Differentiation of Friend leukemia cells was assayed by the benzidine staining method. On the sixth day, 0.2 ml of the cell suspension was mixed with 0.02 ml of freshly prepared benzidine solution (a 10:1 mixture of 2% 3,3'-dimethoxy-benzidine in 0.5 M acetic acid and 30% hydrogen peroxide). The cells that were stained blue, indicating the presence of hemoglobin, were scored in a hemacytometer as the percentage of the total cell number.

Colony formation assay (Ikawa et al., 1979) To examine the cell tumorigenicity, a colony-forming assay was carried out. F5-5 cells were suspended in Ham's F-12 medium supplemented with 10% fetal calf serum to give a cell density of 2×10^3 cells ml^{-1}. The cell suspension (4 ml) was mixed with 16 ml of agar medium consisting of 0.35 ml 7.5% $NaHCO_3$, 11.25 ml Ham's F-12 medium concentrated twofold, 2.5 ml of fetal calf serum, 3.75 ml 3% agar (Difco), and 7.15 ml of water.

FL-657 (Morioka et al., 1985a, 1985b)

Assay method The mycelia from a *Streptomyces* sp. are separated from 10 ml of the culture broth and extracted with 10 ml of chloroform:methanol (1:1). The extracts are evaporated and again dissolved in 1 ml of methanol. Serial 1:2 dilutions of the mycelium extracts and supernatant solutions are made in a 96-well plate. F5-5 cells are cultured with 0.02 ml of the serially diluted test samples at a final concentration of 10% for 6 days. The titer of the sample expressed as units per milliliter of the extract or supernatant is defined as the reciprocal of the maximal dilution necessary to induce benzidine-positive cells by at least 1%. The supernatant solutions are sterilized with filter papers (pore size, 0.45 μm) and subjected to the benzidine staining method and colony-forming assay.

FL-657 Morioka et al. (1985a) have found that the initiation of hemoglobin biosynthesis by F5-5 cells was induced by the action of isolate FL-657, which was produced by *Streptomyces sioyaensis* and consisted of more than two active compounds. FL-657B (8) (Figure 6.4), which was identified as trichostatic acid, caused approximately 90% of the cells to be benzidine positive and reduced the growth to approximately 30–70% of the control at 2.42 μg ml^{-1}. FL-657B also showed strong inhibition of the colony formation of the cells in the soft agar medium at a concentration that did not cause growth inhibition in liquid medium or the induction of hemoglobin biosynthesis. FL-657B was active as an inducer not only for F5-5 but also for the C9-6 and 1745A cell lines.

Other differentiation inducers Other differentiation inducers for leukemic cells are summarized in Table 6.1.

6.3 Recently Developed Characteristic Screening Methods

Screening by the use of a neuroblastoma cell line Most cell differentiators have been screened by the use of several leukemia cell lines.

Some human neuroblastoma cell lines differentiate to ganglionic or neu-

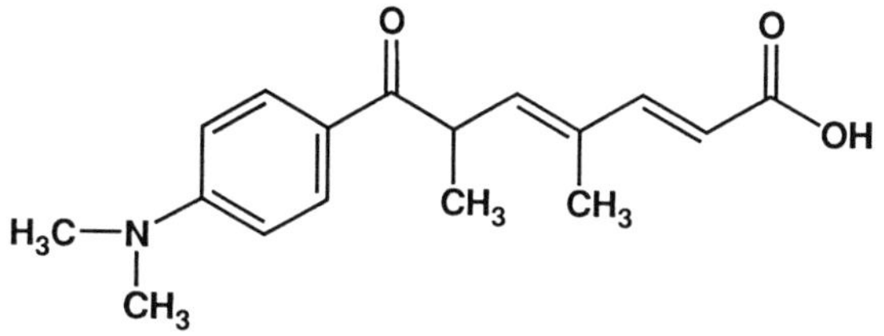

FL-657B (8)

Figure 6.4 Structure of FL-657B (8).

Table 6.1 Other differentiation inducers for leukemic cells

Compound	Cells	Screening method (concentration)	Reference
Capoamycin	Mouse myeloid leukemia cells (M1)	Phagocytosis (20–320 ng ml^{-1})	Hayakawa et al., 1987
Citrinin	Mouse erythroleukemia cells (B8)	Benzidine staining (10–20 µg ml^{-1})	Kawashima et al., 1983
	M1	Phagocytosis (10–20 µg ml^{-1})	Kawashima et al., 1983
Cosmomycin A	Friend leukemia cells (F5-5)	Benzidine staining (1.25 µg ml^{-1})	Morioka et al., 1985b
B	F5-5	Benzidine staining (1.25 µg ml^{-1})	
C	F5-5	Benzidine staining (7.81 ng ml^{-1})	
D	F5-5	Benzidine staining (15.6 ng ml^{-1})	
Diastovaricins I and II	Friend erythroleukemia cells	Benzidine staining (125 µg ml^{-1})	Hotta et al., 1986
Differenol A	B8	Benzidine staining (20–40 µg ml^{-1})	Asahi et al., 1981
	M1	Lysozome activity (10–60 µg ml^{-1})	
Differanisol A	B8	Benzidine staining (75 µg ml^{-1})	Oka et al., 1985
Trichostatin C	Friend erythroleukemia cells (DS-19)	Benzidine staining (100–400 ng ml^{-1})	Yoshida et al., 1985

roblastoma cells when treated with dibutyryl-cyclic adenosine monophosphate (But$_2$ cAMP) (Kitamura and Miyake, 1975; Prasad and kummar, 1975). Therefore, this cell line is also available for the screening of cell differentiators. A human neuroblastoma cell line, NB-1, established by Miyake et al. (1973), was used for the screening. NB-1 was cultured in RPMI-1640 supplemented with 10% fetal calf serum at 37°C for 3–4 days in a 5% CO$_2$ incubator.

Screening method Differentiation induced by test solution is screened by morphological changes such as elongation of neurites. NB-1 (2 × 10^3 cells well^{-1} of a 96-well plate) in 0.2 ml of the medium was incubated for 3 days with a test solution. The degree of differentiation is expressed as a percentage of the cells having elongated neurites of the total number of cells. Viable cells are determined by the trypane blue dye exclusion method.

The test solutions from culture fluids are sterilized by passage through a membrane filter (pore size, 0.45 µm) and assayed with NB-1 at a concentration of 1%.

Staurosporine Morioka et al. (1985c) employed culture fluids of *Streptomyces* species in the differentiation screening system using the human neuroblastoma cell line, NB-1, and found that staurosporine [cf. Chapter 7 (p.

141)], produced by *Streptomyces actuosus*, induced differentiation of NB-1 in a short incubation time at an extremely low concentration. Staurosporine has first been found as a microbial alkaloid by chemical screening (see Chapter 14; also Ōmura et al., 1977). Staurosporine strongly induced differentiation in NB-1, its effect being greater than that of But$_2$ cAMP. Staurosporine induced elongation of neurites and cell enlargement at concentrations as low as 0.02 μM and cell enlargement 1 h after treatment of NB-1. Staurosporine is also active as a potent inhibitor of protein kinase C (Nakanishi et al., 1986) and a potent platelet aggregation inhibitor (Oka et al., 1986).

Ōmura et al. (1991) have developed a screening method for differentiation inducers of the mouse neuroblastoma cell line Neuro 2A from microbial metabolites, and found that *Streptomyces sp.* OM-6519 produced a novel compound lactacystin (9) (Figure 6.5). The compound was revealed to induce neuritogenesis of these neuroblastoma cells.

Screening by the use of Rous sarcoma virus-infected rat kidney cells The search for compounds that inhibit the activity of multifunctioning oncogene(s) may be rewarding. As a model for this type of study, Uehara et al. (1985) screened microbial products having the ability to convert the transformed morphorogy of Rous sarcoma virus-infected rat kidney cells to normal morphology. They have used cells of a rat kidney line infected with ts25, a T-class mutant of Rous sarcoma virus Prague strain (ts/NRK) originally isolated by Chen et al. (1977).

Screening method (Uehara et al., 1985) The cells are cultured either at a permissive temperature (33°C) for transformation or, where specified, at a nonpermissive temperature (39°C) in Dulbecco's modified Eagle medium sup-

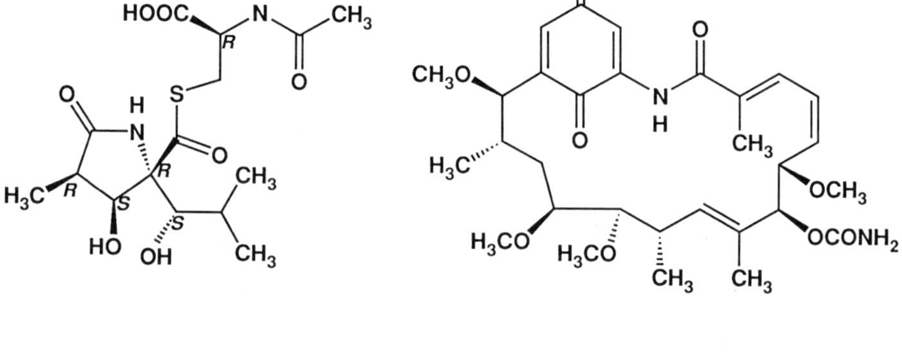

Lactacystin (9)   Herbimycin (10)

Figure 6.5 Structures of lactacystin (9) and herbimycin (10).

plemented with 10% heat-inactivated calf serum in humidified air with 5% CO_2. For screening of microbial products, the cells maintained at 33°C are seeded at 3×10^4 cells 2 ml^{-1} per 35-mm dish and grown overnight at 33°C. Test samples are added to the dishes (usually 10 µl of a fermentation broth per dish after sterilization by mixing with ether), incubation is continued at 33°C, and the cell morphology observed under a phase contrast microscope the next day.

The transformed morphology, characterized by small, densely stuffed, spindle-shaped cells in 33°C cultures, is observed. These cells can reach high cell densities. At 39°C, however, the cells show normal (or essentially normal) morphology, being flattened out and sensitive to contact inhibition. These morphological changes are reversible in either direction at 12–16 h after the temperature shift.

Herbimycin By the use of this screening method, Uehara et al. (1985) found a fermentation product of *Streptomyces* sp. MH237-CF8 as an active substance that is identified as herbimycin (**10**) (see Figure 6.5). Herbimycin had first been isolated during the course of a screening program for herbicidal antibiotics from *Actinomycetes* by Ōmura et al. (1979). When herbimycin was added to a 33°C culture, the cells changed their transformed morphology to that of normal cells. The morphological changes were observed at concentrations ranging from 0.1 to 1.0 µg ml^{-1}, where cell growth was inhibited from 10% to 90%, respectively. At more than 0.5 µg ml^{-1}, all cells were converted to the normal morphology. When the cells were washed free of herbimycin and allowed to grow in fresh medium at 33°C, the transformed morphology returned within 16 h.

No morphological changes were observed with the cells (ts/NRK) maintained at 39°C or with uninfected NRK cells maintained at either temperature. Therefore, it seemed likely that the effect of herbimycin on the cell morphology was a reflection of some interference with *Src* gene expression by the antibiotic. Uehara et al. (1985) also determined the effect of herbimycin on the synthesis and the phosphokinase activity of p60^{src}, the src gene product, in 33°C cells. The results suggested that herbimycin has no direct effect on *src* kinase but destroys its intracellular environment, resulting in an irreversible alteration of the enzyme that leads to loss of catalytic activity. Morphological alteration of cultured cells is an important characteristic of transformed phenotypes in a variety of cell systems. The assay based on microscopic inspection of cell morphology is rapid and simple. Thus, this assay method should be useful in finding and developing antitumor agents which act on oncogene function.

6.4 Concluding Remarks

Although cell differentiators include known antitumor drugs that have potent cytotoxicity, a cell differentiator that is specific for target cells generally shows low toxicity. Therefore, an antitumor drug as a specific cell differentiation inducer is expected to have low cytotoxicity compared with known antitumor drugs. However, the reactivity and side effects of cell differentiators against host cells also have to be considered in clinical use.

Cell differentiators have been screened from natural and synthetic products during the past 10 years. However, clinically effective cell differentiators have not yet been found because most known examples were weak or have no antitumor activity in vivo. Cell differentiation is a complicated phenomenon, and to find clinically useful cell differentiation inducers complete elucidation of the basic principles of cell differentiation is very important. Uehara et al. (1985) screened for compounds that inhibit the activity of multifunctioning oncogenes, and demonstrated that herbimycin changes Rouse sarcoma-transformed cell morphology to that of normal cells and affects oncogene function. This approach is expected to be a good screening system for finding cell differentiation inducers that have antitumor activity.

Retinoid, the synthetic and natural analogs of vitamin A, frequently blocks the phenotypic expression of cancer in vitro, and also induces differentiation in many animal and human malignant cell types (Lippman et al., 1987). Retinoid affects the protein kinase-C cascade system and influences the synthesis and distribution of membrane glycoproteins and other glycoconjugates (Lippman et al., 1987). These biochemical processes also may be used as targets for screening for specific cell differentiators.

References

Abe, E. and Miyamura, C. 1981. Differentiation of mouse myeloid leukemia cells induced by 1α, 25-dihydroxyvitamin D3 *Proceedings of National Academy of Sciences of the United States of America* 78:4990–4994.

Asahi, K., Ono, I., Kusakabe, H., Nakamura, G. and Isono, K. 1981. Studies on differentiation inducing substances of animal cells. I Differenol A, a differentiation inducing substance against mouse leukemia cells. *Journal of Antibiotics* (Tokyo) 34:919–920.

Chen, Y. C., Hayaman, M. S. and Vogt, P. K. 1977. Properties of mammalian cells transformed by temperature-sensitive mutants of avian sarcoma virus. *Cell* 11:513–521.

Friend, C. and Patuleia, M. C. 1966. Erythrocytic maturation *in vitro* of murine (Friend) virus-induced leukemic cells. *National Cancer Institute Monograph* 22:505–522.

Hayakawa, Y., Nakagawa, M., Ando, T., Shimazu, A., Seto, H. and Otake, N. 1982. Studies on the differentiation inducers of myeloid leukemic cells. I. Lactosillan, a new inducer of the differentiation of M1 cells. *Journal of Antibiotics* (Tokyo) 35:1252–1254.

Hayakawa, Y., Ando, T., Shimazu, A., Seto, H. and Otake, N. 1985a. Lactosillan, a new differentiation inducer of mouse myeloid leukemia cells (M1): Taxonomy, fermentation, isolation, physicochemical properties and biological properties. *Agricultural and Biological Chemistry* 49:2437–2442.

Hayakawa, Y., Nakagawa, M., Kawai, H., Tanabe, K., Nakayama, H., Shimazu, A., Seto, H. and Otake, N. 1985b. Spicamycin, a new differentiation inducer of mouse

myeloid leukemia cells (M1) and human promyelocytic leukemia cells (HL-60). *Agricultural and Biological Chemistry* 49:2685–2691.
Hayakawa, Y., Adachi, K., Iwakiri, T., Imamura, K., Furuhata, K., Seto, H. and Otake, N. 1987. Capoamycin, a new isotetracenone antibiotic. *Agricultural and Biological Chemistry* 51:2237–2243.
Hotta, M., Hayakawa, Y., Furuhata, K., Shimazu, A., Seto, H. and Otake, N. 1986. New ansamycin antibiotics, diastovaricins I and II. *Journal of Antibiotics* (Tokyo) 39:311–313.
Hozumi, M. 1982. A new approach to chemotherapy of myeloid leukemia: Control of leukemogenecity of myeloid leukemia cells by inducer of normal differentiation. *Cancer Biology Reviews* 3:153–211.
Ichikawa, Y. 1969. Differentiation of a cell line of myeloid leukemia. *Journal of Cell Physiology* 74:223–234.
Ikawa, Y., Obinata, M. and Sugano, H. 1979. Friend leukemia system as a model for decancerization and erythro differentiation. *Gann Monograph on Cancer Research* 24:263–278.
Kawashima, A., Nakagawa, M., Hayakawa, Y., Kawai, H., Seto, H. and Otake, N. 1983. Studies on the differentiation inducers of myeloid leukemic cells. II. Citrinin, a new inducer of the differentiation of M1 cells. *Journal of Antibiotics* (Tokyo) 36:173–174.
Kitamura, Y. and Miyake, S. 1975. Neuroblastoma. In: Oboshi, S. and Sugano, H. (editors), *Cultures of Human Tumor Cells*, pp.220–225 Asakura Shoten, Tokyo.
Lippman, S. M., Kessler, J. F. and Meyskens, Jr., F. L. 1987. Retinoids as preventive and therapeutic anticancer agents (Part 1), *Cancer Treatment Reports* 71:391–405.
Marks, P. A. and Rifkind, R. A. 1978. Erythrolekemic differentiation. *Annual Review of Biochemistry* 47:419–448.
Miyake, S., Shimo, T., Kitamura, Y., Nojyo, T., Nakamura, S., Imashuku, S. and Abe, T. 1973. Characteristics of continuous and functional cell line NB-1, derived from a human neuroblastoma, *Autonomic Nervous System* 10:115–120.
Morin, M. J. and Sartorelli, A. C. 1984. Inhibition of glycoprotein biosynthesis by the inducers of HL-60 cell differentiation, aclacinomycin A and marcellomycin. *Cancer Research* 44:2807–2812.
Morioka, H., Ishihara, M., Takezawa, M., Hirayama, K., Suzuki, E., Komoda, Y. and Shibai, H. 1985a. A new differentiation inducer of Friend leukemia cells, trichostatic acid. *Agricultural and Biological Chemistry* 49:1365–1370.
Morioka, H., Etoh, Y., Hirono, I., Takezawa, M., Ando, T., Hirayama, K., Kano, H. and Shibai, H. 1985b. Production and isolation of cosmomycins A, B, C and D: New differentiation inducers of Friend cell F5-5. *Agricultural and Biological Chemistry* 49:1951–1958.
Morioka, H., Ishihara, M., Shibai, H. and Suzuki, T. 1985c. Staurosporine-induced differentiation in a human neuroblastoma cell line, NB-1. *Agricultural and Biological Chemistry* 49:1959–1963.
Morioka, H., Takezawa, M. and Shibai, H. 1985d. Actinomycin V as a potent differentiation inducer of F5-5 Friend leukemia cells. *Agricultural and Biological Chemistry* 49:2835–2942.
Nakanishi, S., Matsuda, Y., Iwahashi, K. and Kase, H. 1986. K-252b, c and d, potent inhibitiors of protein kinase C from microbial origin, *Journal of Antibiotics* (Tokyo) 39:1066–1071.
Oka, S., Kodama, M., Takeda, H., Tomizawa, N. and Suzuki, H. 1986. Staurosporine, a potent platelet aggregation inhibitor from a *Streptomyces* species. *Agricultural and Biological Chemistry* 50:2723–2727.
Oka, H., Asahi, K., Morishima, H., Sanada, M., Shiratori, K., Iimura, Y., Sakurai, T., Uzawa, J., Iwadare, S. and Takahashi, N. 1985. Differanisole A, a new differentiation inducing substance. *Journal of Antibiotics* (Tokyo) 38:1100–1102.
Ōmura, S., Iwai, Y., Hirano, A., Nakagawa, A., Awaya, H., Tsuchiya, H., Takahashi,

Y. and Masuma, R. 1977. A New alkaloid AM-2282 of *Streptomyces* origin taxonomy, fermentation, isolation and preliminary characterization. *Journal of Antibiotics* (Tokyo) 30:275-282.

Ōmura, S., Iwai, Y., Takahashi, Y., Sadakane, N. and Nakagawa, A. 1979. Herbimycin, a new antibiotic produced by a strain of *Streptomyces*. *Journal of Antibiotics* (Tokyo) 32:255-261.

Ōmura, S., Fujimoto, T., Otoguro, K., Matsuzaki, K., Moriguchi, R., Tanaka, H. and Sasaki, Y. 1991. Lactacystin, a novel microbial metabolite, induced neuritogenesis of neuroblastoma cells. *Journal of Antibiotics* (Tokyo) 44:113-116.

Orkin, S. H., Harosi, E. I. and Leder, P. 1975. Differentiation in erythroleukemic cells and their somatic hybrids. *Proceedings of National Academy of Sciences of the United States of America* 72:98-102.

Osserman, E. F. and Lawlor, D. P. 1966. Serum and urinary lysozyme (muramidase) in monocytic and monomyelocytic leukemia. *Journal of Experimental Medicine* 124:921-951.

Prasad, K. N. and Kummar, S. 1975. Role of cyclic AMP in differentiation of human neuroblastoma cells in culture. *Cancer* 36:1338-1343.

Sachs, L. 1978. Control of normal cell differentiation in leukemic white blood cells. *Annual Symposium on Fundamental Cancer Research* 30:223-237.

Sartorelli, A. C., Morin, M. J. and Ishiguro, K. 1987. Cancer chemotherapeutic agents as inducers of leukemia cell differentiation. In: *New Avenues in Developmental Cancer Chemotherapy*, pp.229-244. Academic Press, London.

Schwartz, E. L. and Sartorelli, A. C. 1982. Structure-activity relationship for the induction of differentiation of HL-60 human acute promyelocytic leukemia cells by anthracyclines. *Cancer Research* 42:2651-2655.

Schwartz, E. L., Brown, B. J., Nierenberg, M., Marsh, J. C. and Sartorelli, A. C. 1983. Evaluation of some anthracycline antibiotics in an *in vivo* model for studying drug-induced human leukemia cell differentiation. *Cancer Research* 43:2725-2730.

Sugano, H., Kawaguchi, T., Furusawa, M. and Ikawa, I. 1975. Differentiation of Friend virus-induced leukemia cells. In: Ito, Y. and Dutcher, R. M. (editors), *Comparative Leukemia Research 1973*, pp.221-228. University of Tokyo Press, Tokyo.

Uehara, Y., Hori, M., Takeuchi, T. and Umezawa, H. 1985. Screening of agents which convert "transformed morphology" of Rous sarcoma virus-infected rat kidney cells to "normal morphology": Identification of an active agent as herbimycin and its inhibition of intracellular *src* kinase. *Japanese Journal of Cancer Research* 76:672-675.

Yoshida, M., Iwamoto, Y., Uozumi, T. and Beppu, T. 1985. Trichostatin C, a new inducer of differentiation of Friend leukemic cells. *Agricultural and Biological Chemistry* 49:563-565.

Part 3
Enzyme Inhibitors

7

General Screening of Enzyme Inhibitors

Haruo Tanaka, Kazuhito Kawakita, Nobutaka Imamura, Kazuo Tsuzuki, and Kazuro Shiomi

7.1 Introduction

In the past several decades, microorganisms have provided us with many antibiotics useful as chemotherapeutic agents. Screening for new enzyme inhibitors from microorganisms has led us to new knowledge about secondary metabolites that are important as pharmacologically active drugs.

It has been considered that many enzymes have important roles in the maintenance of homeostasis in living organisms and that disease results from a breakdown of homeostasis. The correlation between the biological functions of several enzymes and disease processes has been clarified. However, it remains to be clarified with many other diseases.

Some inhibitors of enzymes in which the correlation with disease processes has been established are very important as pharmacologically active drugs. Other inhibitors are also useful in elucidating the biochemical and regulatory aspects of living organisms. Many analogs of the normal substrates of various enzymes have been synthesized chemically and subjected to the investigation of structure–activity relationships (Jung, 1978; Krenitsky and Elin, 1982). Some of the antagonists synthesized, such as captopril, an inhibitor of angiotensin converting enzyme, have been used as drugs (Ondetti and Cushman, 1982).

In the latter half of the 1960s, Umezawa's and Murao's groups independently started the screening of microorganisms for new enzyme inhibitors. From the results of studies on antibiotics and enzyme inhibitors of animal and plant origin, they speculated that microorganisms also produce inhibitors of various enzymes. In fact, Hoyem and Skulberg (1962) had reported that *Clostridium botulinum* produces trypsin inhibitors. Shimada and Matsushima (1967) reported the production of a protease inhibitor by *Penicillium cyclopium* and isolated the inhibitor, which was characterized as poly-L-malic acid (Shimada and Matsushima, 1969; Shimada et al., 1969). In the screening work by Umezawa's and Murao's groups, Aoyagi et al. (1969) first reported the plasmin inhibitor leupeptin, and Murao and Satoi (1970) discovered the pepsin inhibitor S-PI. Following this, Umezawa and his coworkers (Umezawa, 1972) discovered many new metabolites inhibiting proteases, membrane-associated enzymes, catecholamine-synthesizing enzymes, etc. After that, many other research groups including our laboratory have also been responsible for screening studies, and to date about 150 enzyme inhibitors of microbial origin have been discovered. Table 7.1 shows the enzyme inhibitors of microbial origin that have been found in screening systems targeting specific enzymes. Among them, the inhibitors shown in Table 7.2 are under clinical use or in trial.

The screening search for enzyme inhibitors useful as drugs needs a conventional method for routine assay as the primary screening system, followed by an evaluation system in vitro and in vivo using cell cultures, organs, or intact animals. We believe that whether a new useful inhibitor can be discovered depends very much on the screening method itself.

This chapter describes the concrete strategies and methods used in screening for typical enzyme inhibitors for which the pharmacological activities have been established. The emphasis is given to screening systems for enzyme inhibitors in fields in which industry has a great interest in development. Enzyme inhibitors functioning as chemotherapeutic agents, such as β-lactamase inhibitors and reverse transcriptase inhibitors, are described in Chapters 1 and 3.

A number of reviews concerning enzyme inhibitors of microbial origin have been published (Aoyagi et al., 1977; Umezawa, 1977; Frommer et al., 1979; Schindler, 1980; Fleck, 1981; Umezawa, 1982). Brannon and Fuller (1974) also have discussed the rationale, general principles, and techniques of screening microbial cultures for pharmacologically active compounds.

7.2 Angiotensin-Converting Enzyme

Rationale Two important enzyme systems, the renin-angiotensin (hypertensive) system and the kallikrein-kinin (hypotensive) system, are involved in the regulation of blood pressure in mammals (Figure 7.1). Renin (EC 3.4.4.15) cleaves one peptide bond in angiotensinogen to release the biologically inactive decapeptide angiotensin I. Angiotensin-I-converting enzyme (dipeptidyl carboxypeptidase, ACE, EC 3.4.15.1) causes removal of the C-terminal histidyl leucine moiety from angiotensin I to give the octapeptide angiotensin II, which is a potent vasoconstrictor and stimulant of adrenocortical aldosterone secretion. The same enzyme (another name, kininase II) inactivates the vasodilator bradykinin by the hydrolytic release of one or more carboxyl-terminal dipeptide residues. Thus, ACE inhibitors can reduce hypertension either by the suppression of angiotensin II biosynthesis or by the stimulation of bradykinin breakdown.

Chapter 7 General Screening of Enzyme Inhibitors   119

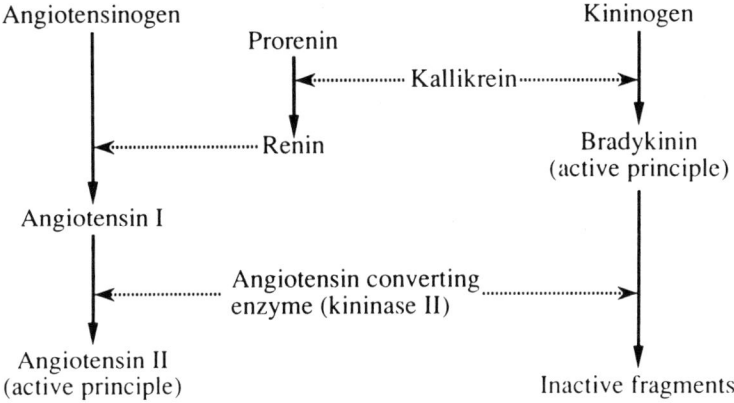

Figure 7.1 Renin–angiotensin and kallikrein–kinin systems.

Methods

General assay

Preparation of the enzyme (Cheung et al., 1980)   Rabbit lung acetone powder is extracted with 10 vol of 50 mM sodium borate buffer, pH 8.3, and centrifuged for 40 min at 40,000 x g. The supernatant, containing about 0.2 U ml^{-1} of ACE, is used for the assay.

Enzyme assay   A spectrophotometric procedure employing specific amino-substituted tripeptide substrates is usually used to detect activity (Ondetti and Cushman, 1982). The hippuric acid (Bz-Gly) produced from Hip-His-Leu or Hip-Gly-Gly can be quantitated by colorimetric determination of the reaction product with 2,4,6-trichloro-s-triazine or p-dimethylaminobenzaldehyde. Further, direct spectrophotometric assays are available using Boc-Phe(NO$_2$)-Phe-Gly, Z-Phe(NO$_2$)-Gly-Gly, Z-Phe(NO$_2$)-His-Leu, Hip(NO$_2$)-Gly-Gly, hippurylthioglycolyl-glycine, 2-aminobenzoyl-Gly-Phe(NO$_2$)-Pro, and FA-Phe-Gly-Gly as substrates.

Plate assay (O'Conner and Somers, 1985)   The agar plate assay involves (1) applying the sample to be tested to an enzyme-containing layer, and (2) overlaying with a substrate-containing layer. The lower layer contains 1% agarose, 20 mM HEPES buffer, pH 6.5, 100 mM NaCl, 200 µM CoCl$_2$ and ACE of 400 µg protein ml^{-1}. Fermentation broths are placed on the agar plate after being dried on 7-mm filter paper disks. After a 2-h diffusion at room temperature, the enzyme layers are overlaid with the second layer containing

Table 7.1 Enzyme inhibitors of microbial origin[1]

Target enzyme	EC Number	Inhibitor	Reference
Proteases			
Leucine aminopeptidase	3.4.11.1	Bu-2743E	Kobaru et al., 1983
Aminopeptidase M	3.4.11.2	Probestin	Aoyagi et al., 1990
		Leuhistin	Aoyagi et al., 1991a
Aminopeptidase B	3.4.11.6	Bestatin (ubenimex)	Umezawa et al., 1976a
		Arphamenines A and B	Umezawa et al., 1983
		OF4949-I, II, III, and IV	Sano et al., 1986
Aminopeptidase A	3.4.11.7	Amastatin	Aoyagi et al., 1978b
Dipeptidyl aminopeptidase IV	3.4.14.5	Diprotins A and B	Umezawa et al., 1984a
Angiotensin-converting enzyme (ACE)	3.4.15.1	Ancovenin	Kido et al., 1983
		L-681176 (murasmine)	Huang et al., 1984a
		Muraceins A, B, and C	Bush et al., 1984a
		15B2 and 15B1 (K-4)	Kido et al., 1984
		Phenacein	Bush et al., 1984b
		A58365 A and B	Mynderse et al., 1985
		Foroxymithine	Umezawa et al., 1985
		K-26	Yamato et al., 1986
		K-4	Koguchi et al., 1986
		K-13	Kase et al., 1987b
		WF-10129	Ando et al., 1987
Carboxypeptidase A	3.4.17.1	(S)-α-Benzylmalic acid	Tanaka et al., 1984
Carboxypeptidase B	3.4.17.2	Histargin	Umezawa et al., 1984b
α-Chymotrypsin	3.4.21.1	Chymostatin	Umezawa et al., 1970a
Plasmin	3.4.21.7	Leupeptin	Aoyagi et al., 1969
Alkaline proteinase (Microbial serine proteinases)	3.4.21.14	S-SI	Murao and Sato, 1972
		API-2b and 2c	Uyeda et al., 1978
Prolyl endopeptidase	3.4.21.26	Poststatin	Aoyagi et al., 1991b
Elastase	3.4.21.36	Elastatinal	Umezawa et al., 1973c
		Elasnin	Ōmura et al., 1978

Table 7.1 (cont.)

Target enzyme	EC Number	Inhibitor	Reference
Thiol proteinase	3.4.22	E-64	Hanada et al., 1978
		Prohisin	Kimura et al., 1984
		Thiolstatins A, B, C, and D	Murao et al., 1985
		Strepin P-1	Ogura et al., 1985
		Staccopins P1 and P2	Saito et al., 1987
		Caricastatin	Murao et al., 1987
		Estatins A and B	Yaginuma et al., 1989
Papain	3.4.22.2	Antipain	Suda et al., 1972
Acid proteinase	3.4.23	Poly-L-malic acid	Shimada and Matsushma, 1969
		Tyrostatin	Oda et al., 1989
Pepsin	3.4.23.1	S-PI (pepstatin Ac)	Murao and Satoi, 1970
		Pepstatin A	Umezawa et al., 1970c
		Pepstatins B-H	Miyano et al., 1972
		Pepstanone A	
		Pepstatins Bu, Pr, and Ac	Aoyagi et al., 1973
		Hydroxypepstatin	Umezawa et al., 1973b
Renin	3.4.23.15	Ahpatinins A, B, D, E, F, and G	Ōmura et al., 1986a
		Cyclothiazomycin	Aoki et al., 1991
Metalloproteinase	3.4.24	S-MPI	Murao et al., 1978
		FMPI	Murao et al., 1982
Collagenase	3.4.24.3	Actinonin	Faucher et al., 1987
Thermolysin	3.4.24.4	Phosphoramidon	Suda et al., 1973
		MK-I (talopeptin)	Fukuhara et al., 1982a
Enkephalinase B	3.4.24.11	Propioxatins A and B	Inaoka et al., 1986
HIV-I protease		α-MAPI	Stella et al., 1991

Continued next page

Table 7.1 (cont.)

Target enzyme	EC Number	Inhibitor	Reference
Enzymes related to sugar			
Aldose reductase	1.1.1.21	WF-3681	Nishikawa et al., 1987
		Aldostatin	Yaginuma et al., 1988
		WF-2421	Nishikawa et al., 1991
		Thiazocins A and B	Ozasa et al., 1991
Glyceraldehyde 3-phosphate dehydrogenase	1.2.1.12	Koningic acid (heptelidic acid)	Endo et al., 1985d
Glucosyltransferase	2.4.1	Ribocitrin	Takashio and Okami, 1982
		Mutastein	Endo et al., 1983
α-Glucoside hydrolase	3.2.1	Adiposins 1 and 2	Namiki et al., 1982
α-Amylase	3.2.1.1	S-AI	Murao and Ohyama, 1975
		BAYg5421 (acarbose)	Schmidt et al., 1977
		Haim I and II	Murao et al., 1980
		Trestatins A, B, and C	Yokose et al., 1983
		(Ro09-0766, 0767, and 0768)	
		Paim I and II	Oouchi et al., 1985
α-Amylase	3.2.1.2	Deoxynojirimycin (S-GI)	Arai et al., 1983
		Glucosyl S-GI	
Glucoamylase	3.2.1.3	Amylase inhibitor	Ueda and Koba, 1973
		Amylostatins	Fukuhara et al., 1982b
Chitinase	3.2.1.14	Allosamidin	Koga et al., 1987
		Methylallosamidin	
		A82516 (allosamidin)	Somers et al., 1987
		Glucoallosamidins A and B	Nishimoto et al., 1991
Sialidase	3.2.1.18	Panosialin	Aoyagi et al., 1971
		Siastatins A and B	Umezawa et al., 1974b
		Neuraminin	Lin et al., 1975
β-Glucosidase	3.2.1.21	Nojirimycin B	Niwa et al., 1984
		Cyclophellitol	Atsumi et al., 1990

Chapter 7   General Screening of Enzyme Inhibitors   123

Table 7.1 (cont.)

Target enzyme	EC Number	Inhibitor	Reference
β-Galactosidase	3.2.1.23	Isoflavonoids I, II, III-2, III-3, and IV	Aoyagi et al., 1975a
		Pyridindolol	Aoyagi et al., 1975b
		p-Hydroxyphenylacetaloxime	Hazato et al., 1979
		Galactostatin	Miyake and Ebata, 1987
β-Mannosidase	3.2.1.24	Mannostatins A and B	Aoyagi et al., 1989a
Trehalase	3.2.1.28	S-GI	Murao and Miyata, 1980
		Trehazoline (Trehalostatin)	Ando et al., 1991
β-Glucuronidase	3.2.1.31	M-GCI	Uyeda et al., 1984
Enzymes related to lipid			
Hydroxymethylglutaryl-CoA reductase	1.1.1.88	ML-236A, B (compactin), and C	Endo et al., 1976
		Dihydromonacolin L	Endo 1979
		Mevinolin (monacolin K)	Endo 1979
			Alberts et al., 1980
			Albers-Schönberg et al., 1981
		Dihydromevinolin	Endo et al., 1985a
		Monacolin X	Endo et al., 1985b
		Monacolins J and L	Endo et al., 1986
		Monacolin M	Hasumi et al., 1987
		Phenicin (phoenicine)	Ogawa et al., 1991
		Pannorin	Nakayama et al., 1989
Testosterone 5-reductase	1.3.1.22	WS-9659 A and B	Nakayama et al., 1990
		Riboflavin	Kitamura et al., 1986b
12-Lipoxygenase	1.13.11.12	MY3-469	Kitamura et al., 1986a
5-Lipoxygenase	1.13.11.12	KF8940	Hook et al., 1990
		MY12-62a and 62c	
		Carbazomycins B and C	
Steroid 11α-hydroxylase	1.14	Diplodialide A	Ishida and Wada, 1975
Prostagrandin synthase	1.14.99.1	Thielavins A and B	Kitahara et al., 1981
Acyl-CoA:cholesterol acyltransferase	2.3.1.26	Purpactins A, B, and C	Tomoda et al., 1991

Continued next page

Table 7.1 (cont.)

Target enzyme	EC Number	Inhibitor	Reference
Pancreatic lipase	3.1.1.3	Lipstatin	Weibel et al., 1987
Phospholipase A_2	3.1.1.4	Luteosporin	Singh et al., 1985
		Plastatin	
		Plipastatins A1, A2, B1, and B2	Umezawa et al., 1986b
		Duramycins B and C	Fredenhagen et al., 1990
		Thielocins A1α and A1β	Yoshida et al., 1991b
Phospholipase C	3.1.4.3	(S,S)-N,N'-Ethylenediamine disuccinic acid	Nishikiori et al., 1984
Hydroxymethylglutaryl-CoA synthase	4.1.3.5	1233A (F-244)	Ōmura et al., 1987
		Dihydroxerulin	Kuhnt et al., 1990
Acyl-CoA synthetase	6.2.1.3	Triacsins A and B	Ōmura et al., 1986c
Acetyl-CoA carboxylase	6.4.1.2	Octylpentanedioic acid	Endo et al., 1985c
		1-Decenyl-1-pentenedioic acid	
		Decanyl-1-pentenedioic acid	
		Decanyl-2-pentenedioic acid	
Aromatase		TAN-931	Ishii et al., 1991
Enzymes related to nucleic acids			
Xanthine oxidase	1.2.3.2	5-Formyluracil	Umezawa et al., 1972b
Thymidylate synthase	2.1.1.45	Vanoxonin	Kanai et al., 1983
		Diazaquinomycins A and B	Ōmura et al., 1985b
		HMPGG	Murata et al., 1987
Poly(ADP-ribose) synthetase	2.4.99.-	Benadrostin	Aoyagi et al., 1988
		2-Methyl-4[3H]quinazolinone	Yoshida et al., 1991
Reverse transcriptase	2.7.7.49	Revistin	Numata et al., 1975
		Retrostatin	Nishio et al., 1983
		Chromostin	Inouye et al., 1985
		Limocrocin	Hanajima et al., 1985
5'-Nucleotidase	3.1.3.5	Nucleoticidin	Ogawara et al., 1985a
		Nelanocidins A and B	Ogawara et al., 1985b

Table 7.1 (cont.)

Target enzyme	EC Number	Inhibitor	Reference
DNase	3.1.21.1	Dotriacolide	Ogawara et al., 1982
Adenosine deaminase	3.5.4.4	Deoxycoformycin	Woo et al., 1974
		Adechlorin	Ōmura et al., 1985a
		(2'-Chloropentostatin)	Tunac and Underhill 1985
		Adecypenol	Ōmura et al., 1986b
DNA-Topoisomerase I	5.99.1.2	Topostins A1, A2, and B	Ikegami et al., 1990
		BE-13793C	Kojiri et al., 1991
DNA-Topoisomerase II	5.99.1.3	BE-10988	Oka et al., 1991
		BE-13793C	Kojiri et al., 1991
Phosphatases and kinases			
Protein kinase C	2.7.1.37	K252a	Kase et al., 1986
		K252b, c, and d	Nakanishi et al., 1986
		Staurosporine	Ōmura et al., 1977
		UCN-01 and 02	Takahashi et al., 1987
		Calphostins A, B, C, D, and I	Kobayashi et al., 1989
		RK-286C	Osada et al., 1990
Phosphatidylinositol kinase	2.7.1.67	Toyocamycin	Nishioka et al., 1990
Tyrosine protein kinase	2.7.1.112	Erbstatin	Umezawa et al.,1986a
		Orobol	Ogawara et al., 1986
		Genistein	
Alkaline phosphatase	3.1.3.1	Forphenicine	Aoyagi et al., 1978a
		Alphostatin	Aoyagi et al., 1989a
		Reticulol	Furutani et al., 1975
cAMP phosphodiesterase	3.1.4.17	PDE-I and II	Enomoto et al., 1978
		APD-I, II, and III	Hosono and Suzuki, 1983
		Terferol	Nakagawa et al., 1984
		Griseolic acid	Nakagawa al., 1985

Continued next page

Table 7.1 (cont.)

Target enzyme	EC Number	Inhibitor	Reference
Ca^{2+}/Calmodulin-dependent cyclic nucleotide phosphodiesterase	3.1.4.17	K-254-I (genistein)	Goto et al., 1987
		K-259-2	Matsuda et al., 1987
		KS-619-1	Matsuda and Kase, 1987
		KS-501 and 502	Nakanishi et al., 1989a
		KS-504a, b, d, and e	Nakanishi et al., 1989b
ATPase	3.6.1.3	L-681110	Huang et al., 1984b
Enzymes related to neurotransmitter			
Monoamine oxidase	1.4.3.4	Pimprinine	Takeuchi et al., 1973
		trans-Cinnamic acid amide	
Tyrosine hydroxylase	1.14.16.2	Oudenone	Umezawa et al., 1970b
Tryptophan hydroxylase	1.14.16.4	2,5-Dihydro-L-phenylalanine	Okabayashi et al., 1977
Dopamine β-hydroxylase	1.14.17.1	Fusaric acid	Hidaka et al., 1969
		Dopastin	Iinuma et al., 1972
		Oosponol	Umezawa et al., 1972a
		Phenopicolinic acid	Nakamura et al., 1975
Tyrosinase	1.14.18.1	Kojic acid	Saruno et al., 1979
Catechol-O-methyltransferase	2.1.1.6	Methylspinazarin	Chimura et al., 1973a
		6,7-Dihydromethylspinazarin	
		7-O-Methylspinochrome B	
		6-(3-Hydroxybutyl)-7-O-methylspinochrome B	Chimura et al., 1973a
		New Ifv II and III	
		Dehydrodicaffeic acid dilactone	Chimura et al., 1975
Acetylcholinesterase	1.7	I-6123	Kumada et al., 1976
Cholinesterase	3.1.1.8	Mycelianamide	Ogata et al., 1974
Histidine decarboxylase	4.1.1.22	Lecanoric acid	Naganawa et al., 1976
DOPA decarboxylase	4.1.1.28	Psi-tectorigenin	Umezawa et al., 1974a
		Genistein	Umezawa et al., 1975
		New Ifv I	
		Orobol	
		8-Hydroxygenistein	

Table 7.1 (cont.)

Target enzyme	EC Number	Inhibitor	Reference
Prolyl hydroxylase	1.14.11.2	P-1894B (vineomycin A1)	Okazaki et al., 1981
		Fibrostatins A, B, C, D, E, and F	Ishimaru et al., 1987
Others			
N-Methyltransferase	2.1.1	1-[2-(3,4,5,6-Tetrahydropyridyl)]-1,3-pentadiene	Kumada et al., 1974
Spermidine synthase	2.5.1.16	Juglorin	Hamaguchi et al., 1987
Aspartate aminotransferase	2.6.1.1	Gostatin	Murao et al., 1981
Esterase	3.1.1	Esterastin	Umezawa et al., 1978
		Ebelactones A and B	Umezawa et al., 1980
		Valielactone	Kitahara et al., 1987
Dehydropeptidase	3.5.1.14	WS1358A1 and B1	Hashimoto et al., 1990
β-Lactamase	3.5.2.6	KA-107	Iwai et al., 1973
		MC696-SY2-A and B	Umezawa et al., 1973a
		MM 4550	Brown et al., 1976
		Izumenolide	Liu et al., 1980
		Dotriacolide	Ikeda et al., 1981
Ornithine decarboxylase	4.1.1.17	Sarcomycin	Fujiwara et al., 1978
		Dihydrosarcomycin	
		Ciotrinin	Yamato et al., 1985
Isocitrate lyase	4.1.3.1	Mycenon	Hautzel et al., 1990
Glyoxylase I	4.4.1.5	Glyo-I and -II	Takeuchi et al., 1975
Glutamine synthetase	6.3.1.2	Phosalasine	Ōmura et al., 1984a
		Oxetin	Ōmura et al., 1984b

[1] Enzyme an arranges in ascending order of EC numbers in each group.

Table 7.2 Pharmacologically active enzyme inhibitors from microorganisms that are in clinical use or trial

Inhibitor	Target enzyme	Clinical entity
Acarbose (BAYg 5421)	α-Glucosidase	Diabetes, obesity
Bestatin (ubenimex)	Aminopeptidase B	Tumors
Deoxycoformycin	Adenosine deaminase	Leukemia
Lovastatin (monacolin K, mevinolin)[1]	Hydroxymethylglutaryl CoA reductase	Hypercholesterolemia
Mutastein	Glucosyltransferase	Decayed teeth

[1] The modified compounds pravastatin and simvastatin are also used.

1% agarose, 20 mM HEPES buffer, pH 6.5, 100 mM NaCl, 200 μM CoCl$_2$, and p-nitrobenzyl oxycarbonylglycyl-(S-4-nitrobenzo-2-oxa-1,3-diazole)-L-cysteinylglycine (NBGCG) at 200 μg ml^{-1}. The plates are incubated at 37°C for a further 2 h and flooded with 0.1 N NaOH for 10 min after which the solution is poured off. Inhibition zones, initially seen as colorless on an amber background, turn pink after 30 min and fade completely in 2 h.

The measurement and recording of zones of inhibition is facilitated by viewing plates through a blue filter. Under these conditions, black-and-white photographs show white zones on a black background.

Inhibitors Many potent and highly specific synthetic inhibitors of ACE have proliferated since the 1970s (Soffer et al., 1976; Cushman and Ondetti, 1980). These inhibitors fall into two major classes: (1) snake venom oligopeptides and their analogs, and (2) nonpeptidic inhibitors designed for effective interactions with ACE, especially with its associated zinc ion.

On the other hand, as shown in Table 7.3 and Figure 7.2, there are not as many ACE inhibitors derived from microorganisms. Among them, A58365A and A58365B were discovered using the agar plate assay just described. Phenacein acts as a pure competitive inhibitor (K_i, 0.58 μM), while K-4, K-13, and K-26 inhibit noncompetitively. ACE inhibition by phenacein could be reversed by Zn^{2+}, indicating that phenacein may chelate the active site zinc atom of ACE.

7.3 Hydroxymethylglutaryl-CoA Reductase

Rationale High levels of plasma cholesterol (hypercholesterolemia) are known to cause atherosclerosis. The greater part of cholesterol in the human body is synthesized in the liver. The biosynthesis from acetyl-CoA consists of more than 20 enzymatic reactions, and the rate-limiting enzyme is 3-hydroxy-3-methylglutaryl (HMG)-CoA reductase (EC 1.1.1.88), which catalyzes the reduction of HMG-CoA to mevalonate. Therefore, an inhibitor of this enzyme would be expected to control the level of cholesterol in plasma and to prevent athreosclerosis (Endo, 1987).

Table 7.3 ACE inhibitors of microbial origin

Inhibitor	Molecular formula	Substrate	Activity		Reference
			In vitro (IC_{50})	In vivo[1]	
Ancovenin	$C_{85}H_{121}N_{23}O_{25}S_3$	p-Nitrobenzoyl-Gly-Gly-Gly	85 nM		Kido et al., 1983
Aspergillomarasmine A	$C_{10}H_{17}N_3O_8$	Hip-His-Leu	1.2 μM	75	Mikami and Suzuki 1983
Aspergillomarasmine B	$C_9H_{14}N_2O_8$	Hip-His-Leu	1.0 μM	50	
L-681,176	$C_{12}H_{23}N_5O_7$	Hip-His-Leu	1.3 μg ml^{-1}	142	Huang et al., 1984a
Muracein A	$C_{26}H_{44}N_6O_{14}$	NBGCG[2]	280 nM		Bush et al., 1984a
		FAPGG[3]			
Muracein B	$C_{32}H_{54}N_8O_{16}$		12 μM		Kido et al., 1984
Muracein C	$C_{32}H_{54}N_8O_{17}$		170 μM		
I5B2	$C_{23}H_{32}N_3O_7P$	Hip-His-Leu	91 nM		Bush et al., 1984b
Phenacein	$C_{13}H_8N_2O_4$	NBGCG[2]	390 nM		
		FAPGG[3]			
A58365A	$C_{12}H_{13}NO_6$	NBGCG[2]	1.8 nM		O'Connor and Somers 1985
A58365B	$C_{13}H_{15}NO_6$	NBGCG[2]	4.0 nM		
K-26	$C_{25}H_{34}N_3O_8P$	Hip-His-Leu	6.7 ng ml^{-1}	0.1	Yamato et al., 1986
Foroxymithine	$C_{22}H_{37}N_7O_{11}$		7 μg ml^{-1}		Umezawa et al., 1985
K-4 (I5B1)	$C_{23}H_{32}N_3O_6P$	Hip-His-Leu	80 ng ml^{-1}	0.21	Koguchi et al., 1986
K-13	$C_{29}H_{29}N_3O_8$	Hip-His-Leu	177 ng ml^{-1}	3	Kase et al., 1987b
WF-10129	$C_{20}H_{28}N_2O_8$	ABGPP[4]	14 nM	0.3	Ando et al., 1987D

[1] Inhibition of blood pressure response to angiotensin I in rats (ID_{50}, mg kg^{-1}, iv).
[2] NBGCG, p-Nitrobenzyl-oxycarbonylglycyl-(S-4-nitrobenzo-2-oxa-1,3-diazole)-L-cysteinylglycine.
[3] FAPGG, Furanacryloyl-phenylalanylglycylglycine.
[4] ABCPP, O-Aminobenzoylglycyl-p-nitro-L-phenylalanyl-L-proline.

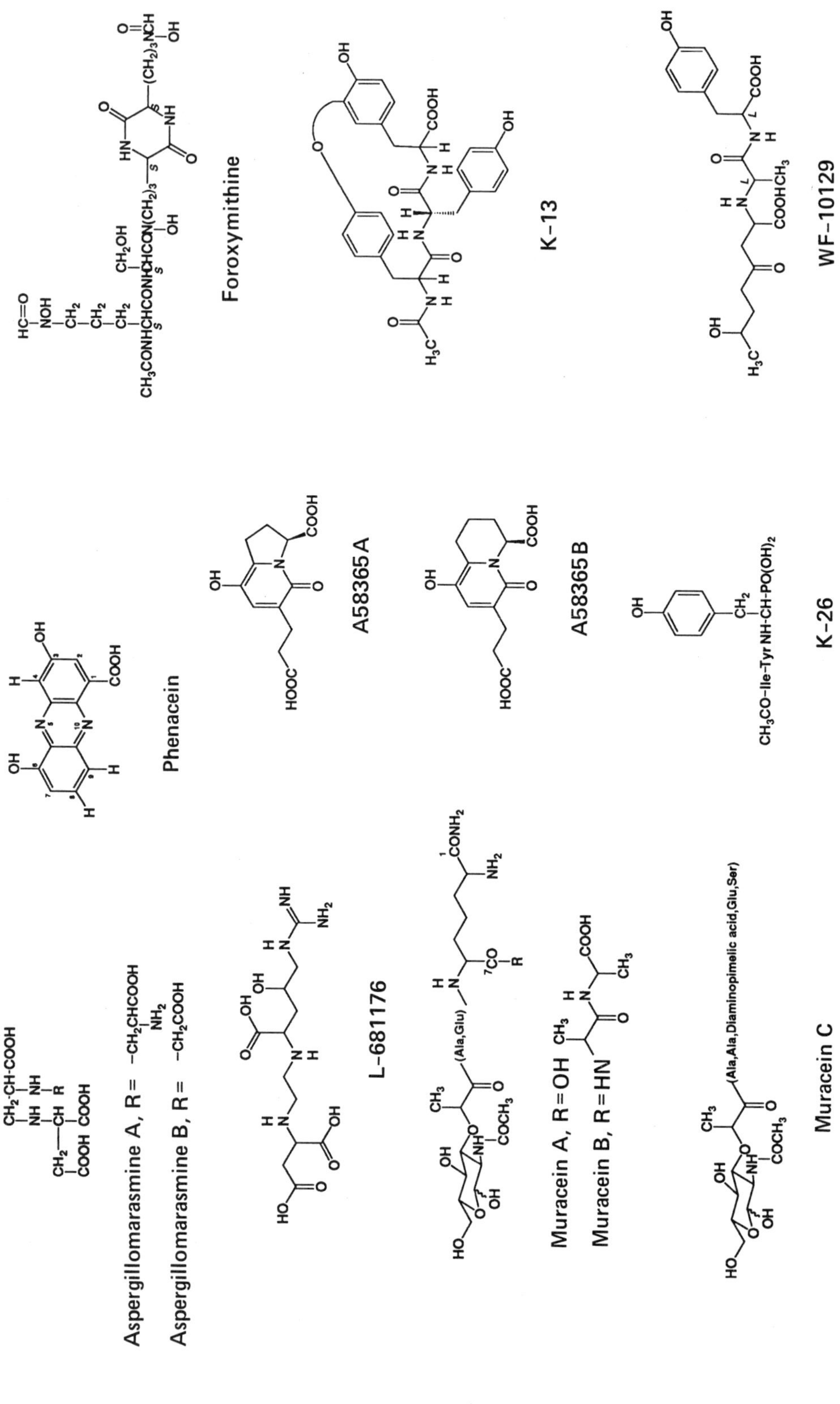

Figure 7.2 Structures of angiotensin-converting enzyme inhibitors from microorganisms.

Method (Knauss et al., 1959; Endo, 1987)

Preparation of the enzyme Male Wistar-Imamichi rats are fed a 2% cholestyramine diet with 8% added corn oil for 5 days to elevate the hepatic HMG-CoA reductase activity. The livers are homogenized in 2 vol of a cold buffer containing 100 mM potassium phosphate, pH 7.4, 15 mM nicotinamide, and 2 mM MgCl$_2$, and centrifuged at 12,000 × g for 30 min. The supernatant is further fractionated by centrifugation at 105,000 × g for 60 min, resulting in supernatant (cytosolic) and pellet (microsomal) fractions. The cytosolic fraction is subjected to 40%–80% (NH$_4$)$_2$SO$_4$ precipitation, dissolved in 100 mM potassium phosphate buffer, pH 7.4, and dialyzed against the same buffer at 4°C for 6 h.

Enzyme assay Incubation mixture in a final volume of 0.2 ml is as follows: 1 mM ATP, 10 mM glucose-1-phosphate, 6 mM glutathione, 6 mM MgCl$_2$, 40 μM CoA, 0.25 mM NAD$^+$, 0.25 mM NADP$^+$, 100 mM potassium phosphate buffer, pH 7.4, 1 mM [1-^{14}C]acetate (3.0 Ci mol^{-1}), microsomes (0.15–0.20 mg protein), and the (NH$_4$)$_2$SO$_4$ fraction (1.5–2.0 mg protein). After incubation at 37°C for 60 min, they are treated with 1 ml of 15% alcoholic KOH. The reaction mixtures are saponified at 75°C for 60 min and extracted with light petroleum. Aliquots of the extracts (nonsaponifiable fraction) are evaporated at 75°C for 10 min and counted. Sterols isolated from another aliquot of these extracts as digitonides are dissolved in 0.5 ml of Hyamine 10-X hydroxide (Packard) and counted.

Inhibitors In the course of the screening efforts by Endo and his co-workers (Endo, 1979, 1987; Endo et al., 1976, 1985a, 1985b, 1986), Alberts et al. (1980), and Albers-Schönberg et al. (1981), compactin and related compounds (Figure 7.3) were isolated from microorganisms as inhibitors of HMG-CoA reductase. All these metabolites are structurally related to each other and exhibit specific inhibitory activities. In 1987, the Food and Drug Administration in the United States authorized the use of lovastatin (monacolin K, mevinolin) as a drug for the treatment of hypercholesterolemia. After that, its related compounds, pravastatin and simvastatin are also being used in medical fields.

7.4 Glucosyltransferase

Rationale *Streptococcus mutans* is believed to be the primary dental cariogenic bacterium in both humans and animals. The virulency of *S. mutans* depends on its ability to adhere to smooth enamel tooth surfaces, inducing cariogenic plaque formation and the production of high yields of organic acids responsible for demineralization of the enamel tooth surfaces. The adherence of *S. mutans* to smooth tooth surfaces is mediated by the de novo synthesis

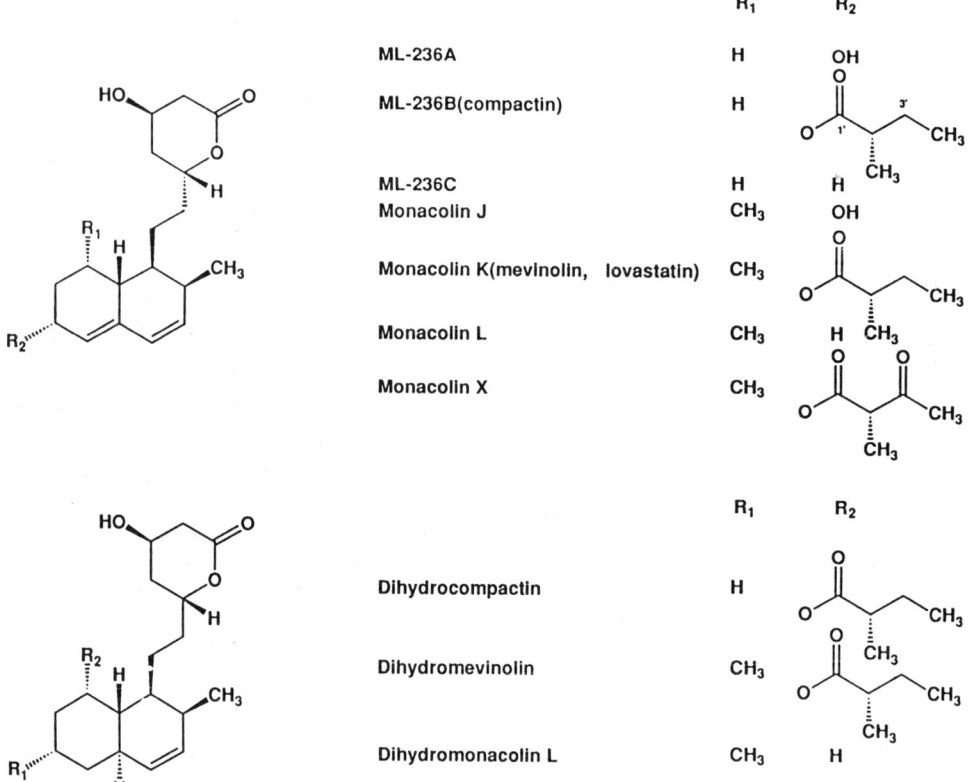

Figure 7.3 Structures of compactin and related compounds.

of adhesive, insoluble glucans from dietary sucrose caused by the action of cell-free or associated forms of glucosyltransferases (GTase, dextransucrase, EC 2.4.1.5) (Hamada and Slade, 1980). Thus, GTase inhibitors may be expected to be useful for preventing plaque formation and caries development.

Methods (Takashio and Okami, 1982)

Preparation of the enzyme Ten milliliters of cariogenic *S. mutans* E49 cultivated in brain-heart Infusion broth (BHI) overnight at 37°C is transferred to 1000 ml of BHI supplemented with 0.05% Tween 80 and incubated for 16–18 h at 37°C. The culture filtrate is precipitated with 50% $(NH_4)_2SO_4$. This precipitate is used for assay as crude enzyme preparation.

Enzyme assay Substrate solution (2.7 ml of 10 mM sucrose dissolved in 50 mM sodium phosphate buffer, pH 6.9, containing 0.04% sodium azide), 0.3 ml of test solution, and 50 μl of enzyme ($\approx$200 μg protein 3 ml^{-1} of reaction mixture) are mixed and incubated at 37°C for 14 h. The reaction is stopped by heating in a boiling water bath for 3 min, and the turbidity of the reaction mixture is measured at 600 nm.

Inhibitors   Among about 7000 cultured microbial broths tested by Takashio and Okami, only one strain, a species of *Streptomyces neyagawaensis*, was found to produce a low molecular weight inhibitor of GTase named ribocitrin (Figure 7.4; Ohnuki et al., 1981; Okami et al., 1981). This strain inhibited noncompetitively with sucrose; K_i was 2.7×10^{-5} M (pH 6.9). Ribocitrin at 8.9×10^{-5} M exhibited 50% inhibition of the adherence of *S. mutans* E49 to a glass surface.

Endo et al. (1983) found a strong inhibitor of GTase isolated from the culture filtrate of a strain of *Aspergillus terreus*. It was named mutastein, and the molecular weight was reported to be more than 20,000. It specifically inhibited the conversion of the soluble glucan into adherent insoluble glucan (Nakano et al., 1987). Dental caries development and plaque accumulation were markedly reduced when rats were fed a high-sucrose diet supplemented with 0.1%–0.4% mutastein. Mutastein is now on the market as an ingredient of chewing gum and other high-sucrose foods.

7.5   Aldose Reductase

Rationale   The administration of insulin to diabetics has prolonged their life, but its use has not prevented the late-onset complications associated with diabetes, such as cataracts, retinopathy, neuropathy, and nephropathy. An

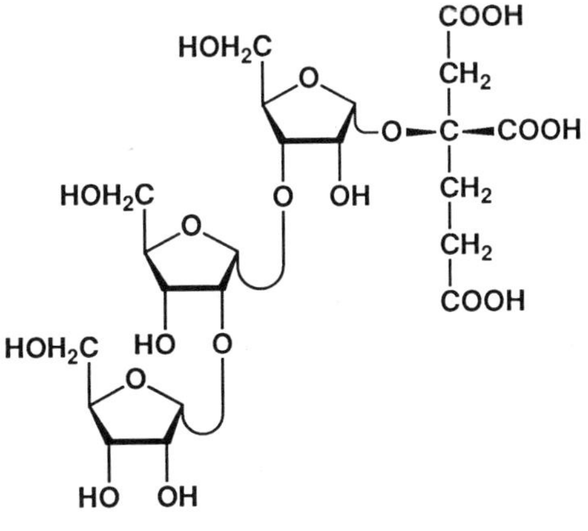

Figure 7.4   Structure of ribocitrin.

increased synthesis of intracellular polyols such as sorbitol is considered to play a critical pathophysiological role in the development of these diabetic complications. The sorbitol pathway (Figure 7.5) has been found in many tissues such as the lens, retina, nerve, and kidney in which diabetic complications appear. Aldose reductase (alditol: $NADP^+$ oxidoreductase, EC 1.1.1.21), which is involved in the conversion of glucose to sorbitol, is a key enzyme in the polyol pathway (Kador et al., 1985).

Under normal physiological conditions in many tissues, aldose reductase and hexokinase compete each other for the utilization of glucose. The affinity of hexokinase for glucose is greater than that of aldose reductase so that glucose is preferentially phosphorylated. On the other hand, under nonphysiological conditions, such as in diabetes, the concentration of sorbitol formed by aldose reductase increases because hexokinase is saturated by high levels of glucose. The intracellular accumulation of sorbitol leads to hyperosmotic conditions, which are associated with the diabetic complications.

Therefore, it is possible that cell function in diabetics can be maintained at a normal level by the inhibition of aldose reductase activity. In fact, aldose reductase inhibitors obtained by chemical synthesis are known to reduce tissue sorbitol content in diabetic animals and are now under clinical trials.

Methods (Hayman and Kinoshita, 1965)

Preparation of the enzyme   The lenses of calf eyes are homogenized and centrifuged at $10,000 \times g$ for 15 min. The supernatant fluid is saturated with 50%–75% $(NH_4)_2SO_4$. The precipitate obtained by centrifugation is dissolved in 0.05 M NaCl and is used as an enzyme preparation.

Enzyme assay   Oxidation of NADPH to $NADP^+$ is assayed by following UV absorption at 340 nm with a spectrophotometer. A reaction mixture (3 ml) containing 100 mM phosphate buffer, pH 6.0, 0.04 mM NADPH, an enzyme solution, and 0.5 mM DL-glyceraldehyde is incubated for 3 min at 37°C and then subjected to determination of the decrease of A_{340}. DL-Glyceraldehyde

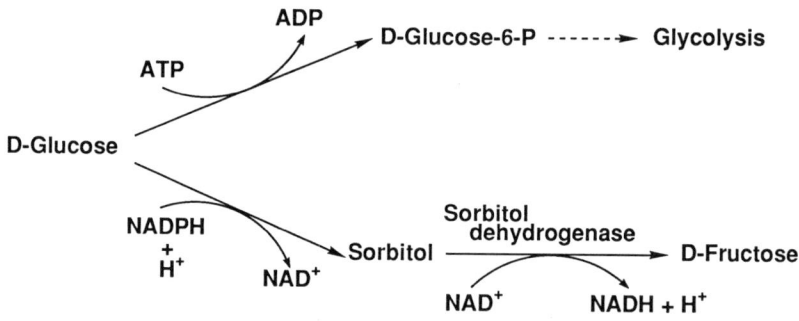

Figure 7.5 Pathways of glucose metabolism.

is usually used as a substrate for the enzyme reaction because the K_m value (3×10^{-5} M) is much smaller than that of glucose (7×10^{-2} M).

Inhibitors Fujisawa's group (Namiki et al., 1987; Nishikawa et al., 1987) found a new inhibitor of aldose reductase, WF-3681 (Figure 7.6), from the culture filtrate of a fungus, *Chaetomella raphigera* No. 3681. WF-3681 inhibits aldose reductase uncompetitively with the substrate DL-glyceraldehyde (IC$_{50}$, 2.5×10^{-7} M). WF-3681 achieved a high level in blood when it was administered to mice p.o. and decreased the sorbitol concentration in the sciatic nerve of mice under diabetic conditions caused by the administration of streptozotocin.

Toyo Jozo's group (Yaginuma et al., 1988) found another new inhibitor of aldose reductase, aldostatin, which was isolated from a culture broth of *Pseudomonas zonatum*. Aldostatin inhibits calf lens aldose reductase uncompetitively with DL-glyceraldehyde (IC$_{50}$, 1.2×10^{-6} M). When they used rat lens prepared as described by Peterson et al. (1979), aldostatin suppressed sorbitol accumulation in the lens in the presence of high concentrations of glucose.

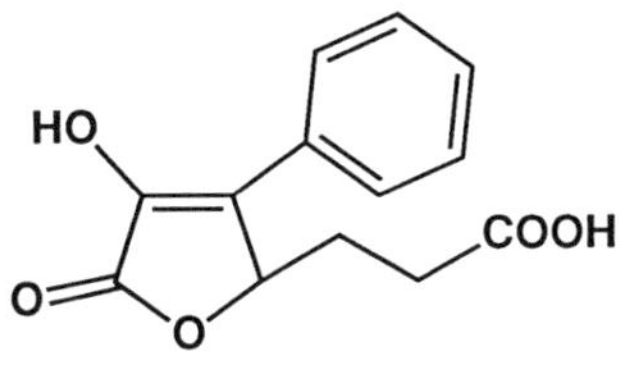

WF-3681

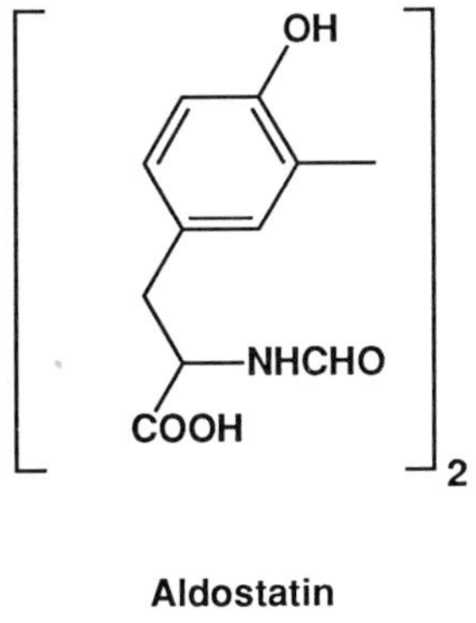

Aldostatin

Figure 7.6 Structures of aldose reductase inhibitors.

7.6 Aminopeptidase B

Rationale Aminopeptidase B (EC 3.4.11.6) hydrolyses an N-terminal peptide bond containing L-arginine or L-lysine. This enzyme is widely distributed among mammalian tissues and cells including rat liver, bovine pituitary gland, tumor cells, polymorphonuclear leucocytes, macrophages, and lymphocytes (Aoyagi and Umezawa, 1980; Aoyagi et al., 1976, 1977, 1978a). It has been suggested that aminopeptidase is located not only in the cytoplasm but also on the cell membrane without being released extracellularly. It is possible that exopeptidases have a strong effect on cell surfaces compared to endopeptidases and are involved in various cellular phenomena. Thus, the screening for inhibitors against aminopeptidase B was started to elucidate the biological role of this enzyme in various functions of cells.

Methods

General assay

Preparation of the enzyme (Hopsu et al., 1966) Aminopeptidase B was purified from rat liver using fractionation by $(NH_4)_2SO_4$, gel filtration, and column chromatography with substituted celluloses and hydroxyapatite.

Enzyme assay (Umezawa et al., 1976a) The reaction mixture consists of 0.25 ml of 2 mM L-arginine-β-naphthylamide, 0.5 ml of 0.1 M Tris-HCl buffer, pH 7.0, and 0.1 ml of water with or without an inhibitor. After preincubation at 37°C for 3 min, the reaction is started by adding 0.15 ml of aminopeptidase B solution. Exactly 30 min later, 1.0 ml of a solution of the stabilized diazonium salt Garnet GBC (1 mg ml^{-1}) in 1.0 M acetic acid buffer, pH 4.2, containing 10% Tween 20, is added. After standing for 15 min at room temperature, absorbance at 525 nm is measured.

Assay using intact Ehrlich ascites carcinoma (EAC) cells (Sano et al., 1986) The reaction mixture contains 0.1 ml of 3 mM L-arginine-β-naphthylamide, 0.7 ml of Hanks' balanced salt solution, and 10 µl of water with or without the inhibitor. After incubation for 3 min at 37°C, 0.2 ml of an EAC cell suspension is added. Exactly 30 min later, incubation is stopped by adding 3 ml of the stabilized diazonium salt Garnet GBC (0.3 mg ml^{-1}) in 1.0 M acetate buffer, pH 4.2, containing 3% Tween 20. The mixture is left for 15 min at room temperature, centrifuged, and its absorbance is measured at 525 nm.

Inhibitors Umezawa et al. (1976a) obtained the aminopeptidase inhibitor bestatin (ubenimex; Figure 7.7) from the culture filtrate of *Streptomyces olivoreticuli*. The inhibitor also has immunomodulatory and antitumor activities.

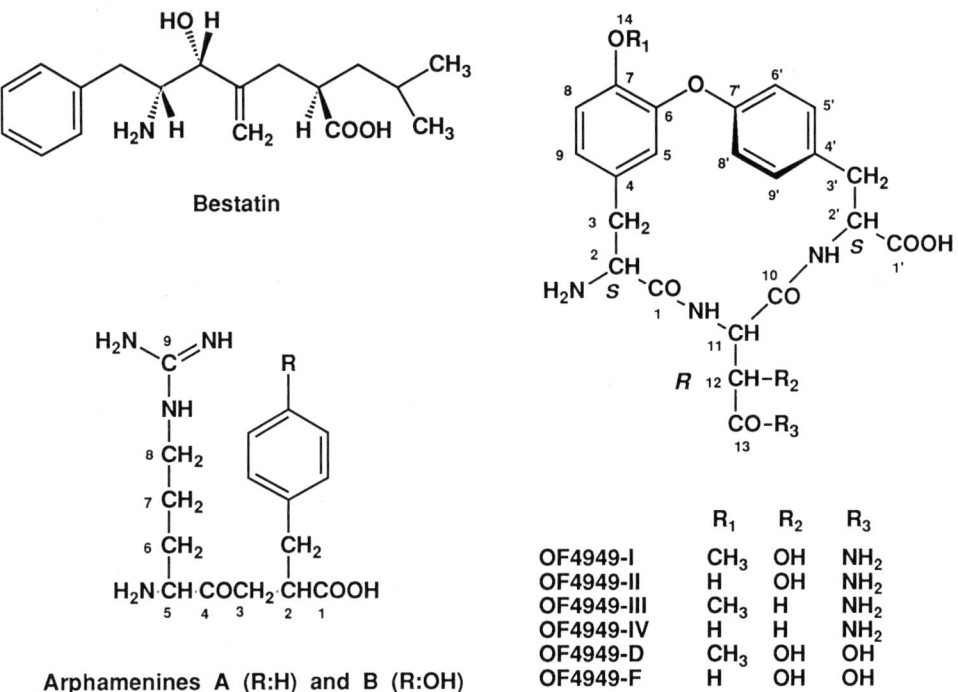

Figure 7.7 Structures of aminopeptidase B inhibitors.

It inhibits aminopeptidase B (K_i, 6×10^{-8} M) and leucine aminopeptidase (K_i, 2×10^{-8} M).

Bestatin has been shown to affect various pathoimmunological processes. (1) Bestatin enhanced the effects in delayed-type hypersensitivity to sheep red blood cells (Umezawa et al., 1976b). (2) It inhibited the enzyme activities in intact spleen cells and peritoneal macrophages from mice (Kuramochi et al., 1987). (3) It exhibited antitumor activity against mouse myeloid leukemia and enhanced the therapeutic effect of some cytotoxic antitumor agents; mitomycin C, 5-fluorouracil, and cisdichlorodiammineplatinum (Abe et al., 1984, 1985). (4) It activated macrophage tumoricidal activity (Schorlemmer, 1983). (5) It stimulated differentiation of polyclonal bone marrow progenitor cells to granulocytes (Ishizuka et al., 1981).

Bestatin in combination with other therapy has been used as a drug for the treatment of nonlymphocytic leukemia in adults since 1987 in Japan. Bestatin is the first enzyme inhibitor of microbial origin used clinically.

Umezawa et al. (1983) discovered another group of aminopeptidase inhibitors that were named arphamenines A and B (see Figure 7.7). The producing strain was classified as *Chromobacterium violaceum*. The arphamenines show strong inhibition against aminopetidase B, but they do not inhibit leucine aminopeptidase and aminopeptidase A. They are competitive with the substrate. The K_i values of arphamenines A and B are 2.5×10^{-9} M and 8.4×10^{-10} M, respectively. They also enhanced immune cellular responses.

Sano et al. (1986) screened culture broths of microorganisms using intact EAC cells and found OF4949-I, -II, -III, and -IV (see Figure 7.7) in the culture broth of *Penicillium rugulosum* OF4949. OF-4949-I and -II inhibit aminopeptidase B from EAC cells and enhance delayed-type hypersensitivity to sheep red blood cells in mice. The K_i values for both OF-4949-I and -II against L-arginine-β-naphthylamide were 8×10^{-9} M. OF4949-I showed antitumor activity against subcutaneous solid IMC carcinoma and protected against pulmonary metastases of Lewis lung carcinoma (Sano et al., 1987).

7.7 Adenosine Deaminase

Rationale Adenosine deaminase (adenosine aminohydrolase, ADA, EC 3.5.4.4), widespread in mammalian tissue, is involved in the regulation of the intracellular levels of adenosine and deoxyadenosine. These substrates control a number of important physiological functions and serve as precursors of nucleic acid biosynthesis. A congenital defect of ADA in lymphocytes and erythrocytes results in severe combined immunodeficiency, suggesting that the presence of ADA is essential for lymphocytic functions. However, many adenosine analogs, which are important in cancer chemotherapy, immunology, and virology, are substrates for ADA and are often inactivated by the enzyme. Thus, ADA inhibitors are responsible for alteration in adenosine and deoxyadenosine levels and lymphocytic growth and functions, and also enhance the chemotherapeutic effects of adenosine analogs (Agarwal, 1982).

Methods

General assay (Meier and Conscience, 1980) The substrate solution contains 1.4 mM adenosine and 1.4 mM dithiothreitol in 0.01 M Tris-HCl buffer, pH 7.5. The enzyme solution contains 1 mU ml^{-1} of calf intestinal ADA (Type III, Sigma Chem. Co.) in 0.01 M Tris-HCl buffer, pH 7.5. The reaction is initiated by the addition of 10 μl of enzyme solution to 280 μl of substrate solution with or without 10 μl of inhibitor. After incubation for 20 min at 20°C, 5 μl of the reaction mixture is spotted on a cellulose TLC plate and developed with 5% Na$_2$HPO$_4$ saturated with isoamyl alcohol. The A$_{265}$ of the spots corresponding to adenosine and inosine are measured using a TLC-UV scanner. Enzyme activity is calculated in terms of the decrease in A$_{265}$ as a result of the conversion of adenosine to inosine.

Assay using a test organism sensitive to the ADA inhibitor (Tunac and Underhill, 1985) *Enterococcus faecalis* PD 05045 is resistant to a number of antibiotics but especially sensitive to ADA inhibitors such as coformycin and deoxycoformycin. The assay plate contains this test organism and adenine sulfate. Samples are applied on the assay plate using 12.7-mm paper disks. After incubation at 37°C for 16–18 h, activity is measured as inhibitory zones.

Visual assay using an agar plate containing a pH indicator (Katsuragi et al., 1985) The agar plate contains 6 µg of adenosine deaminase, 0.5 mmol of adenosine, 0.25 mmol of potassium phosphate buffer, pH 5.9, 20 mg of phenol red, and 1 g of agar in 100 ml of water. Paper disks wetted with solutions of inhibitor are placed on the plate. After a few hours of incubation, yellow color zones remain around the edges of the paper disks containing inhibitor, while the background color turns to a purplish red. Color change by the pH indicator depends on ammonia liberation in the deamination reaction and increased pH.

Inhibitors Coformycin was discovered as a by-product of formycin fermentation by *Nocardia interforma* and *Streptomyces kaniharaensis* (Niida et al., 1967; Sawa et al., 1967a, 1967b). In the studies of antitumor and antimicrobial activities of formycin, it was found that a minor component prevented the inactivation (deamination) of formycin. The component isolated from the crude material was named coformycin.

Deoxycoformycin (Figure 7.8) was isolated from the fermentation broth of a strain of *Streptomyces antibioticus* in the course of screening studies on ADA inhibitors (Woo et al., 1974). The compound possesses a chemotherapeutic effect on hairy cell and adult T-cell leukemia (O'Dwyer et al., 1988). Coformycin and deoxycoformycin were tight-binding-type inhibitors (K_i, 2.1 × 10^{-10} M and 7.6 × 10^{-11} M, respectively).

During screening studies by Ōmura et al. (1985a, 1986a, 1986d), two

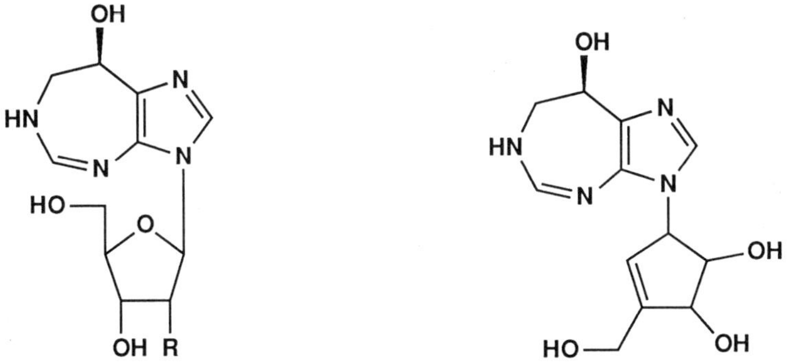

R
OH : Coformycin
H : Deoxycoformycin
Cl : Adechlorin

Adecypenol

Figure 7.8 Structures of adenosine deaminase inhibitors.

new inhibitors, adechlorin and adecypenol (see Figure 7.8), were isolated from the culture broths of *Actinomadura* sp. OMR-37 and *Streptomyces* sp. OM-3223, respectively. Adechlorin is an inhibitor of the tight-binding type (K_i, 5.3×10^{-10} M), while adecypenol is a semitight-binding inhibitor (K_i, 4.7×10^{-9} M). Adechlorin enhanced the antiviral activity of Ara-A against HeLa S3 cells infected with HSV-1. Adecypenol potentiated the antitumor activity of Ara-A against mouse leukemia L-1210 in vivo (Tanaka and Ōmura, 1988).

Tunac and Underhill (1985) independently isolated the same compound (adechlorin), using their screening method mentioned earlier, and named it 2'-chloropentostatin.

7.8 Protein Kinase C

Rationale Cyclic AMP (cAMP), cyclic GMP (cGMP), and Ca^{2+} are well recognized as intracellular second messengers for regulating many cellular functions such as secretion, contraction, phototransduction, cell division and differentiation, and alteration in the transport of ions. The biological effects of cAMP and cGMP have been proposed to be mediated by activation of cAMP-dependent protein kinase (A-PK) and cGMP-dependent protein kinase (C-PK), respectively, and the effects of Ca^{2+} are mediated by Ca^{2+}-binding proteins such as calmodulin. Recently, it became clear that the Ca^{2+} messenger system involves a mediation of Ca^{2+} by activating a phospholipid-sensitive, Ca^{2+}-dependent protein kinase (protein kinase C, C-PK, EC 2.7.1.37) in addition to calmodulin (Nishizuka, 1984a).

C-PK is widespread in a variety of organs and tissues and possesses more potent enzymatic activity than those of the other protein kinases found in many tissues such as brain and platelets. In living cells, diacylglycerol, formed from inositol phospholipid turnover stimulated by the extracellular informational signals, activates C-PK (Nishizuka, 1984b). This kinase is also activated directly by tumor promoters such as phorbol esters, suggesting that C-PK plays key roles in both signal transduction and cellular proliferation. Moreover, C-PK is proposed to be closely associated with secretion, platelet aggregation, and smooth muscle contraction. Thus, many researchers have been interested in inhibitors of C-PK. These may affect cellular proliferation and also may be useful in clarifying its in vivo functions.

Methods

General assay

Preparation of enzymes It is usually necessary to assay the inhibiting activities against not only C-PK but also A-PK and G-PK to determine the specificity of an inhibitor. C-PK is partially purified from rat brain using a DE-52 column (Whatman) (Kikkawa et al., 1982). The holoenzyme of A-PK type I is purified from rabbit skeletal muscle using DEAE-cellulose, CM-Sephadex, and Sephadex G-100 columns (Bachtel et al., 1977). G-PK from pig lung is partially purified using a DEAE-cellulose column (Lincoln et al., 1977).

Enzyme assay (Kase et al., 1987a)   C-PK activity is assayed using a reaction mixture containing, in a final volume of 0.25 ml: 20 mM Tris-HCl buffer, pH 7.5; 10 mM magnesium acetate; 200 µg ml^{-1} histone H-1; 3.2 µg ml^{-1} enzyme; 2.5 mM CaCl$_2$ (or 2.5 mM EGTA); 80 µg ml^{-1} phosphatidyl serine; 3.2 µg ml^{-1} diolein; and 5 µM [γ-^{32}P]ATP. The reaction mixture for the assay of A-PK contains, in a final volume of 0.25 ml: 40 mM phosphate buffer, pH 7.0; 10 mM magnesium acetate; 3.2 µM cAMP; 200 µg ml^{-1} histone H-1; enzyme; and 5 µM [γ-^{32}P]ATP. The reaction mixture of G-PK assay contains, in a final volume of 0.25 ml: 20 mM Tris-HCl buffer, pH 7.5; 100 mM magnesium acetate; 0.1 µM cGMP; 100 µg ml^{-1} histone IIA; enzyme; and 10 µM [γ-^{32}P]ATP. After incubation of the foregoing reaction mixtures for 3–5 min at 30°C, the reaction is terminated by the addition of 1 ml of 25% trichloroacetic acid. Acid-precipitable materials are collected on a nitrocellulose membrane filter and washed four times with 1 ml of 5% trichloroacetic acid. The radioactivity of the filter is determined using a liquid scintillation spectrometer.

Bleb-forming assay (Magae et al., 1988)   When K562, a human chronic myeloid leukemia cell, was treated with phorbol 12,13-dibutyrate or teleocidin, which are activators of C-PK, many blebs appeared on the cell surface of K562 within 10 min. This appearance of blebs is inhibited by staurosporine and H7, C-PK inhibitors. The bleb-forming assay satisfies the criteria (simplicity and specificity) required for preliminary screening of activators or inhibitors of C-PK.

Inhibitors   In the course of the screening studies for C-PK inhibitors by Kase et al. (1986, 1987a), K-252a and K-252b (Figure 7.9) were isolated from a culture broth of *Nocardiopsis* sp. and found to be potent inhibitors (K_i, 25 nM and 20 nM, respectively). Their inhibition types were competitive with ATP. K_i-252a was a nonselective inhibitor for protein kinases (K_i for A-PK, 18 nM; K_i for G-PK, 20 nM).

Staurosporine (see Figure 7.9), a microbial alkaloid discovered by Ōmura et al. (1977) that had been known to have antifungal and hypotensive activities, was found to be the most potent inhibitor of C-PK among the known inhibitors (K_i, 0.7 nM) and to inhibit both A-PK and G-PK (K_i, 7 nM and 8.5 nM, respectively) (Tamaoki et al., 1986). K-252a and staurosporine seriously affect the functions of various cells and tissues; for example, they cause inhibition of platelet aggregation and relaxation of vascular smooth muscle (Yamada et al., 1987).

Takahashi et al. (1987) have isolated UCN-01 (see Figure 7.9), a selective inhibitor of C-PK, from a culture broth of *Streptomyces* sp., and UCN-01

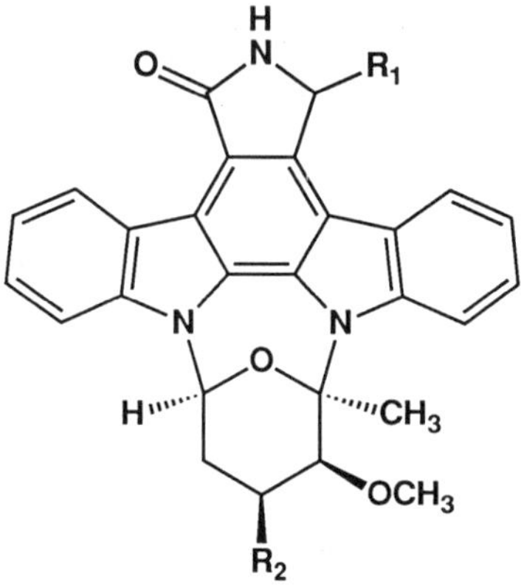

Figure 7.9 Structures of staurosporine-related inhibitors of protein kinase C.

	R_1	R_2
Staurosporine	H	NHCH$_3$
UCN-01 and 02	OH	NHCH$_3$
RK-286C	H	OH

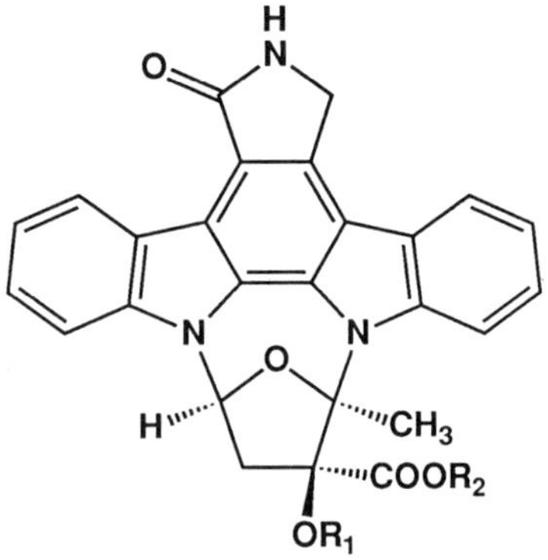

	R_1	R_2
K-252a	H	CH$_3$
K-252b	H	H
KT5720	H	(CH$_2$)$_5$CH$_3$
KT5822	CH$_3$	CH$_3$

exhibited antitumor activity against murine lymphotic leukemia P388 in mice. Magae et al. (1988) found the C-PK activator tautomycin (Figure 7.10) using the bleb-forming assay. It has been revealed that tautomycin is a protein phosphatase inhibitor and causes an accumulation of phosphorylated proteins (Magae et al., 1990).

Staurosporine homologs, UCN-01 and UCN-02 (Takahashi et al., 1989) and PK-286C (Osada et al., 1990) (see Figure 7.9) were isolated from the broths of *Streptomyces* sp. Calphostin A, B, C, D, and I, isolated from a broth of *Cladosporium cladosporiodes* (Kobayashi et al., 1989) are C-PK inhibitors of 3,10-perylenequione skeleton (see Figure 7.10).

7.9 Tyrosine-Specific Protein Kinase

Rationale Epidermal growth factor (EGF) and platelet-derived growth factor (PDGF) can induce DNA synthesis and proliferation of specific target cells. The binding of the growth factor to its receptor stimulates phosphorylation of a tyrosine residue of the receptor and triggers cell proliferation. The receptors of other growth factors such as insulin, insulin-like growth factor, and the transforming growth factors also were found to have tyrosine-specific protein kinase activities (tyrosine kinase, EC 2.7.1.112).

Moreover, the transforming proteins of RNA tumor viruses as well as growth factor receptors have tyrosine kinase activity. The oncogenes such as

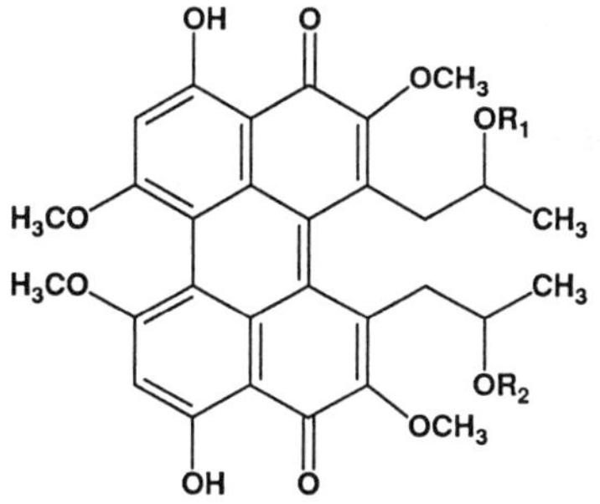

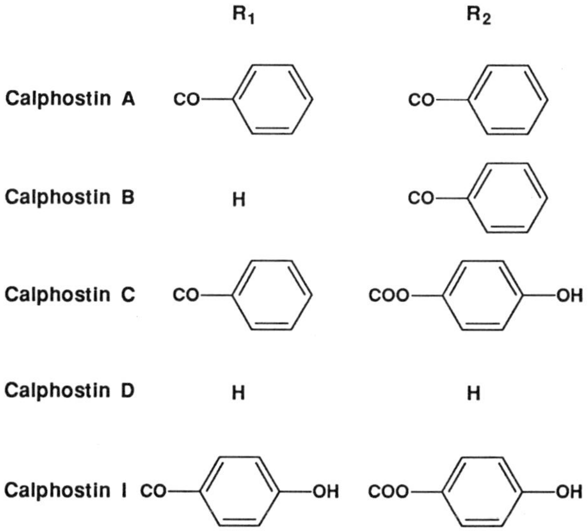

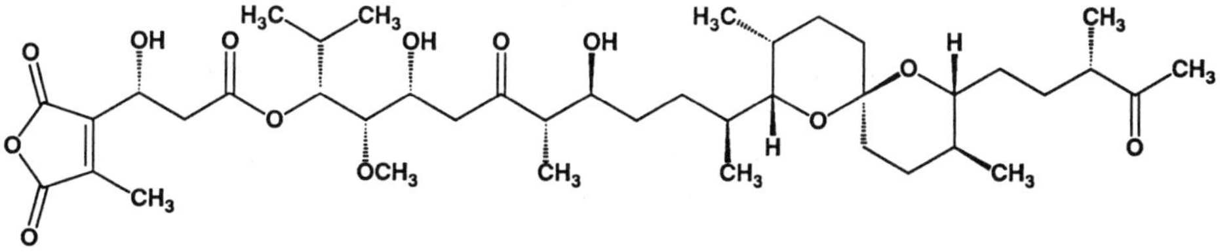

Chapter 7 General Screening of Enzyme Inhibitors 143

	R₁	R₂
Calphostin A	CO—C₆H₅	CO—C₆H₅
Calphostin B	H	CO—C₆H₅
Calphostin C	CO—C₆H₅	COO—C₆H₄—OH
Calphostin D	H	H
Calphostin I	CO—C₆H₄—OH	COO—C₆H₄—OH

Tautomycin

Figure 7.10 Structures of tautomycin and calphostins.

src, *yes*, *fps*, and *abl*, which encode transforming proteins possessing tyrosine kinase activity, transform normal cells and cause various types of tumors in human and animals (Bishop, 1983; Klein and Klein, 1985).

It has been shown that the transforming protein of simian sarcoma virus has a close structural and functional relationship to PDGF, and that the EGF receptor matches very closely to a part of the amino acid sequence of the v-*erbB* transforming protein of avian erythroblastosis virus (Downward et al., 1984).

These findings suggest the possibility that tyrosine kinase activity of growth factor receptors and *src*-related oncogenes is associated with the regulation of normal and malignant cell proliferation. Thus, specific inhibitors against tyrosine kinase may be useful for inhibiting cancer development.

Methods

General assay (Umezawa et al., 1986a)   The membrane fraction of the human epidermoid carcinoma cell line A-431 prepared by the method of Thom et al. (1977) can be used as an enzyme preparation. The reaction mixture, containing 1 mM $MnCl_2$, 10 ng EGF, 40 µg protein of A-431 membrane fraction, 75 µg albumin, 3 µg histone, and 20 mM HEPES buffer, pH 7.4, in a final volume of 50 µl, is preincubated for 10 min in the presence or absence of inhibitor. The reaction is initiated by the addition of 10 µl of [γ-^{32}P]ATP. After incubation for 30 min at 0°C, an aliquot of the mixture (50 µl) is pipetted onto Whatman 3MM filter paper and dropped immediately into a beaker of cold 10% trichloroacetic acid (TCA) containing 0.01 M sodium pyrophosphate. The filter papers are washed with TCA solution, extracted with alcohol, and dried. The radioactivity is measured by a scintillation counter.

Assay using autoradiography (Ogawara et al., 1986)   After a reaction mixture similar to that just described is incubated for 5 min at 0°C, the reaction is terminated by the addition of Laemmli's sample buffer (Laemmli, 1970) and by boiling for 2 min. The sample is analysed by sodium dodecyl sulfate (SDS) polyacrylamide gel electrophoresis (PAGE) and autoradiography.

Inhibitors   Umezawa et al. (1986a) reported a novel compound, erbstain (Figure 7.11), isolated as a tyrosine kinase inhibitor from a strain closely related to *Streptomyces viridosporus*. It exhibited a strong inhibitory activity against tyrosine kinase (IC_{50}, 0.55 µg ml^{-1}) and a very weak activity against cAMP-dependent protein kinase (IC_{50}, 100 µg ml^{-1}).

Two known compounds were also found to exhibit potent inhibitory activity against tyrosine kinase during screening work with microorganisms. An isoflavone compound, orobol (see Figure 7.11), isolated from *Streptomyces neyagawaensis* var. *orobolere*, exhibits inhibitory activity (IC_{50}, 3 g ml^{-1}) (Umezawa et al. 1986a). Another isoflavone compound, genistein (Figure 7.11),

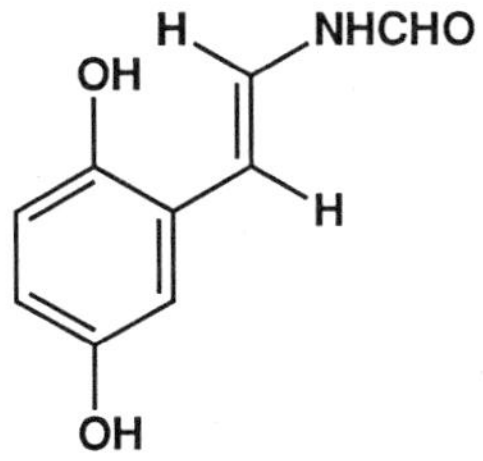

Erbstatin

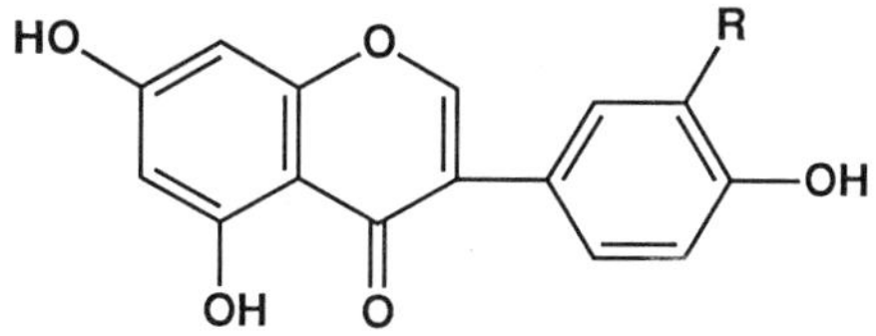

Genistein (R:H)
Orobol (R:OH)

Figure 7.11 Structures of inhibitors of tyrosine-specific protein kinase.

isolated from *Pseudomonas* sp., also inhibited the enzyme activity (IC_{50}, 0.7 µg ml^{-1}) (Ogawara et al., 1986).

7.10 Concluding Remarks

There are three important problems to face when starting to screen for new enzyme inhibitors useful as drugs: (1) The first is whether a correlation between enzyme inhibition and pharmacological effect can be established; (2) The second is whether a conventional and sensitive assay method is available for the screening (the lack of a suitable method prevents efficient screening of large numbers of microbial samples). In general, a chromogenic or radioactive substrate (or cofactor) is used in the enzyme assay. When a substrate is not chromogenic, a chromogenic derivative of the substrate sometimes can be used successfully for the enzyme assay. An example in which the development of a unique conventional method made efficient screening possible can be seen in the angiotensin-converting enzyme assay as described previously. The screening of enzyme inhibitors based on antimicrobial activity or cytotoxicity (see Chapter 8) is also conventional, although such methods are not often applied.

(3) The third problem concerns whether an in vivo evaluation system for pharmacological activity is available. The evaluation needs to be done, if possible, before the active principle is isolated as a pure material. If an isolated compound does not show a strong pharmacological activity when established during in vivo evaluation, the compound often can be subjected to chemical modification to give derivatives or analogs with higher activity. Microorganisms often produces good lead compounds for drug design, with structures that one cannot predict from the structure of the substrates.

As mentioned, the screening of enzyme inhibitors from microorganisms requires the knowledge and techniques obtained in many fields of natural sciences—microbiology, biochemistry, chemistry, and pharmacology. These disciplines concern the isolation and cultivation of microorganisms, measurement of enzyme reactions, isolation and characterization of the inhibitors and their evaluation in animals, etc. During the past three decades, about 300 new bioactive compounds have been discovered during screening for enzyme inhibitors. Among them, bestatin has been used as a drug for the treatment of cancer in Japan, and, also in Japan, mutastein has been for preventing decay of teeth. Since 1987, HMG-CoA reductase inhibitors such as lovastatin, pravastatin, and simvastatin have been used for the treatment of hypercholesterolemia in the world. Several compounds shown in Table 7.2 are under clinical trial. In the near future, the correlation between enzyme function and disease processes will be further clarified and more enzyme inhibitors useful as pharmacologically active drugs will be discovered. Additionally, antagonists and agonists of hormone and autacoid receptors will be discovered and developed as useful drugs for the treatment of various diseases.

References

Abe, F., Shibuya, K., Ashizawa, J., Takahashi, K., Horinishi, H., Matsuda, A., Ishizuka, M., Takeuchi, T. and Umezawa, H. 1985. Enhancement of antitumor effect of cytotoxic agents by bestatin. *Journal of Antibiotics* (Tokyo) 38:411–414.

Abe, F., Shibuya, K., Uchida, M., Takahashi, K., Horinishi, H., Matsuda, A., Ishizuka, M., Takeuchi, T. and Umezawa, H. 1984. Effect of bestatin on syngeneic tumors in mice. *Gann* 75:89–94.

Agarwal, R. P. 1982. Inhibitors of adenosine deaminase. *Pharmacology Therapeutics* 17:399–429.

Albers-Schönberg, G., Joshua, H., Lopez, M. B., Hensens, O. D., Springer, J. P., Chen, J., Ostrove, S., Hoffman, C. H., Alberts, A. W. and Patchett, A. A. 1981. Dihydromevinolin, a potent hypocholesterolemic metabolite produced by *Aspergillus terreus*. *Journal of Antibiotics* (Tokyo) 34:507–512.

Alberts, A. W., Chen, J., Kuron., Hunt, V., Huff, J., Hoffman, C., Rothrock, J., Lopez, M., Joshua, H., Harris, E., Patchett, A., Monaghan, R., Currie, S., Stapley, E., Albers-Schonberg, G., Hensens, O., Hirshfield, J., Hoogsteen, K., Liesch, J. and Springer, J. 1980. Mevinolin. A highly potent competitive inhibitor of hydroxymethylglutaryl-coenzyme A reductase and a cholesterol-lowering agent. *Proceedings of the National Academy of Sciences of the United States of America* 77:3957–3961.

Ando, O., Satake, H., Itoi, K., Sato, A., Nakajima, M., Takahashi, S., Haruyama, H., Ohkuma, Y., Kinoshita, T. and Enokita, R. 1991. Trehazolin, a new trehalase inhibitor. *Journal of Antibiotics* (Tokyo) 44:1165–1168

Ando, T., Okada, S., Uchida, I., Hemmi, K., Nishikawa, M., Tsurumi, Y., Fujie, A., Yoshida, K. and Okuhara, M. 1987. WF- 10129, a novel angiotensin converting enzyme inhibitor produced by a fungus, *Doratomyces putredinis*. *Journal of Antibiotics (Tokyo)* 40:468–475.

Aoki, M., Ohtsuka, T., Yamada, M., Ohba, Y., Yoshizaki, H., Yasuno, H., Sano, T., Watanabe, J., Yokose, K. and Seto, H. 1991. Cyclothiazomycin, a novel polythiazole-containing peptide with renin inhibitory activity. Taxonomy, fermentation, isolation and physico-chemical characterization. *Journal of Antibiotics* (Tokyo) 44:582–588.

Aoyagi, T. and Umezawa, H. 1980. Hydrolytic enzymes on the cellular surface and

their inhibitors found in microorganisms. *Proceedings of the Federation of European Biochemical Society Meeting* 61:89–99.
Aoyagi, T., Ishizuka, M., Takeuchi, T. and Umezawa, H. 1977. Enzyme inhibitors in relation to cancer therapy. *Japanese Journal of Antibiotics* 30 (Suppl.): S121–S132.
Aoyagi, T., Hazato, T., Kumagai, M., Hamada, M., Takeuchi, T. and Umezawa, H. 1975a. Isoflavone rhamunosides, inhibitors of beta- galactosidase produced by actinomycetes. *Journal of Antibiotics* (Tokyo) 28:1006–1008.
Aoyagi, T., Kumagai, M., Hazato, T., Hamada, M., Takeuchi, T. and Umezawa, H. 1975b. Pyridindolol, a new beta-galactosidase inhibitor produced by actinomycetes. *Journal of Antibiotics* (Tokyo) 28:555–557.
Aoyagi, T., Yamamoto, T., Kojiri, K., Kojima, F., Hamada, M., Takeuchi, T. and Umezawa, H. 1978a. Forphenicine, an inhibitor of alkaline phosphatase produced by actinomycetes. *Journal of Antibiotics* (Tokyo) 31:244–246.
Aoyagi, T., Tobe, H., Kojima, F., Hamada, M., Takeuchi, T. and Umezawa, H. 1978b. Amastatin, an inhibitor of aminopeptidase A, produced by actinomycetes. *Journal of Antibiotics* (Tokyo) 31:636–639.
Aoyagi, T., Nagai, M., Iwabuchi, M., Liaw, W. S., Andoh, T. and Umezawa, H. 1978c. Aminopeptidase activities on the surface of mammalian cells and their alterations associated with transformation. *Cancer Research* 38:3505–3508.
Aoyagi, T., Yoshida, S., Nakamura, Y., Shigihara, Y., Hamada, M. and Takeuchi, T. 1990. Probestin, a new inhibitor of aminopeptidase M, produced by *Streptomyces azureus* MH663-2F6. *Journal of Antibiotics* (Tokyo) 43:143–148.
Aoyagi, T., Yagisawa, M., Kumagai, M., Hamada, M., Okami, Y., Takeuchi, T. and Umezawa, H. 1971. An enzyme inhibitor, panosialin, produced by *Streptomyces*. I. Biological activity, isolation and characterization of panosialin. *Journal of Antibiotics* (Tokyo) 24:860–869.
Aoyagi, T., Yagisawa, Y., Kumagai, M., Hamada, M., Morishima, H., Takeuchi, T. and Umezawa, H. 1973. New pepstatins, pepstatins Bu, Pr and Ac produced by *Streptomyces*. *Journal of Antibiotics* (Tokyo) 27:539–541.
Aoyagi, T., Suda, H., Nagai, M., Ogawa, K., Suzuki, J., Takeuchi, T. and Umezawa, H. 1976. Aminopeptidase activities on the surface of mammalian cells. *Biochimica et Biophysica Acta* 452:131–143.
Aoyagi, T., Takeuchi, T., Matsuzaki, A., Kawamura, K., Kondo, S., Mamada, M., Maeda, K. and Umezawa, H. 1969. Leupeptins, new protease inhibitors from actinomycetes. *Journal of Antibiotics* (Tokyo) 22:283–286.
Aoyagi, T., Morishima, H., Kojiri, K., Yamamoto, T., Kojima, F., Nagaoka, K., Hamada, M., Takeuchi, T. and Umezawa, H. 1989a. Alphostatin, an inhibitor of alkaline phosphatase of bovine liver produced by *Bacillus megaterium*. *Journal of Antibiotics* (Tokyo) 42:486–488.
Aoyagi, T., Yamamoto, T., Kojiri, K., Morishima, H., Nagai, M., Hamada, M., Takeuchi, T. and Umezawa, H. 1989b. Mannostatins A and B: New inhibitors of alpha-D-mannosidase, produced by *Streptoverticillium verticillus* var. *zuintum* ME3-AG3: Taxonomy, production, isolation, physico-chemical properties and biological activities. *Journal of Antibiotics* (Tokyo) 42:883–889.
Aoyagi, T., Yoshida, S., Harada, S., Okuyama, A., Nakayama, C., Yoshida, T., Hamada, M., Takeuchi, T. and Umezawa, H. 1988. Benadrostin, new inhibitor of poly(ADP-ribose)synthetase, produced by actinomycetes. I. Taxonomy, production, isolation, physico-chemical properties and biological activities. *Journal of Antibiotics* (Tokyo) 41:1009–1014.
Aoyagi, T., Yoshida, S., Matsuda, N., Ikeda, T., Hamada, M. and Takeuchi, T. 1991a. Leuhistin, a new inhibitor of aminopeptidase M, produced by *Bacillus laterosporus* BMI156-14F1. 1. Taxonomy, production, isolation, physico-chemical properties and biological activities. *Journal of Antibiotics* (Tokyo) 44:573–578.
Aoyagi, T., Nagai, M., Ogawa, K., Kojima, F., Okada, M., Ikeda, T., Hamada, M. and

Takeuchi, T. 1991b. Poststatin, a new inhibitor of proryl endopeptidase, produced by *Streptomyces viridochromogenes* MH534-30F3. 1. Taxonomy, production, isolation, physico-chemical properties and biological activities. *Journal of Antibiotics* (Tokyo) 44:949-955.

Arai, M., Sumida, M., Nakatani, S. and Murao, S. 1983. A novel beta-amylase inhibitor. *Agricultural and Biological Chemistry* 47:183-185.

Atsumi, S., Umezawa, K., Iinuma, H., Naganawa, H., Nakamura, H., Iitaka, Y. and Takeuchi, T. 1990. Production, isolation and structure determination of novel beta-glucosidase inhibitor, cyclophellitol, from *Phellinus* sp. *Journal of Antibiotics* (Tokyo) 43:49-53.

Bachtel, P. J., Beavo, J. A. and Krebs, E. G. 1977. Purification and characterization of catalytic subunit of skeletal muscle adenosine 3':5'-monophosphate-dependent protein kinase. *Journal of Biological Chemistry* 252:2691-2697.

Bishop, J. M. 1983. Cellular oncogenes and retroviruses. *Annual Review of Biochemistry* 52:301-354.

Brannon, D. R. and Fuller, R. W. 1974. Microbial production of pharmacologically active compounds other than antibiotics. *Lloydia* (Cincinnati) 37:134-146.

Brown, A. G., Butterworth, D., Cole, M., Hanscomb, G., Hood, J. D., Reading, C. and Rolinson, G. N. 1976. Naturally-occurring beta-lactamase inhibitors with antibacterial activity. *Journal of Antibiotics* (Tokyo) 29:668-669.

Bush, K., Henry, P. R. and Slusarchyk, D. S. 1984a. Muraceins—muramyl peptides produced by *Nocardia orientalis* as angiotensin-converting enzyme inhibitors. *Journal of Antibiotics* (Tokyo) 37:330-335.

Bush, K., Henry, P. R., Souser-Woehleke, M., Trejo, W. H. and Slusarchyk, D. S. 1984b. Phenacein—an angiotensin converting enzyme inhibitor produced by a *Streptomycete*. *Journal of Antibiotics* (Tokyo) 37:1308-1312.

Chimura, H., Sawa, T., Kumada, Y., Naganawa, H., Matsuzaki, M., Takita, T., Hamada, M., Takeuchi, T. and Umezawa, H. 1975. New isoflavones, inhibiting catechol-O-methyltransferase, produced by *Streptomyces*. *Journal of Antibiotics* (Tokyo), 28:619-626.

Chimura, H., Sawa, T., Kumada, Y., Nakamura, F., Matsuzaki, M., Takita, T., Takeuchi, T. and Umezawa, H. 1973b. 7-O-methylspinochrome B and its 6-(3-hydroxy-N-butyl)-derivative, catechol-O-methyl transferase inhibitors, produced by Fungi imperfecti. Journal of Antibiotics (Tokyo) 26:618-620.

Chimura, H., Sawa, T., Takita, T., Matsuzaki, M., Takeuchi, T., Nagatsu, T. and Umezawa, H. 1973a. Methylspinazarin and dihydromethylspinazarin, catechol-O-methyl transferase inhibitors produced by *Streptomyces*. *Journal of Antibiotics* (Tokyo) 26:112-114.

Cheung, H. S., Wang, F. L., Ondetti, M. A., Sabo, E. F. and Cushman, D. W. 1980. Binding of peptide substrates and inhibitors of angiotensin-converting enzyme. *Journal of Biological Chemistry* 255:401-407.

Cushman, D. W. and Ondetti, M. A. 1980. Inhibitors of angiotensin- converting enzyme. In: Ellis, G. P. and West, G. B. (editors), *Progress in Medicinal Chemistry*, Vol. 17, pp. 42-104. Elsevier/North-Holland, Amsterdam.

Downward, J., Yarden, Y., Mayes, E., Scrace, G., Notty, N., Stockwell, P., Ullrich, A., Schlessinger, J. and Waterfieke, M. D. 1984. Close similarity of epidermal growth factor receptor and v-*erb*-B oncogene protein sequences. *Nature* (London) 307:521-526.

Endo, A. 1979. Monacolin K, a new hypocholesterolemic agent produced by a *Monascus* species. *Journal of Antibiotics* (Tokyo) 32:852-854.

Endo, A. 1987. Drugs inhibiting HMG-CoA reductase. *International Encyclopedia of Pharmacology and Therapeutics* 31:257-267.

Endo, A., Hayashida, O. and Murakawa, S. 1983. Mutastein, a new inhibitor of ad-

hesive-insoluble glucan synthesis by glucosyltransferases of *Streptococcus mutans*. *Journal of Antibiotics* 36 (Tokyo):203–207.
Endo, A., Kuroda, M. and Tsujita, Y. 1976. ML-236a, ML-236b, and ML- 236c, new inhibitors of cholesterogenesis produced by *Penicillium citrinum*. *Journal of Antibiotics* (Tokyo) 29:1346–1348.
Endo, A., Komagata, D. and Shimada, H. 1986. Monacolin M, a new inhibitor of cholesterol biosynthesis. *Journal of Antibiotics* (Tokyo) 39:1670–1673.
Endo, A., Hasumi, K., Nakamura, T., Kunishima, M. and Masuda, M. 1985a. Dihydromonacolin L and monacolin X, new metabolites those inhibit cholesterol biosynthesis. *Journal of Antibiotics* (Tokyo) 38:321–327.
Endo, A., Hasumi, K. and Negishi, S. 1985b. Monacolins J and L, new inhibitors of cholesterol biosynthesis produced by *Monascus ruber*. *Journal of Antibiotics* (Tokyo) 38:420–422.
Endo, A., Takeshima, H. and Kuwabara, K. 1985c. Acetyl-CoA carboxylase inhibitors from the fungus *Gongronella butreli*. *Journal of Antibiotics* (Tokyo) 38:599–604.
Endo, A., Hasumi, K., Sakai, K. and Kanbe, T. 1985d. Specific inhibition of glyceraldehyde-3-phosphate dehydrogenase by koningic acid (heptelidic acid). *Journal of Antibiotics* (Tokyo) 38:920–925.
Enomoto, Y., Furutani, Y., Naganawa, H., Hamada, M., Takeuchi, T. and Umezawa, H. 1978. Isolation and characterization of pDE-I and II, the inhibitors of cyclic adenosine-3',5'-monophosphate phosphodiesterase. *Agricultural and Biological Chemistry* 42:1331–1336.
Faucher, D. C., Lelievre, Y. and Cartwright, T. 1987. An inhibitor of mammalian collagenase active at micromolar concentrations from an actinomycete culture broth. *Journal of Antibiotics* (Tokyo) 40:1757–1761.
Fleck, W. F. 1981. Enzym-Inhibitoren mikrobieller Herkunft. *Biologische Rundschau* 19:78–88.
Fredenhagen, A., Fendrich, G., Marki, F., Marki, W., Gruner, J., Raschdorf, F. and Peter, H. H. 1990. Duramycins B and C, two new lanthionine containing antibiotics as inhibitors of phospholipase A_2. Structural revision of duramycin and cinnamycin. *Journal of Antibiotics* (Tokyo) 43:1403–1412.
Frommer, W., Junge, B., Muller, L., Schmidt, D. and Truscheit, E. 1979. Neue enzyminhibitoren aus Mikroorganismen. *Planta Medica* 35:195–217.
Fujiwara, A., Shiomi, Y., Suzuki, K. and Fujiwara, M. 1978. Ornithine decarboxylase inhibitors of microbial origin, a possible approach for screening of antitumor agents. *Agricultural and Biological Chemistry* 42:1435–1436.
Fukuhara, K., Katsura, M. and Murao, S. 1982a. Purification and some properties of talopeptin(MK-I), a novel proteinase inhibitor produced by *Streptomyces mozunensis* MK-23. *Agricultural and Biological Chemistry* 46:1707–1710.
Fukuhara, K., Murai, H. and Murao, S. 1982b. Isolation and structure-activity relationship of some amylostatins (f-1b fraction) produced by *Streptomyces diastaticus* subsp. *Amylostaticus* No. 9410. *Agricultural and Biological Chemistry* 46:1941–1945.
Furutani, Y., Shimada, M., Hamada, M., Takeuchi, T. and Umezawa, H. 1975. Reticulol, an inhibitor of cyclic adenosine 3',5'- monophosphate phosphodiesterase. *Journal of Antibiotics* (Tokyo) 28:558–560.
Goto, J., Matsuda, Y., Asano, K., Kawamoto, I., Yasuzawa, T., Shirahata, K., Sano, H. and Kase, H. 1987. K-254-I (genistein), a new inhibitor of Ca^{2+} and calmodulin-dependent cyclic nucleotide phosphodiesterase from *Streptosporangium vulgare*. *Agricultural and Biological Chemistry* 51:3003–3009.
Hamada, S. and Slade, H. D. 1980. Biology, immunology, and cariogenicity of *Streptococcus mutans*. *Microbiological Reviews* 44:331–384.
Hamaguchi, K., Iwakiri, T., Imamura, K., Furihata, K., Seto, H. and Otake, N. 1987. Juglorin, a new spermidine synthase inhibitor. *Journal of Antibiotics* (Tokyo) 40:717–719.

Hanada, K., Tamai, M., Yamagishi, M., Ohmura, S., Sawada, J. and Tanaka, I. 1978. Isolation and characterization of E-64, a new thiol protease inhibitor. *Agricultural and Biological Chemistry* 42:523–528.

Hanajima, S., Ishimaru, K., Dakano, K., Roy, S. K., Inouye, Y. and Nakamura, S. 1985. Inhibition of reverse transcriptase by limocrocin. *Journal of Antibiotics* (Tokyo) 38:803–805.

Hashimoto, S., Murai, H., Ezaki, M., Morikawa, N., Hatanaka, H., Okuhara, M., Kohsaka, M. and Imanaka, H. 1990. Studies on new dehydropeptidase inhibitor. I. Taxonomy, fermentation, isolation and physicochemical properties. *Journal of Antibiotics* (Tokyo) 43:29–37.

Hasumi, K., Arahira, M., Sakai, K. and Endo, A. 1987. Irreversible inhibition of 3-hydroxy-3-methylglutaryl coenzyme A reductase by phenicin (phoenicine). *Journal of Antibiotics* (Tokyo) 40:224–226.

Hautzel, R., Anke, H. and Sheldrick, W. S. 1990. Mycenon, a new metabolite from a *Mycena* species TA87202 (basidiomycetes) as an inhibitor of isocitrate lyase. *Journal of Antibiotics* (Tokyo) 43:1240–1244.

Hayman, S. and Kinoshita, J. H. 1965. Isolation and properties of lens aldose reductase. *Journal of Biological Chemistry* 240:877–882.

Hazato, T., Kumagai, M., Naganawa, H., Aoyagi, T. and Umezawa, H. 1979. p-Hydroxyphenylacetaldoxime, an inhibitor of beta-galactosidase, produced by actinomycetes. *Journal of Antibiotics* (Tokyo) 32:91–93.

Hidaka, H., Nagatsu, T., Takeya, K., Takeuchi, T., Suda, H., Kojiri, K., Matsuzaki, M. and Umezawa, H. 1969. Fusaric acid, a hypotensive agent produced by fungi. *Journal of Antibiotics* (Tokyo) 22:228–230.

Hook, D. J., Yacobucci, J. J., O'Connor, S., Lee, M., Kerns, E., Krishnan, B., Matson, J. and Hesler, G. 1990. Identification of the inhibitory activity of carbazomycins B and C against 5-lipoxygenase, a new activity for these compounds. *Journal of Antibiotics* (Tokyo) 43:1347–1348.

Hopsu, V. K., Makinen, K. K. and Glenner, G. G. 1966. Purification of a mammalian peptidase selective for N-terminal arginine and lysine residues: Aminopeptidase B. *Archives of Biochemistry and Biophysics* 114:557–566.

Hosono, K. and Suzuki, H. 1983. Acylpeptides, the inhibitors of cyclic adenosine 3',5'-monophosphate phosphodiesterase. *Journal of Antibiotics* (Tokyo) 36:194–196.

Hoyem, T. and Skulberg, A. 1962. Trypsin inhibitors produced by *Clostridium botulinum* cultures. *Nature* (London) 195:922–923.

Huang, L., Rowin, G., Dunn, J., Sykes, R., Dobna, R., Mayles, B. A., Gross, D. M. and Burg, R. W. 1984a. Discovery, purification and characterization of the angiotensin converting enzyme inhibitor, L-681,176, produced by *Streptomyces* sp. MA 5143a. *Journal of Antibiotics* (Tokyo) 37:462–465.

Huang, L., Albers-Schonberg, G., Monaghan, R. L., Jakubas, K., Pong, S. S., Hensens, P. D., Burg, R. W., Ostlind, D. A., Conroy, J. and Stapley, E. O. 1984b. Discovery, production and purification of the Na^+, K^+ activated ATPase inhibitor, L-681,110 from the fermentation broth of *Streptomyces* sp. Ma-5038. *Journal of Antibiotics* (Tokyo) 37:970–975.

Iinuma, H., Takeuchi, T., Kondo, S., Matsuzaki, M., Umezawa, H. and Ohno, M. 1972. Dopastin, an inhibitor of dopamine beta-hydroxylase. *Journal of Antibiotics* (Tokyo) 25:497–500.

Ikeda, Y., Kondo, S., Sawa, T., Tsuchiya, M., Ikeda, D., Hamada, M., Takeuchi, T. and Umezawa, H. 1981. Dotriacolide, a new beta-lactamase inhibitor. *Journal of Antibiotics* (Tokyo) 34:1628–1630.

Ikegami, Y., Takeuchi, N., Hanada, M., Hasegawa, Y., Ishii, K., Andoh, T., Saito, T., Suzuki, K., Yamaguchi, H., Miyazaki, S., Nagai, K., Watanabe, S. and Saito, T. 1990. Topostin, a novel inhibitor of mammalian DNA topoisomerase I from *Flexibacter topostinus* sp. nov. II. Purification and some properties of Topostin. *Journal of Antibiotics* (Tokyo) 43:158–162.

Inaoka, Y., Tamaoki, H., Takahashi, S., Enokita, R. and Okazaki, T. 1986. Propioxatins A and B, new enkephalinase B inhibitors. I. Taxonomy, fermentation, isolation and biological properties. *Journal of Antibiotics* (Tokyo) 39:1368–1377.

Inouye, Y., Manabe, N., Mukai, H., Nakamura, S., Matsugi, T., Amanuma, H. and Ikawa, Y. 1985. Chromostin, a novel specific inhibitor against reverse transcriptase. *Journal of Antibiotics* (Tokyo) 38:519–521.

Ishida, T. and Wada, K. 1975. A steroid hydroxylase inhibitor, diplodialide A from *Diplodia pinea*. *Journal of the Chemical Society: Chemical Communications* 209–210.

Ishii, T., Hida, T., Ishimaru, T., Iinuma, S., Sudo, K., Muroi, M., Kanamaru, T. and Okazaki, H. 1991. TAN-931, a novel nonsteroidal aromatase inhibitor produced by *Penicillium funiculosum* No. 8974. I. Taxonomy, fermentation, isolation, characterization and biological activities. *Jounal of Antibiotics* (Tokyo) 44:589–599.

Ishimaru, T., Kanamaru, T., Ohta, K. and Okazaki, H. 1987. Fibrostatins, new inhibitors of prolyl hydroxylase. I. Taxonomy, isolation and characterization. *Journal of Antibiotics* (Tokyo) 40:1231–1238.

Ishizuka, M., Aoyagi, T., Takeuchi, T. and Umezawa, H. 1981. Activity of bestatin: Enhancement of immune responses and antitumor effect. In: Umezawa, H. (editor), *Small Molecular Immunomodifiers of Microbial Origin: Fundamental and Clinical Studies of Bestatin*, pp. 17–38 Japan Scientific Societies Press, Tokyo.

Iwai, Y., Ohno, H., Takeshima, H., Yamaguchi, N., Ōmura, S. and Hata, T. 1973. Screening and isolation of penicillinase inhibitor, KA-107. *Antimicrobial Agents and Chemotherapy* 4:222–225.

Jung, M. J. 1978. Selective enzyme inhibitors in medicinal chemistry. *Annual Reports in Medicinal Chemistry* 13:249–260.

Kador, P. F., Kinoshita, J. H. and Sharpless, N. E. 1985. Aldose reductase inhibitors. A potential new class of agents for the pharmacological control of certain diabetic complications. *Journal of Medicinal Chemistry* 28:841–849.

Kanai, F., Sawa, T., Hamada, M., Naganawa, H., Takeuchi, T. and Umezawa, H. 1983. Vanoxonin, a new inhibitor of thymidylate synthetase. *Journal of Antibiotics* (Tokyo) 36:656–660.

Kase, H., Iwahashi, K. and Matsuda, Y. 1986. K-252a, a potent inhibitor of protein kinase C from microbial origin. *Journal of Antibiotics* (Tokyo) 39:1059–1065.

Kase, H., Iwahashi, K., Nakanishi, S., Matsuda, Y., Yamada, K., Takahashi, M., Murakata, C., Sato, A. and Kaneko, M. 1987a. K-252 compounds, novel and potent inhibitors of protein kinase C and cyclic nucleotide-dependent protein kinases. *Biochemical and Biophysical Research Communications* 142:436–440.

Kase, H., Kaneko, M. and Yamada, K. 1987b. K-13, a novel inhibitor of angiotensin I converting enzyme produced by *Micromonospora halophytica* subsp. *exilisia*. *Journal of Antibiotics* (Tokyo) 40:450–454.

Katsuragi, T., Sakai, T. and Tonomura, K. 1985. Visual assay for specific inhibitors of adenosine deaminase with agar plates containing pH indicator. *Journal of Fermentation Technology* 63:431–436.

Kido, Y., Hamakado, T., Yoshida, T., Anno, M. and Motoki, Y. 1983. Isolation and characterization of ancovenin, a new inhibitor of angiotensin converting enzyme, produced by actinomycetes. *Journal of Antibiotics* (Tokyo) 36:1295–1299.

Kido, Y., Hamakado, T., Anno, M., Miyagawa, E. and Motoki, Y. 1984. Isolation and characterization of I5B2, a new phosphorus containing inhibitor of angiotensin I converting enzyme produced by *Actinomadura* sp. *Journal of Antibiotics* (Tokyo) 37:965–969.

Kikkawa, U., Takai, Y., Minakuchi, R., Inohara, S. and Nishizuka, Y. 1982. Calcium-activated, phospholipid-dependent protein kinase from rat brain. *Journal of Biological Chemistry* 257:13341–13348.

Kimura, T., Tsuchiya, K. and Ōmura, S. 1984. Prohisin, new thiol protease inhibitor produced by *Cephalosporium* sp. KM 388. *Agricultural and Biological Chemistry* 48:1685–1686.

Kitahara, M., Asano, M., Naganawa, H., Maeda, K., Hamada, M., Aoyagi, T., Umezawa, H., Iitaka, Y. and Nakamura, H. 1987. Valilactone, an inhibitor of esterase, produced by actinomycetes. *Journal of Antibiotics* (Tokyo) 40:1647–1650.

Kitahara, N., Endo, A., Furuya, K. and Takahashi, S. 1981. Thielavins A and B, new inhibitors of prostagrandin biosynthesis produced by *Thielavia terricola*. *Journal of Antibiotics* (Tokyo) 34:1562–1568.

Kitamura, S., Iida, T., Shirahata, K. and Kase, H. 1986a. Studies on lipoxygenase inhibitors, I. MY3-469 (3-methoxytropolone), a potent and selective inhibitor of 12-lipoxygenase, produced by *Streptoverticillium hadanonense* KY11449. *Journal of Antibiotics* (Tokyo) 39:589–593.

Kitamura, S., Hashizume, K., Iida, T., Miyashita, E., Shirahata, K. and Kase, H. 1986b. Studies on lipoxygenase inhibitors, II. KF-8940 (2-n-heptyl-4-hydroxyquinoline-N-oxide), a potent and selective inhibitor of 5-lipoxygenase, produced by *Pseudomonas methanica*. *Journal of Antibiotics* (Tokyo) 39:1160–1166.

Klein, G. and Klein, E. 1985. Evolution of tumors and the impact of molecular oncology. *Nature* (London) 315:190–195.

Knauss, H. J., Peter, J. W. and Wasson, G. 1959. The biosynthesis of mevalonic acid from 1-^{14}C-acetate by a rat liver enzyme system. *Journal of Biological Chemistry* 234:2835–2840.

Kobaru, S., Tsunakawa, M., Hanada, M., Konishi, M., Tomita, K. and Kawaguchi, H. 1983. Bu-2743E, a leucine aminopeptidase inhibitor, produced by *Bacillus circulans*. *Journal of Antibiotics* (Tokyo) 36:1396–1399.

Kobayashi, E., Ando, K., Nakano, H., Iida, T., Ohno, H., Morimoto, M. and Tamaoki, T. 1989. Calphostins (UCN-1028), novel an specific inhibitors of protein kinase C. I. Fermentation, isolation, physicochemical properties and biological activities. *Journal of Antibiotics* (Tokyo) 42:1470–1474.

Koga, D., Isogai, A., Sakuda, S., Matsumoto, S., Suzuki, A., Kimura, S. and Ide, A. 1987. Specific inhibition of *Bombyx mori* chitinase by allodsamidin. *Agricultural and Biological Chemistry* 51:471–476.

Koguchi, T., Yamada, K., Yamato, M., Okachi, R., Nakayama, K. and Kase, H. 1986. K-4, a novel inhibitor of angiotensin I converting enzyme produced by *Actinomadura spiculosospora*. *Journal of Antibiotics* (Tokyo) 39:364–371.

Kojiri, K., Kondo, H., Yoshinari, T., Arakawa, H., Nakajima, S., Satoh, F., Kawamura, K., Okura, A., Suda, H. and Okanishi, M. 1991. A new antitumor substance, BE-13793C, produced by a streptomycete. Taxonomy, fermentation, isolation structure determination and biological activity. *Journal of Antibiotics* (Tokyo) 44:723–728.

Krenitsky, T. A. and Elin, G. B. 1982. Enzymes as tools and targets in drug research. In: Bunisman, K. (editor), *Strategy in Drug Research*, pp. 65–87. Elsevier, The Netherlands.

Kuhnt, D., Anke, T., Besl, H., Bross, M., Herrmann, R., Mocek, U., Steffan, B. and Steglich, W. 1990. Antibiotics from basidiomycetes. XXXVII. New inhibitors of cholesterol biosynthesis from cultures of *Xerula melanotricha* Dorfelt. *Journal of Antibiotics* (Tokyo) 43:1413–1420.

Kumada, Y., Naganawa, H., Hamada, M., Takeuchi, T. and Umezawa, H. 1974. 1-[2-(3,4,5,6-Tetrahydropyridyl)]-1,3-pentadiene, an N-methyltransferase inhibitor produced by actinomycetes. *Journal of Antibiotics* (Tokyo) 27:726–728.

Kumada, Y., Naganawa, H., Iinuma, H., Matsuzaki, M., Takeuchi, T. and Umezawa, H. 1976. Dehydrodicaffeic acid dilactone, an inhibitor of catechol-O-methyl transferase. *Journal of Antibiotics* (Tokyo) 29:882–889.

Kuramochi, H., Motegi, A., Iwabuchi, M., Takahashi, A., Horinishi, H. and Umezawa, H. 1987. Action of ubenimex on aminopeptidase activities in spleen cells and peritoneal macrophages from mice. *Journal of Antibiotics* (Tokyo) 40:1605–1611.

Laemmli, U. K. 1970. Cleavage of structural proteins during the assembly of the head of bacteriophage T4. *Nature* (London) 227:680–685.
Lin, W., Oishi, K. and Aida, K. 1975. A new screening method of viral-neuraminidase inhibitors. *Agricultural and Biological Chemistry* 39:759–765.
Lincoln, T. M., Dills, W. L., Jr. and Corbin, J. D. 1977. Purification and subunit composition of guanosine 3':5'- monophosphate-dependent protein kinase from bovine lung.
Journal of Biological Chemistry 252:4269–4275.
Liu, W-C., Astle, G., Wells, J. S., Jr., Trejo, W. H., Principe, P. A., Rathnum, M. L., Parker, W. L., Kocy, O. R. and Sykes, R. 1980. Izumenolide—a novel beta-lactamase inhibitor produced by *Micromonospora*. *Journal of Antibiotics* (Tokyo) 33:1256–1261.
Magae, J., Watanabe, C., Osada, H., Cheng, X.-C. and Isono, K. 1988. Induction of morphological change of human myeloid leukemia and activation of protein kinase C by a novel antibiotic, tautomycin. *Journal of Antibiotics* (Tokyo) 41:932–937.
Magae, J., Osada, H., Fujiki, H. Saido, T. C., Suzuki, K., Nagai, K., Yamasaki, M. and Isono, K. 1990. Morphological changes of human myeloid leukemia K562 cells by a protein phosphatase inhibitor, tautomycin. *Proceedings of The Japan Academy* 66:209–212.
Matsuda, Y. and Kase, H. 1987. KS-619-1, a new inhibitor of Ca^{2+} and calmodulin-dependent cyclic nucleotide phosphodiesterase from *Streptomyces californicus*. *Journal of Antibiotics* (Tokyo) 40:1104–1110.
Matsuda, Y., Asano, K., Kawamoto, I. and Kase, H. 1987. K-259-2, a new inhibitor of Ca^{2+} and calmodurin-dependent cyclic nucleotide phosphodiesterase from *Micromonospora olivasterospora*. *Journal of Antibiotics* (Tokyo) 40:1092–1100.
Meier, W. and Conscience, J.-F. 1980. A fast and simple radiometric assay for adenosine deaminase using reverse-phase thin-layer chromatography. *Analytical Biochemistry* 105:334–339.
Mikami, Y. and Suzuki, T. 1983. Novel microbial inhibitors of angiotensin-converting enzyme, aspergillomarasmines A and B. *Agricultural and Biological Chemistry* 47:2693–2695.
Miyake, Y. and Ebata, M. 1987. Galactostatin, a new beta-galactosidase inhibitor from *Streptomyces lydicus*. *Journal of Antibiotics* (Tokyo) 40:122–123.
Miyano, T., Tomiyasu, M., Iizuka, H., Tomisaka, S., Takita, T., Aoyagi, T. and Umezawa, H. 1972. New pepstatins, pepstatins B and C, and pepstanone A, produced by *Streptomyces*. *Journal of Antibiotics* (Tokyo) 25:489–491.
Murao, S. and Miyata, S. 1980. Isolation and characterization of a new trehalase inhibitor, S-GI. *Agricultural an Biological Chemistry* 44:219–221.
Murao, S. and Ohyama, K. 1975. New amylase inhibitor (S-AI) from *Streptomyces diastaticus* var. *amylostaticus* No. 2476. *Agricultural and Biological Chemistry* 39:2271–2273.
Murao, S. and Satoi, S. 1970. New pepsin inhibitor (S-PI) from *Streptomyces* EF-44-201. *Agricultural and Biological Chemistry* 34:1265–1267.
Murao, S. and Sato, S. 1972. S-SI, a new alkaline protease inhibitor from *Streptomyces albogriseolus* S-3253. *Agricultural and Biological Chemistry* 36:160–163.
Murao, S., Goto, A., Matsui, Y. and Ohyama, K. 1980. New proteinous inhibitor (Haim) of animal alpha-amylase from *Streptomyces griseosporeus* YM-25. *Agricultural and Biological Chemistry* 44:1679–1681.
Murao, S., Nishino, T., Katayama, N. and Nagano, H. 1981. New aspartate aminotransferase inhibitor (gostatin) produced by *Streptomyces sumanensis* nov. sp. NK-23. *Agricultural and Biological Chemistry* 45:1039–1041.
Murao, S., Kasai, N., Kimura, Y., Oda, K. and Fukuhara, K. 1982. A new metalloproteinase inhibitor (fMpI) produced by *Streptomyces rishiriensis* NK-122. *Agricultural and Biological Chemistry* 46:855–857.
Murao, S., Shin, T., Katsu, Y., Nakatani, S. and Hirayama, K. 1985. Novel thiol pro-

teinase inhibitor, thiolstatin, produced by a strain of *Bacillus cereus*. *Agricultural and Biological Chemistry* 49:895–897.

Murao, S., Shin, T., Katsu, Y., Iwahara, M. and Hirayama, K. 1987. Thiol proteinase inhibitor, caricastatin, produced by a strain of *Nigrosabulum* novosp. *Agricultural and Biological Chemistry* 51:2029–2031.

Murata, M., Tanaka, H. and Ōmura, S. 1987. 7-Hydro-8- methylpteroylglutamylglutamic acid, a new anti-folate from an actinomycete. *Journal of Antibiotics* (Tokyo) 40:251–257.

Mynderse, J. S., Samlaska, S. K., Fukuda, D. S., Dubus, R. H. and Baker, P. J. 1985. Isolation of A58365a and A58365b, angiotensin converting enzyme inhibitors produced by *Streptomyces chromofuscus*. *Journal of Antibiotics* (Tokyo) 38:1003–1007.

Naganawa, T., Mori, N., Tani, Y. and Ogata, K. 1976. A butyrylcholinesterase inhibitor produced by *Penicillium* sp. No. C-81 and its identity with mycelianamide. *Journal of Antibiotics* (Tokyo) 29:526–531.

Nakagawa, F., Enokita, R. and Naito, A. 1984. Terferol, an inhibitor of cyclic adenosine 3′,5′-monophosphate phosphodiesterase. *Journal of Antibiotics* (Tokyo) 37:6–9.

Nakagawa, F., Okazaki, T., Naito, A., Iijima, Y. and Yamazaki, M. 1985. Griseolic acid, an inhibitor of cyclic adenosine 3′,5′-monophosphate phosphodiesterase. *Journal of Antibiotics* (Tokyo) 38:824–829.

Nakamura, T., Yasuda, H., Obayashi, A., Tanabe, O., Matsumura, S., Ueda, F. and Ohata, K. 1975. Phenopicolinic acid, a new microbial product inhibiting dopamine beta-hydroxylase. *Journal of Antibiotics* (Tokyo) 28:477–478.

Nakanishi, S., Matsuda, Y., Iwahashi, K. and Kase, H. 1986. K- 252b, c and d, potent inhibitors of protein kinase C from microbial origin. *Journal of Antibiotics* (Tokyo) 39:1066–1071.

Nakanishi, S., Ando, K., Kawamoto, I. and Kase, H. 1989a. KS-501 and KS-502, new inhibitors of Ca^{2+} and calmodulin-dependent cyclic-nucleotide phosphodiesterase from *Sporothrix* sp. *Journal of Antibiotics* (Tokyo) 42:1049–1055.

Nakanishi, S., Ando, K., Kawamoto, I., Yasuzawa, T., Sano, H. and Kase, H. 1989b. KS-504 compounds, novel inhibitors of Ca^{2+} and calmodulin-dependent cyclic nucleotide phosphodiesterase from *Mollisia ventosa*. *Journal of Antibiotics* (Tokyo) 42:1775–1783.

Nakano, Y., Murakawa, S. and Endo, A. 1987. Inhibitory effect of mutastein on the synthesis of artificial dental plaque by *Streptococcus mutans*. *Journal of Antibiotics* (Tokyo) 40:226–227.

Nakayama, O., Yagi, M., Tanaka, M., Kiyoto, S., Okuhara, M. and Kohsaka, M. 1989. WS-9659 A and B, novel testosterone 5-alpha-reductase inhibitors isolated from a *Streptomyces*. I. Taxonomy, fermentation, isolation, physicochemical characteristics. *Journal of Antibiotics* (Tokyo) 42:1221–1229.

Nakayama, O., Yagi, M., Kiyoto, S., Okuhara, M. and Kohsaka, M. 1990. Riboflavin, a testosterone 5-α-reductase inhibitor. *Journal of Antibiotics* (Tokyo) 43:1615–1616.

Namiki, S., Kangouri, K., Nagate, T., Hara, H., Sugita, K. and Ōmura, S. 1982. Studies on the alpha-glucoside hydrolase inhibitor, adiposin. I. Isolation and physicochemical properties. *Journal of Antibiotics* (Tokyo) 35:1234–1236.

Namiki, T., Nishikawa, M., Itoh, Y., Uchida, I. and Hashimoto, M. 1987. Studies of WF-3681, a novel aldose reductase inhibitor. II. Structure determination and synthesis. *Journal of Antibiotics* (Tokyo) 40:1400–1407.

Niida, T., Niwa, T., Tsuruoka, T., Ezaki, N., Shomura, T. and Umezawa, H. 1967. Isolation and characteristics of coformycin. *153rd Scientific Meeting of the Japanese Antibiotic Research Association Jan.*, p. 27.

Nishikawa, M., Tsurumi, Y., Namiki, T., Yoshida, K. and Okuhara, M. 1987. Studies on WF-3681, a novel aldose reductase inhibitor. I. Taxonomy, fermentation, isolation and characterization. *Journal of Antibiotics* (Tokyo) 40:1394–1399.

Nishikawa, M., Tsurumi, Y., Murai, H., Yoshida, K., Okamoto, M., Takase, S., Tanaka,

H., Hirota, H., Hashimoto, M. and Kohsaka, M. 1991 WF-2421, a new aldose reductase inhibitor produced from a fungus, Humicola grisea. *Journal of Antibiotics* (Tokyo) 44:130-135.

Nishikiori, T., Okuyama, A., Naganawa, H., Takita, T., Hamada, M., Takeuchi, T., Aoyagi, T. and Umezawa, H. 1984. Production by actinomycetes of (S, S)-N, N'-ethylenediaminedisuccinic acid, an inhibitor of phospholipase C. *Journal of Antibiotics* (Tokyo) 37:426-427.

Nishimoto, Y., Sakuda, S., Takayama, S. and Yamada, Y. 1991 Isolation and characterization of new allosamidins. *Journal of Antibiotics* (Tokyo) 44:716-722.

Nishio, M., Kuroda, M., Suzuki, M., Ishimaru, K., Nakamura, S. and Nomi, R. 1983. Retrostatin, a new specific enzyme inhibitor against avian myeloblastosis virus reverse transcriptase. *Journal of Antibiotics* (Tokyo) 36:761-769.

Nishioka, H., Sawa, T., Hamada, M., Shimura, M., Imoto, M. and Umezawa, K. 1990. Inhibition of phosphatidylinositol kinase by toyocamycin. *Journal of Antibiotics* (Tokyo) 43:1586-1589.

Nishizuka, Y. 1984a. The role of protein kinase C in cell surface signal transduction and tumor promotion. *Nature* (London) 308:693-698.

Nishizuka, Y. 1984b. Turnover of inositol phospholipids and signal transduction. *Science* 225:1365-1370.

Niwa, T., Tsuruoka, T., Goi, H., Kodama, Y., Itoh, J., Inouye, S., Yamada, Y., Niida, T., Nobe, M. and Ogawa, Y. 1984. Novel glycosidase inhibitors, nojirimycin-B and D-mannonic-delta-lactam. *Journal of Antibiotics* (Tokyo) 37:1579-1586.

Numata, M., Nitta, K., Utahara, R., Maeda, K. and Umezawa, H. 1975. Revistin found by screening for inhibitors of reverse transcriptase of an oncogenic virus. *Journal of Antibiotics* (Tokyo) 28:757-763.

O'Connor, S. and Somers, P. 1985. Methods for the detection and quantitation of angiotensin converting enzyme inhibitors in fermentation broths. *Journal of Antibiotics* (Tokyo) 38:993-996.

Oda, K., Fukuda, Y., Murao, S., Uchida, K. and Kainosho, M. 1989. A novel proteinase inhibitor, tyrostatin, inhibiting some pepstatin-insensitive carboxyl proteinases. *Agricultural and Biological Chemistry* 53:405-415.

O'Dwyer, P. J., Wagner, B., Leyland-Jones, B., Wittes, R. E., Cheson, B. D. and Hoth, D. F. 1988. 2'-Deoxycoformycin (pentostatin) for lymphoid malignanancies. *Annals of Internal Medicine* 108:733-743.

Ogata, K., Ueda, K., Naganawa, T. and Tani, Y. 1974. A cholinesterase inhibitor produced by *Aspergillus terreus*. *Journal of Antibiotics* (Tokyo) 27:343-345.

Ogawa, H., Hasumi, K., Sakai, K., Murakawa, S. and Endo, A. 1991. Pannorin, a new 3-hydroxy-3-methylglutaryl coenzyme A reductase inhibitor produced by *Chrysosporium pannorum*. *Journal of Antibiotics* (Tokyo) 44:762-767.

Ogawara, H., Uchino, K., Akiyama, T. and Watanabe, S. 1985a. A new 5'-nucleotidase inhibitor, nucleoticidin. *Journal of Antibiotics* (Tokyo) 38:153-156.

Ogawara, H., Uchino, K., Akiyama, T. and Watanabe, S. 1985b. New 5'-nucleotidase inhibitors, melanocidin A and melanocidin B. *Journal of Antibiotics* (Tokyo) 38:587-591.

Ogawara, H., Akiyama, T., Ishida, J., Watanabe, S. and Suzuki, K. 1986. A specific inhibitor for tyrosine protein kinase from *Pseudomonas*. *Journal of Antibiotics* (Tokyo) 39:606-608.

Ogawara, H., Horikawa, S., Yanagida, T., Nakano, M. M., Andoh, T., Ishii, K., Hori, M., Goto, T., Hamada, M. and Umezawa, H. 1982. A novel deoxyribonuclease inhibitor from *Micromonospora*. *Journal of Antibiotics* F(Tokyo) 35:248-250.

Ogura, K., Maeda, M., Nagai, M., Tanaka, T., Nomoto, K. and Murachi, T. 1985. Purification and structure of a novel cysteine proteinase inhibitor, strepin P-1. *Agricultural and Biological Chemistry* 49:799-805.

Ohnuki, T., Takashio, M., Okami, Y. and Umezawa, H. 1981. The structure of a novel inhibitor of dextransucrase. *Tetrahedron Letters* 22:1267–1270.

Oka, H., Yoshinari, T., Murai, T., Kawamura, K., Satoh, F. Funaishi, K., Okura, A., Suda, H., Okanishi, M. and Shizuri, Y. 1991. A new topoisomerase-II inhibitor, BE-10988, produced by a streptomycete. I. Taxonomy, fermentation, isolation and characterization. *Journal of Antibiotics* (Tokyo) 44:582–588.

Okabayashi, K., Morishima, H., Hamada, M., Takeuchi, T. and Umezawa, H. 1977. A tryptophan hydroxylase inhibitor produced by a streptomycete. 2,5-dihydro-L-phenylalanine. *Journal of Antibiotics* (Tokyo) 30:675–677.

Okami, Y., Takashio, M. and Umezawa, H. 1981. Ribocitrin, a new inhibitor of dextransucrase. *Journal of Antibiotics* (Tokyo) 34:344–345.

Okazaki, H., Ohta, K., Kanamaru, T., Ishimaru, T. and Kishi, T. 1981. A potent prolyl hydroxylase inhibitor, P-1894B, produced by a strain of *Streptomyces*. *Journal of Antibiotics* (Tokyo) 34: 1355–1356.

Ōmura, S., Ohno, H., Saheki, T., Yoshida, M. and Nakagawa, A. 1978. Elasnin, a new human granulocyte elastase inhibitor produced by a strain of *Streptomyces*. *Biochemical Biophysical Research Communication* 83:704–709.

Ōmura, S., Tanaka, H., Kuga, H. and Imamura, N. 1986a. Adecypenol, a unique adenosine deaminase inhibitor containing homopurine and cyclopentene rings. *Journal of Antibiotics* (Tokyo) 39:309–310.

Ōmura, S., Imamura, N., Kawakita, K., Mori, Y., Yamazaki, Y., Masuma, R., Takahashi, Y., Tanaka, H., Hang, L-Y. and Woodruff, H. B. 1986b. Ahpatinins, new acid protease inhibitors containing 4-amino-3-hydroxy-5-phenylpentanoic acid. *Journal of Antibiotics* (Tokyo) 39:1079–1085.

Ōmura, S., Tomoda, H., Xu, Q. M., Takahashi, Y. and Iwai, Y. 1986c. Triacsins, new inhibitors of acyl-CoA synthetase produced by *Streptomyces* sp. *Journal of Antibiotics* (Tokyo) 39:1211–1218.

Ōmura, S., Ishikawa, H., Kuga, H., Imamura, N., Taga, S., Takahashi, Y. and Tanaka, H. 1986d. Adecypenol, a unique adenosine deaminase inhibitor containing homopurine and cyclopentene rings. Taxonomy, production and enzyme inhibition. *Journal of Antibiotics* (Tokyo) 39:1219–1224.

Ōmura, S., Murata, M., Hanaki, H., Hinotozawa, K., Oiwa, R. and Tanaka, H. 1984a. Phosalacin, a new herbicidal antibiotic containing phosphinothricin. Fermentation, isolation, biological activity and mechanism of action. *Journal of Antibiotics* (Tokyo) 37:829–835.

Ōmura, S., Murata, M., Imamura, N., Iwai, Y., Tanaka, H., Furusaki, A. and Matsumoto, T. 1984b. Oxetin, a new antimetabolite from an actinomycete. *Journal of Antibiotics* (Tokyo) 37:1324–1332.

Ōmura, S., Imamura, N., Kuga, H., Ishikawa, H., Yamazaki, Y., Okano, K., Kimura, K., Takahashi, Y. and Tanaka, H. 1985a. Adechlorin, a new adenosine deaminase inhibitor containing chlorine. Production, isolation and properties. *Journal of Antibiotics* (Tokyo) 38:1008–1015.

Ōmura, S., Murata, M., Kimura, K., Matsukura, S., Nishihara, T. and Tanaka, H. 1985b. Screening for new antifolates of microbial origin and a new antifolate AM-8402. *Journal of Antibiotics* (Tokyo) 38:1016–1024.

Ōmura, S., Iwai, Y., Hirano, A., Nakagawa, A., Awaya, J., Tsuchiya, H., Takahashi, Y. and Masuma, R. 1977. A new alkaloid AM-2282 of *Streptomyces* origin. Taxonomy, fermentation, isolation and preliminary characterization. *Journal of Antibiotics* (Tokyo) 30:275–282.

Ōmura, S., Tomoda, H., Kumagai, H., Greenspan, M. D., Yodokovitz, J. B., Chen, J. S., Alberts, A. W., Martin, I., Mochales, S., Monaghan, R. L., Chabala, J. C., Schwartz, R. E. and Patchett, A. A. 1987. Potent inhibitory effect of antibiotic 1233A on cholesterol biosynthesis which specifically blocks 3-hydroxy-3-methylglutaryl coenzyme A synthase. *Journal of Antibiotics* 40:1356–1357.

Ondetti, M. A. and Cushman, D. W. 1982. Enzymes of the renin- angiotensin system and their inhibitors. *Annual Reviews of Biochemistry* 51:283–308.
Oouchi, N., Arai, M. and Murao, S. 1985. Purification and some properties of an alpha-amylase inhibitor (Paim) from *Streptomyces corchorushii*. *Agricultural and Biological Chemistry* 49:793–797.
Osada, H., Takahashi, H., Tsunoda, K., Kusakabe, H. and Isono, K. 1990. A new inhibitor of protein kinase C, RK-286C (4'-demethylamino-4'-hydroxystaurosporine). I. Screening, taxonomy, fermentation and biological activity. *Journal of Antibiotics* (Tokyo) 43:163–167.
Ozasa, T., Yoneda, T., Hirasawa, M., Suzuki, K., Tanaka, K., Kadota, S. and Iwanami, M. 1991. Thiazocins, new aldose reductase inhibitors from *Actinosynnema* sp. 1. Fermentation, isolation and characterization. *Journal of Antibiotics* (Tokyo)44:768–773.
Peterson, M. J., Sarges, S., Aldinger, C. E. and MacDonald, D. P. 1979. CP-45,634: A novel aldose reductase inhibitor that inhibits polyol pathway activity in diabetic and galactosemic rats. *Metabolism. Clinical and Experimental* 28:456–461.
Saito, M., Kawaguchi, N., Hashimoto, N., Kodama, T., Higuchi, N., Tanaka, T., Nomoto, K. and Murachi, T. 1987. Purification and structure of novel cysteine proteinase inhibitors, staccopins P-1 and P-2, from *Staphylococcus tanabeensis*. *Agricultural and Biological Chemistry* 51:861–868.
Sano, S., Ikai, K., Kuroda, H., Nakamura, T., Obayashi, A., Ezure, Y. and Enomoto, H. 1986. OF4949, new inhibitors of aminopeptidase B. I. Taxonomy, fermentation, isolation and characterization. *Journal of Antibiotics* (Tokyo) 39:1674–1684.
Sano, S., Kuroda, H., Ueno, M., Yoshikawa, Y., Nakamura, T. and Obayashi, A. 1987. OF4949, new inhibitors of aminopeptidase B. V. Effect on the murine immune system. *Journal of Antibiotics* (Tokyo) 40:519–525.
Saruno, R., Kato, F. and Ikeno, T. 1979. Kojic acid, a tyrosinase inhibitor from *Aspergillus albus*. *Agricultural and Biological Chemistry* 43:1337–1338.
Sawa, T., Fukagawa, Y., Homma, I., Takeuchi, T. and Umezawa, H. 1967a. Mode of inhibition of coformycin on adenosine deaminase. *Journal of Antibiotics* (Tokyo) Series A 20:227–231.
Sawa, T., Fukagawa, Y., Homma, I., Takeuchi, T. and Umezawa, H. 1967b. Formycin-deaminating activity of microorganisms. *Journal of Antibiotics* (Tokyo) Series A 20:317–321.
Schindler, P. 1980. Enzyme inhibitors of microbial origin. *Philosophical Transactions of the Royal Society of London* B290:291–301.
Schmidt, D. D., Frommer, W., Junge, B., Muller, L., Wingender, W. and Trusheit, E. 1977. Alpha-glucosidase inhibitors, new complex oligosaccharides of microbial origin. *Naturwissenschaften* 64:535–536.
Schorlemmer, H. U., Bosslet, K. and Sedlacek, H. H. 1983. Ability of the immunomodulating dipeptide bestatin to active cytotoxic mononuclear phagocytes. *Cancer Research* 43:4148–4153.
Shimada, K. and Matsushima, K. 1967. Studies on the production of protease inhibitor by molds. Production of protease inhibitor by *Penicillium cyclopium*. *Nippon Nogeikagaku Kaishi* 41:454–458.
Shimada, K. and Matsushima, K. 1969. A protease inhibitor from *Penicillium cyclopium*. Part I. Purification and partial characterization. *Agricultural and Biological Chemistry* 33:544–548.
Shimada, K., Matsushima, K., Fukumoto, J. and Yamamoto, T. 1969. Poly-(L)-malic acid; a new protease inhibitor from *Penicillium cyclopium*. *Biochemical and Biophysical Research Communications* 35:619–624.
Singh, P. D., Johnson, J. H., Aklonis, K. B., Fisher, S. M. and O'Sullivan, J. 1985. Two new inhibitors of phospholipase A_2 produced by *Penicillium chermesinum*. *Journal of Antibiotics* (Tokyo) 38:706–712.

Soffer, R. L. 1976. Angiotensin-converting enzyme and the regulation of vasoactive peptides. *Annual Review of Biochemistry* 45:73–94.

Somers, P. J. B., Yao, R. C., Doolin, L. E., McGowan, M. J., Fukuda, D. S. and Mynderse, J. S. 1987. Method for the detection and quantitation of chitinase inhibitors in fermentation broths; Isolation and insect life cycle effect of A82516. *Journal of Antibiotics* (Tokyo) 40:1751–1756.

Stella, S., Saddler, G., Sarubbi, E., Colombo, L., Stefanelli, S., Denaro, M. and Selva, E. 1991. Isolation of α-MAPI from fermentation broths during a screening program for HIV-1 protease inhibitors. *Journal of Antibiotics* (Tokyo) 44:1019–1022.

Suda, H., Aoyagi, T., Hamada, M., Takeuchi, T. and Umezawa, H. 1972. Antipain, a new protease inhibitor isolated from actinomycetes. *Journal of Antibiotics* (Tokyo) 25:263–266.

Suda, H., Aoyagi, T., Takeuchi, T. and Umezawa, H. 1973. A thermolysin inhibitor produced by actinomycetes. Phosphoramidon. *Journal of Antibiotics* (Tokyo) 26:621–623.

Takahashi, I., Kobayashi, E., Asono, K., Yoshida, M. and Nakano, H. 1987. UCN-01, a selective inhibitor of protein kinase C from *Streptomyces*. *Journal of Antibiotics* (Tokyo) 40:1782–1784.

Takahashi, I., Saitoh, Y., Yoshida, M., Sano, H., Nakano, H., Morimoto, M. and Tamaoki, T. 1989. UCN-01 and UCN-02, new selective inhibitors of protein kinase C. II. Purification, physico-chemical properties, structural determination and biological activities. *Journal of Antibiotics* (Tokyo) 42:571–576.

Takashio, M. and Okami, Y. 1982. Screening of a dextransucrase inhibitor. *Agricultural and Biological Chemistry* 46:1457–1464.

Takeuchi, T., Chimura, H., Hamada, M., Umezawa, H., Yoshioka, O., Oguchi, N., Takahashi, Y. and Matsuda, A. 1975. A glyoxalase I inhibitor of a new structural type produced by *Streptomyces*. *Journal of Antibiotics* (Tokyo) 28:737–742.

Takeuchi, T., Ogawa, K., Iinuma, H., Suda, H., Ukita, K., Nagatsu, T., Kato, M., Umezawa, H. and Tanabe, O. 1973. Monoamine oxidase inhibitors isolated from fermented broths. *Journal of Antibiotics* (Tokyo) 26:162–167.

Tamaoki, T., Nomoto, H., Takahashi, I., Kato, Y., Morimoto, M. and Tomita, F. 1986. Staurosporine, a potent inhibitor of phospholipid/Ca^{++} dependent protein kinase. *Biochemical Biophysical Research Communications* 135:397–402.

Tanaka, H. and Ōmura, S. 1988. New adenosine deaminase inhibitors, adechlorin and adecypenol. In: *Abstracts of the First International Conference on the Biotechnology of Microbial Products: Novel Pharmacological and Agrobiological Activities*, S-19. Elsevier, Amsterdam.

Tanaka, T., Suda, H., Naganawa, H., Hamada, M., Takeuchi, T., Aoyagi, T. and Umezawa, H. 1984. Production of (S)-alpha-benzylmalic acid, inhibitor of carboxypeptidase A by actinomycetes. *Journal of Antibiotics* (Tokyo) 37:682–684.

Thom, D., Powell, A. J., Lloyd, C. W. and Rees, D. A. 1977. Rapid isolation of plasma membranes in high yield from cultured fibroblasts. *Biochemical Journal* 168:187–194.

Tomoda, H., Nishida, H., Masuma, R., Cao, J., Okuda, S. and Ōmura, S. 1991 Purpactins, new inhibitors of acyl-CoA:cholesterol acyltransferase produced by *Penicillium purpurogenum*. I. Production, isolation and physico-chemical and biological properties. *Journal of Antibiotics* (Tokyo) 44:136–143.

Tunac, J. B. and Underhill, M. 1985. 2'-Chloropentostatin: Discovery, fermentation and biological activity. *Journal of Antibiotics* (Tokyo) 38:1344–1349.

Ueda, S. and Koba, Y. 1973. Some properties of amylase inhibitor produced by *Streptomyces* sp. No. 280. *Agricultural and Biological Chemistry* 37:2025–2030.

Umezawa, H. 1972. *Enzyme Inhibitors of Microbial Origin*, pp. 1–114. University of Tokyo Press, Tokyo.

Umezawa, H. 1977. Recent advances in bioactive microbial secondary metabolites. *Japanese Journal of Antibiotics* 30 (Suppl.):S138–S173.

Umezawa, H. 1982. Low-molecular-weight enzyme inhibitors of microbial origin. *Annual Review of Microbiology* 36:75–99.

Umezawa, H., Aoyagi, T., Morishima, H., Kunimoto, S., Matsuzaki, M., Hamada, M. and Takeuchi, T. 1970a. Chimostatin, a new chymotrypsin inhibitor produced by actinomycetes. *Journal of Antibiotics* (Tokyo) 23:425–427.

Umezawa, H., Takeuchi, T., Iinuma, H., Suzuki, K., Ito, M. and Matsuzaki, M. 1970b. A new microbial product, oudenone, inhibiting tyrosine hydroxylase. *Journal of Antibiotics* (Tokyo) 23:514–518.

Umezawa, H., Aoyagi, T., Morishima, H., Matsuzaki, M., Hamada, H. and Takeuchi, T. 1970c. Pepstatin, a new pepsin inhibitor produced by actinomyces. *Journal of Antibiotics* (Tokyo) 23:259–262.

Umezawa, H., Iinuma, H., Ito, M., Matsuzaki, M., Takeuchi, T. and Tanabe, O. 1972a. Dopamine beta-hydroxylase inhibitor produced by *Gloeophyllum striatum* and its identity with oosponol. *Journal of Antibiotics* (Tokyo) 25:239–242.

Umezawa, H., Takeuchi, T., Yasuda, S. and Murase, M. 1972b. Fermentative production of 5-formyluracil, a xanthine-oxidase inhibitor. *Japan Kokai* 72:28189.

Umezawa, H., Mitsuhashi, S., Hamada, M., Iyobe, S., Takahashi, S., Osato, Y., Yamazaki, S., Ogawara, H. and Maeda, K. 1973a. Two beta-lactamase inhibitors produced by a *Streptomyces*. *Journal of Antibiotics* (Tokyo) 26:51–54.

Umezawa, H., Miyano, T., Murakami, T., Takita, T., Aoyagi, T., Takeuchi, T., Naganawa, H. and Morishima, H. 1973b. Hydroxypepstatin, a new pepstatin produced by *Streptomyces*. *Journal of Antibiotics* (Tokyo) 26:615–617.

Umezawa, H., Aoyagi, T., Okura, A., Morishima, H., Takeuchi, T. and Okami, Y. 1973c. Elastatinal, a new elastase inhibitor produced by actinomycetes. *Journal of Antibiotics* (Tokyo) 26:787–789.

Umezawa, H., Shibamoto, N., Naganawa, H., Ayukawa, S., Matsuzaki, M., Takeuchi, T., Kono, K. and Sakamoto, T. 1974a. Isolation of lecanoric acid, an inhibitor of histidine decarboxylase from a fungus. *Journal of Antibiotics* (Tokyo) 27:587–596.

Umezawa, H., Aoyagi, T., Komiyama, T., Morishima, H., Hamada, M. and Takeuchi, T. 1974b. Purification and characterization of a sialidase inhibitor, siastatin, produced by *Streptomyces*. *Journal of Antibiotics* (Tokyo) 27:963–969.

Umezawa, H., Tobe, H., Shibamoto, N., Nakamura, F., Nakamura, K., Matsuzaki, M. and Takeuchi, T. 1975. Isolation of isoflavones inhibiting DOPA decarboxylase from fungi and streptomyces. *Journal of Antibiotics* (Tokyo) 28:947–952.

Umezawa, H., Aoyagi, T., Suda, H., Hamada, M. and Takeuchi, T. 1976a. Bestatin, an inhibitor of aminopeptidase B, produced by actinomycetes. *Journal of Antibiotics* (Tokyo) 29:97–99.

Umezawa, H., Ishizuka, M., Aoyagi, T. and Takeuchi, T. 1976b. Enhancement of delayed-type hypersensitivity by bestatin, an inhibitor of aminopeptidase B and leucine aminopeptidase. *Journal of Antibiotics* (Tokyo) 29:857–859.

Umezawa, H., Aoyagi, T., Hazato, T., Uotani, K., Kojima, F., Hamada, M. and Takeuchi, T. 1978. Esterastin, an inhibitor of esterase, produced by actinomycetes. *Journal of Antibiotics* (Tokyo) 31:639–641.

Umezawa, H., Aoyagi, T., Uotani, K., Hamada, M., Takeuchi, T. and Takahashi, S. 1980. Ebelactone, an inhibitor of esterase, produced by actinomycetes. *Journal of Antibiotics* (Tokyo) 33:1594–1596.

Umezawa, H., Aoyagi, T., Ohuchi, S., Okuyama, A., Suda, H., Takita, T., Hamada, M. and Takeuchi, T. 1983. Arphamenines A and B, new inhibitors of aminopeptidase B, produced by bacteria. *Journal of Antibiotics* (Tokyo) 36:1572–1575.

Umezawa, H., Aoyagi, T., Ogawa, K., Naganawa, H., Hamada, M. and Takeuchi, T. 1984a. Diprotins A and B, inhibitors of dipeptidyl aminopeptidase IV, produced by bacteria. *Journal of Antibiotics* (Tokyo) 37:422–425.

Umezawa, H., Aoyagi, T., Ogawa, K., Iinuma, H., Naganawa, H., Hamada, M. and Takeuchi, T. 1984b. Histargin, a new inhibitor of carboxypeptidase B, produced by actinomycetes. *Journal of Antibiotics* (Tokyo) 37:1088–1090.

Umezawa, H., Aoyagi, T., Ogawa, K., Obata, T., Iinuma, H., Naganawa, H., Hamada, M. and Takeuchi, T. 1985. Foroxymithine, a new inhibitor of angiotensin-converting enzyme, produced by actinomycetes. *Journal of Antibiotics* (Tokyo) 38:1813–1815.

Umezawa, H., Imoto, M., Sawa, T., Isshiki, K., Matsuda, N., Uchida, T., Iinuma, H., Hamada, M. and Takeuchi, T. 1986a. Studies on a new epidermal growth factor-receptor kinase inhibitor, erbstatin, produced by MH435-hF3. *Journal of Antibiotics* (Tokyo) 39:170–173.

Umezawa, H., Aoyagi, T., Nishikiori, T., Okuyama, A., Yamagishi, Y., Hamada, M. and Takeuchi, T. 1986b. Plipastatins. New inhibitors of phospholipase A_2, produced by *Bacillus cereus* BMG302-fF67. *Journal of Antibiotics* (Tokyo) 39:737–744.

Uyeda, M., Suzuki, K. and Shibata, M. 1984. M-GCI, a novel beta-glucuronidase inhibitor produced by *Micromonospora* sp. strain No. BR-1613. *Agricultural and Biological Chemistry* 48:29–35.

Uyeda, M., Suzuki, K., Uwatoko, H. and Shibata, M. 1978. API-2b, a new alkaline protease inhibitor produced by *Streptomyces griseoincarnatus* strain No. KTo-250. *Agricultural and Biological Chemistry* 42:49–54.

Weibel, E. K., Hadvary, P., Hochuli, E., Kupfer, E. and Lingsfeld, H. 1987. Lipstatin, an inhibitor of pancreatic lipase, produced by *Streptomyces toxytricini*. *Journal of Antibiotics* (Tokyo) 40:1081–1085.

Woo, P. W. K., Dion, H. W., Lange, S. M., Dahl, L. F. and Durham, L. J. 1974. A novel adenosine and Ara-A deaminase inhibitor, (R)-3-(2-deoxy-β-D-*erythro*-pentofuranosyl)-3,6,7,8-tetrahydroimidazo[4,5-*d*][1,3]diazepin-8-ol. *Journal of Heterocyclic Chemistry* 11:641–643.

Yaginuma, S., Asahi, A., Takada, M., Hayashi, M., Tsujino, M., Mizuno, K. and Murase, J. 1988. Aldostatin, a novel aldose reductase inhibitor. In: *Abstracts of the First International Conference on the Biotechnology of Microbial Products: Novel Pharmacological and Agrobiological Activities*, p. S-19.

Yaginuma, S., Asahi, A., Morishita, A., Hayashi, M., Tsujino, M. and Tokoda, M. 1989. Isolation and characterization of new thiol protease inhibitors estatins A and B. *Journal of Antibiotics* (Tokyo) 42:1362–1369.

Yamada, K., Tanaka, H., Kubo, K. and Kase, H. 1987. Inhibition by K252a, a microbial product, of contraction of isolated rabbit arteries. *Japanese Journal of Pharmacology* 43(Supplement): 284p.

Yamato, M., Hamada, M., Naganawa, H., Maeda, K., Takeuchi, T. and Umezawa, H. 1985. Screening of microbial inhibitors of mammalian ornithine decarboxylase. *Journal of Antibiotics* 38:442–443.

Yamato, M., Koguchi, T., Okachi, R., Yamada, K., Nakayama, K. and Kase, H. 1986. K-26, a novel inhibitor of angiotensin I converting enzyme produced by an actinomycete K-26. *Journal of Antibiotics* (Tokyo) 39:44–52.

Yokose, K., Ogawa, K., Sano, T., Watanabe, H., Maruyama, B. and Suhara, Y. 1983. New alpha-amylase inhibitor, trestatins. I. Isolation, characterization and biological activities of trestatins A, B and C. *Journal of Antibiotics* (Tokyo) 36:1157–1165.

Yoshida, S., Aoyagi, T., Harada, S., Matsuda, N., Ikeda, T., Naganawa, H., Hamada, M. and Takeuchi, T. 1991a. Production of 2-methyl-4[3*H*]-quinazolinone, an inhibitor of poly(ADP-ribose) synthetase, by bacterium. *Journal of Antibiotics* (Tokyo) 44:111–112.

Yoshida, T., Nakamoto, S., Sakazaki, R., Matsumoto, K., Terui, Y., Sato, T., Arita, H., Matsutani, S., Inoue, K. and Kudo, I. 1991b. Thielocins A1α and A1β, novel phospholipase A_2 inhibitors from ascomycetes. *Journal of Antibiotics* (Tokyo) 44:1467–1470.

8

New Strategy for Search of Enzyme Inhibitors: Screening with Animal Cells or Microorganisms with Special Functions

Hiroshi Tomoda and Satoshi Ōmura

8.1 Introduction

A large number of bioactive compounds have been discovered among the metabolites of microorganisms. Researchers have revealed the mechanism of action of several that have practical use, as well as of some that are merely interesting. Most compounds were found to be specific inhibitors of certain enzymes or enzyme systems. Such enzymes or enzyme systems provide targets for selective cytotoxicity or for pharmacological intervention.

On the basis of these results, screening for enzyme inhibitors, as described in Chapter 7, has been carried out extensively with assays using the target enzymes or enzyme systems themselves. In some cases purified enzymes were used for the assay of enzyme reactions, and in other cases partially purified enzyme preparations, such as rat liver microsomal fractions or cell homogenates, were used. Thus, conventional methods for routine assays in the primary screening systems have been established. However, with such a screening strategy one may get a lot of false-positive inhibitors because the assays are too closely linked to enzyme function. Enzymes require their own environment in cells to exhibit their normal activity in vivo. Some enzymes are in the cytosol, some are associated with membranes, and others are localized in certain organelles. Therefore, utilization of living cells in assays may be much better as a strategy in searching for enzyme inhibitors.

This strategy is expected to have the following advantages. (1) The environment where the enzyme exists in vivo is maintained and normal physiological behavior is mimicked. (2) False-positive inhibitors can be omitted. It has been a problem to know whether an enzyme inhibitor can reach the place where the enzyme exists in the cell because of both the membrane permeability and the other properties of an inhibitor. (3) A prodrug type of inhibitor might be picked up. Some inhibitors may be activated only after incorporation into cells by enzymes other than target ones. Concerning these points, there has often been a big gap between an in vitro enzyme assay and an in vivo assay.

Therefore, (4) an inhibitor selected by the method of assaying with living cells will provide a higher possibility for full in vivo efficacy. The assay established according to this strategy lies between an in vitro and an in vivo assay and is much closer to an in vivo case. (5) It should be noted that a large amount of radioactive substrate is often needed when an in vitro assay, as mentioned in Chapter 7, is used as a primary screening. However, radioactive compounds are not needed, if necessary, the amount can be reduced in the whole cell assays. In opposition to this advantage, it must be said that it is often difficult to keep the characteristics of animal cells constant during subculturing. Also, the running cost for animal cell cultures is quite high for routine screening work. This chapter presents some of our recent experiences with the application of this new strategy to the discovery of enzyme inhibitors of lipid metabolism (Tomoda and Ōmura, 1990).

Lipid metabolism is elegantly balanced between synthesis and degradation and closely cooperates with other metabolism to maintain homeostasis. When the balance of lipid metabolism is lost, a variety of serious diseases will develop, including arteriosclerosis (atherosclerosis), hypertension, obesity, diabetes, functional depression of some organs, and so on. To control lipid metabolism by drugs could lead to the treatment of these diseases. Lipid metabolic pathways, such as fatty acid degradation where acyl-CoA synthetase and β-oxidation are involved, fatty acid synthesis where acetyl-CoA carboxylase and fatty acid synthase are involved, triacylglycerol synthesis, cholesterol synthesis, etc., might be possible target sites for treatment. Screening for enzyme inhibitors among these pathways is described here.

8.2 Mevalonate Biosynthesis

Rationale Much attention has been paid to inhibitors of cholesterol biosynthesis as potential hypocholesteremic agents as described (see Chapter 7). The fungal metabolites compactin (ML-236B) (Brown et al., 1976; Endo et al., 1976) and mevinolin (monacolin K) (Endo, 1979; Alberts et al., 1980) were discovered as inhibitors of 3-hydroxy-3-methylglutaryl coenzyme A (HMG-CoA) reductase, the rate-limiting enzyme in cholesterol biosynthesis (Figure 8.1). Recently, their analogs pravastatin (CS-514) (Tsujita et al., 1986) and simvastatin have been developed as clinical drugs. The enzymatic product, mevalonate, is a key intermediate in the cholesterol biosynthetic pathway.

Chapter 8 New Strategy for Search of Enzyme Inhibitors

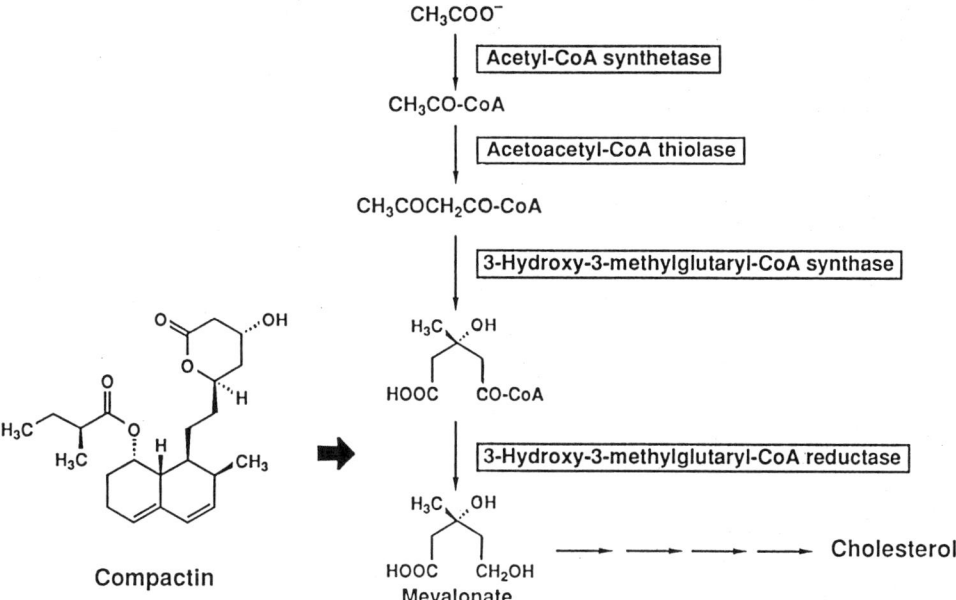

Figure 8.1 Biosynthetic pathway of cholesterol.

Mevalonate is produced from acetyl-CoA by three enzymes, namely, acetoacetyl-CoA thiolase, HMG-CoA synthase and HMG-CoA reductase (see Figure 8.1). These enzymes provide promising targets for pharmaceutical intervention by a hypocholesteremic agent. Kaneko et al. (1978) reported that the growth of cultured animal cells was inhibited by compactin and that the inhibition was overcome by the addition of mevalonate to the medium. Because it has been thought that most microorganisms cannot incorporate mevalonate, cultured animal cells are expected to be an ideal test organism for discovering inhibitors of mevalonate biosynthesis.

Method (Tomoda et al., 1987b; Kumagai et al., 1990) Vero cells (an established cell line from the kidney cells of the African green monkey) were seeded in each well of a 96-well microplate (Corning Co.) at a concentration of 3.0×10^4 cells in 100 µl of Eagle's minimum essential medium supplemented with 2% calf serum. After a 1-h incubation, a paper disk (6 mm, Toyo Roshi Co.) containing various concentrations of a test sample (usually cultured broth of microorganisms) with and without 1 mM mevalonate (final concentration) was put into each well. After a 24-h incubation in a humidified incubator (5% CO_2) at 37°C, the paper disks were removed carefully, and the cell growth was examined with a microscope or microplate photometer at 550 nm after methylrosaniline staining by the method of Armstrong (1971). If a test sample causes inhibition of Vero cell growth and the inhibition is overcome by the addition of mevalonate, the microbial strain producing such a sample should be picked up as a producer of an inhibitor.

Inhibitors (Ōmura et al. 1987; Tomoda et al., 1987b, 1988) A cultured broth of *Scopulariopsis* sp. F-244 was found to show the described phenomena with Vero cells as an inhibitor. As shown in Figure 8.2, the inhibition and the morphological changes of Vero cell growth were observed only in the presence of the broth. The inhibition was reversed by the addition of mevalonate to the media and the cells grew normally. The active principle was isolated and the structure was identified as antibiotic 1233A (Figure 8.3), originally reported by Aldridge et al. (1970, 1971). Detailed studies of 1233A on the site of inhibition in mevalonate biosynthesis demonstrated that 1233A is the first naturally occurring compound that inhibits HMG-CoA synthase strongly and specifically (IC_{50}, 0.2 μM).

Using this screening program, compactin, monacolin K, and their derivatives were also detected and identified. Kuroda and Endo (1977) reported that several long-chain fatty acids inhibit cholesterol synthesis in vitro. These include tridecanoate, which inhibits acetoacetyl-CoA thiolase; highly unsaturated fatty acids, e.g., arachidonate and linoleate; inhibitors of HMG-CoA synthase; and ricinolate and phytanate, which diminish the conversion of mevalonate to sterol. None of these fatty acids showed any effect on the growth of Vero cells.

8.3 Fatty Acid Metabolism

Rationale Fatty acid metabolism in *Candida lipolytica* has been studied extensively by Numa and his coworkers (Kamiryo et al., 1977, 1979; Mishina et al., 1978, 1979; Hosaka et al., 1979). The scheme of lipid metabolism in *Candida lipolytica* is shown in Figure 8.4. This yeast possesses two distinct acyl-CoA synthetases that activate a free long-chain fatty acid by converting it to the corresponding acyl-CoA. The one, designated acyl-CoA synthetase I, is responsible for the synthesis of cellular lipids and the other, acyl-CoA synthetase II, provides an acyl-CoA that is exclusively degraded via β-oxidation to yield acetyl-CoA. Acyl-CoA synthetase I is distributed among different subcellular fractions, including microsomes and mitochondria. On the other hand, acyl-CoA synthetase II is localized in peroxisomes where an acyl-CoA-oxidizing system exists. Acyl-CoA to be utilized for cellular lipid synthesis is also provided via fatty acid synthase.

Target sites of inhibitors of fatty acid metabolism were defined as acyl-CoA synthetase, a β-oxidation system, and fatty acid synthase. For the screening, two kinds of mutant yeast strains, L-7 and A-1, were used as test organisms. The mutant strain L-7 is defective in acyl-CoA synthetase I (Kamiryo et al., 1979), and the mutant strain A-1 lacks a fatty acid synthase (Miyakawa et al., 1984) (Figure 8.4). It is essential to produce both acyl-CoA for lipid synthesis and acetyl-CoA so that these yeasts can grow. These mutant strains were grown on two different media, one containing a long-chain fatty acid as the sole carbon source and the other containing glucose and a small amount

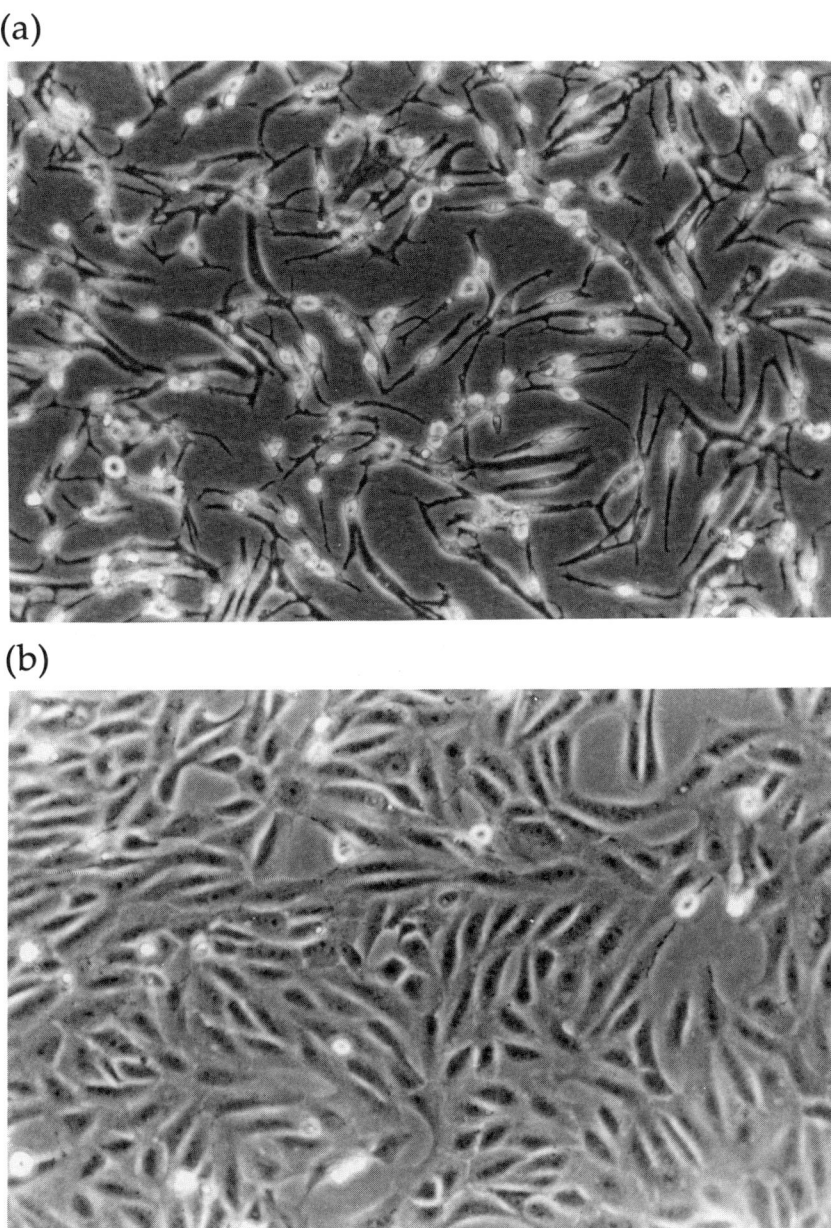

Figure 8.2 Vero cell growth in the presence of a cultured broth of *Scopulariopsis* sp. F-244 alone (a) or with 1 mM mevalonate (b).

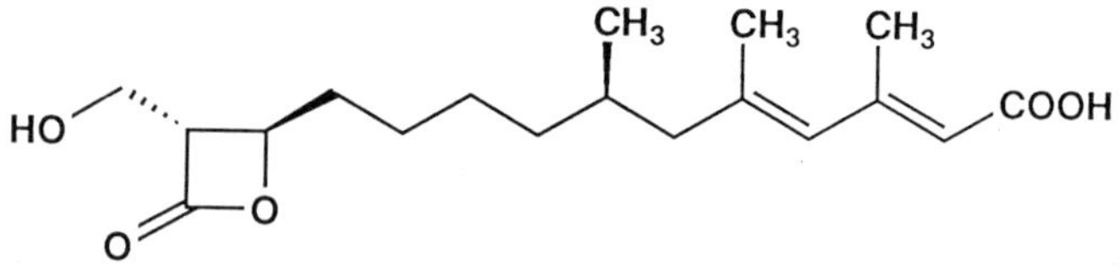

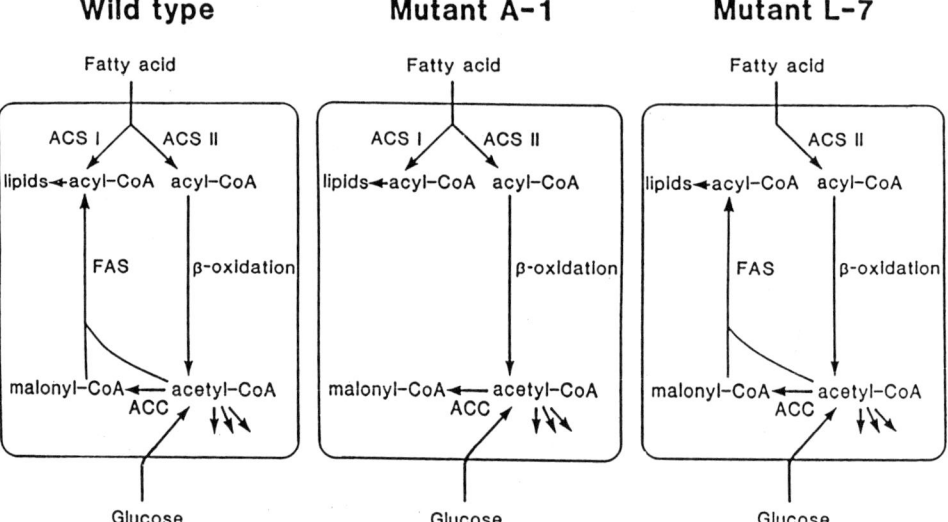

Figure 8.3 Structure of 1233A

Figure 8.4 Fatty acid metabolism in *Candida lipolytica*. Wild type and mutant strains A-1 and L-7. ACS ; acyl-CoA synthetase, ACC; acetyl-CoA carboxylase, FAS; fatty acid synthase.

of fatty acid. One can understand the inhibition target of unknown compounds from microbial sources by evaluating their pattern of inhibitory activity against the two mutant strains in the two different media. The possible inhibition site predicted from the inhibitory pattern of unknown samples are shown in Table 8.1.

Method The mutant strains L-7 and A-1 were incubated for 2–3 days at 27°C in B medium, comprising 0.1% Bacto-yeast, 0.5% $(NH_4)_2SO_4$, 0.25% KH_2PO_4, 0.01% NaOH, 0.1% $MgSO_4 \cdot 7H_2O$, and 0.0001% $FeCl_3 \cdot 6H_2O$(wt/vol), supplemented with 1.0% oleate (dissolved in 1.0% Brij 58) (wt/vol) to give the seed cultures. The mycelia were harvested by centrifugation, washed twice with saline, and resuspended with the same volume of B medium supplemented with 1.0% oleate (in 1.0% Brij 58). The four different agar plates

Chapter 8 New Strategy for Search of Enzyme Inhibitors

Table 8.1 Possible inhibition site in fatty acid metabolism expected from inhibitory patterns against *Candida lipolytica* mutant strains L-7 and A-1 in a medium containing fatty acid or glucose as sole carbon source

Possible inhibition site	Inhibitory pattern			
	Mutant L-7		Mutant A-1	
	Fatty acid	Glucose[1]	Fatty acid	Glucose[1]
Acyl-CoA synthetase I	−[2]	−	+	+
Acyl-CoA synthetase II or β-oxidation	+	−	+	−
Fatty acid synthase	+	+	−	−

[1] A small amount of fatty acid (0.01 %) was supplemented
[2] Symbols: growth inhibition, +; growth; −.

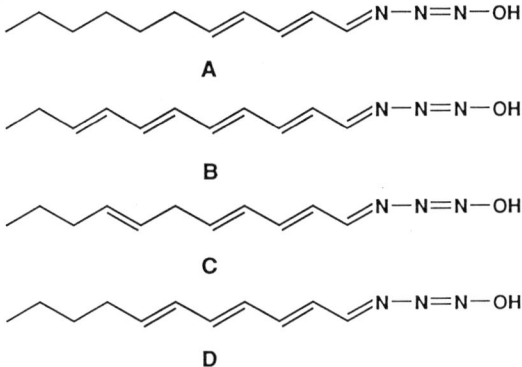

Figure 8.5 Structures of triacsins A, B, C and D.

containing B medium supplemented with 1.0% carbon source (oleate dissolved in Brij 58 or glucose), 1.0%–1.5% Bacto-agar (wt/vol) and 1.0% of the seed culture (mutant strain L-7 or A-1) (vol/vol) were prepared. The agar plates on which were set paper disks containing unknown samples were incubated for 12 h at 27°C.

Inhibitors (Ōmura et al., 1986; Tomoda, et al., 1987a, 1991) A fermented broth of *Streptomyces* sp. SK-1894 showed the inhibitory pattern of an acyl-CoA synthetase I inhibitor. Four active compounds, named triacsins A, B, C, and D (Figure 8.5), were isolated. As expected triacsins inhibited long chain acyl-CoA synthetases from widely different sources and the IC_{50} values are summarized in Table 8.2.

A fermented broth of an unidentified fungus showed the inhibitory pattern of a fatty acid synthase, resulting in the identification of cerulenin, a well-known inhibitor of fatty acid synthase (Ōmura, 1976).

Table 8.2 Inhibition of acyl-CoA synthetases by triacsins

Acyl-CoA synthetases	IC_{50} (μM)[1]			
	Triacsin A	Triacsin B	Triacsin C	Triacsin D
Pseudomonas sp.	17	>200	3.6	>200
Rat liver	18	>200	8.7	>200
Raji cells	5.3	>100	3.2	>100

[1] >100: inhibited by 40–45% at 100 μM of triacsins, >200: inhibited by 40–45% at 200 μM of triacsins.

8.4 Concluding Remarks

Examples that depend on the strategy of using animal cells or microorganisms with special functions to discover new enzyme inhibitors other than those with targets in lipid metabolism include inhibitors of glutamine synthetase (using *Bacillus subtilis* grown on a minimal medium) (Ōmura et al., 1984a, 1984b) (see Chapter 7) and inhibitors of thymidylate synthetase (using *Enterococcus faecium*, which requires folate-related compounds for growth) (Ōmura et al., 1985, and others).

As expected, some advantages of this strategy were demonstrated. For example, Kuroda and Endo (1977) reported fatty acids that appeared to be inhibitors were discovered by the in vitro cholesterol synthesis assay, but such false-positives showed no effect on the Vero cell assay and neither of them was picked up. In another case, phosalacine, an inhibitor of glutamine synthetase that exhibits a potent herbicidal activity, was discovered as a glutamine antagonist by using a bacterium. If an in vitro assay using isolated glutamine synthetase had been used, this compound would not have been discovered, because phosalacine itself is a very weak inhibitor of the enzyme. However, with the whole cell assay, phosalacine could enter the bacterium cell easily, be hydrolyzed to an active form, phosphinothricin, and reach the place where the enzyme reaction takes place.

Further successful applications of this strategy for detecting inhibitors of enzymes are anticipated.

References

Alberts, A. W., Chen, J., Kuron, G., Hunt, V., Huff, J., Hoffman, C., Rothrock, J., Lopez, M., Joshua, H., Harris, E., Patchett, A., Monaghan, R., Currie, S., Stapley, E., Albers-Schonberg, G., Hensens, O., Hirshfield, J., Hoogsteen, K., Liesch, J. and Springer, J. 1980. Mevinolin: A highly potent competitive inhibitor of hydroxymethylglutaryl-coenzyme A reductase and a cholesterol-lowering agent. *Proceedings of the National Academy of Sciences of the United States of America* 77:3957–3961.

Aldridge, D. C., Gile, D. and Turner, W. B. 1970. Antibiotic 1233A, a fungal β-lactone. *Journal of the Chemical Society Chemical Communications* 1970:639.

Aldridge, D. C., Gile, D. and Turner, W. B. 1971. Antibiotic 1233A, a fungal β-lactone. *Journal of the Chemical Society Chemical Communications* 1971:3888–3891.
Armstrong, J. A. 1971. Semi-micro, dye-binding assay for rabbit interferon. *Applied Microbiology* 21:723–725.
Brown, A. G., Smale, T. C., King, T. J., Hasenkamp, R. and Thompson, R. H. 1976. Crystal and molecular structure of compactin, a new antifungal metabolite from *Penicillium brevicompactum*. *Journal of the Chemical Socciety Perkin Transactions*. 1:1165–1170.
Endo, A., Kuroda, M. and Tsujita, Y. 1976. ML-236a, ML-236b, and ML-236c, new inhibitors of cholesterogenesis produced by *Penicillium citrium*. *Journal of Antibiotics* (Tokyo) 29:1346–1348.
Endo, A. 1979. Monacolin K, a new hypocholesterolemic agent produced by a *Monascus* species. *Journal of Antibiotics* (Tokyo) 32:852–854.
Hosaka, K., Mishina, T., Tanaka, T., Kamiryo, T. and Numa, S. 1979. Acyl-coenzyme A synthetase I from *Candida lipolytica*. *European Journal of Biochemistry* 93:197–203.
Kamiryo, T., Mishina, M., Tashiro, S. and Numa, S. 1977. *Candida lipolytica* mutants defective in an acyl-coenzyme A synthetase. Isolation and fatty acid metabolism. *Proceedings of the National Academy of Sciences of the United States of America* 74:4947–4950.
Kamiryo, T., Nishikawa, Y., Mishina, M., Terao, M. and Numa, S. 1979. Involvement of long-chain acyl coenzyme A for lipid synthesis in repression on acetyl-coenzyme A carboxylase in *Candida lipolytica*. *Proceeding of the National Academy of Sciences of the United States of America* 76:4390–4394.
Kaneko, I., Hazama-Shimada, Y. and Endo, A. 1978. Inhibitory effects on lipid metabolism in cultured cells on ML-236B, a potent inhibitor of 3-hydroxy-3-methylglutaryl-coenzyme-A reductase. *European Journal of Biochemistry* 87:313–321.
Kumagai, H., Tomoda, H. and Ōmura, S. 1990. Method of search for microbial inhibitors of mevalonate biosynthesis using animal cells. *Journal of Antibiotics* (Tokyo) 43:397–402.
Kuroda, M. and Endo, A. 1977. Inhibition of *in vitro* cholesterol synthesis by fatty acids. *Biochimica et Biophysica Acta* 486:70–81.
Mishina, M., Kamiryo, T., Tashiro, S. and Numa, S. 1978. Separation and characterization of two long-chain acyl-CoA synthetases from *Candida lipolytica*. *European Journal of Biochemistry* 82:347–354.
Mishina, M., Kamiryo, T., Tashiro, S., Hagihara, T., Tanaka, A., Fukui, S., Osumi, M. and Numa, S. 1979. Subcellular localization of two long-chain acyl-coenzyme A synthetases in *Candida lipolytica*. *European Journal of Biochemistry* 89:321–328.
Miyakawa, T., Nakajima, H., Hamada, K., Tsuchiya, E., Kamiryo, T. and Fukui, S. 1984. Isolation and characterization of a mutant of *Candida lipolytica* which excretes long-chain fatty acids. *Agricultural and Biological Chemistry* 48:499–503.
Ōmura, S. 1976. The antibiotic cerulenin, a novel tool for biochemistry as an inhibitor of fatty acid synthesis. *Bacteriological Reviews* 40:681–697.
Ōmura, S., Murata, M., Hanaki, H., Hinotozawa, K., Oiwa, R. and Tanaka, H. 1984a. Phosalacin, a new herbicidal antibiotic containing phosphinothricin. Fermentation, isolation, biological activity and mechanism of action. *Journal of Antibiotics* (Tokyo) 37:1324–1332.
Ōmura, S., Hinotozawa, K., Imamura, N. and Murata, M. 1984b. The structure of phosalacine, a new herbicidal antibiotic containing posphinothricin. *Journal of Antibiotics* (Tokyo) 37:939–940
Ōmura, S., Murata, M., Kimura, K., Matsukura, S., Nishihara, T. and Tanaka, H. 1985. Screening for new antifolates of microbial origin and a new antifolate AM-8402. *Journal of Antibiotics* (Tokyo) 38:1016–1024.
Ōmura, S., Tomoda, H., Xu, Q. M., Takahashi, Y. and Iwai, Y. 1986. Triacsins, new inhibitors of acyl-CoA synthetase produced by *Streptomyces* sp. *Journal of Antibiotics* (Tokyo) 39:1211–1218.

Ōmura, S., Tomoda, H., Kumagai, H., Greenspan, M. D., Yodkovitz, J. B., Chen, J. S., Alberts, A. W., Martin, I., Mochales, R. L., Monaghan, R. L., Chabala, J. C., Schwartz, R. E. and Patchett, A. A. 1987. Potent inhibitory effect of antibiotic 1233A on cholesterol biosynthesis which specifically blocks 3-hydroxy-3-methylglutaryl coenzyme A synthase. *Journal of Antibiotics* (Tokyo) 40:1356–1357.

Tomoda, H., Igarashi, K. and Ōmura, S. 1987a. Inhibition of acyl-CoA synthetase by triacsins. *Biochimica et Biophysica Acta* 921:595–598

Tomoda, H., Kumagai, H., Tanaka, H. and Ōmura, S. 1987b. F-244 specifically inhibits 3-hydroxy-3-methylglutaryl coenzyme A synthase. *Biochimica et Biophysica Acta* 922:351–356.

Tomoda, H., Kumagai, H., Takahashi, Y., Tanaka, Y., Iwai, Y. and Ōmura, S. 1988. F-244 (1233A), a specific inhibitor of 3-hydroxy-3-methylglutaryl coenzyme A synthase. Taxonomy of producing strain, fermentation, isolation and biological properties. *The Journal of Antibiotics* 41:247–249.

Tomoda, H. and Ōmura, S. 1990. New strategy for discovery of enzyme inhibitors: Screening with intact mamalian cells or intact microorganisms having special functions. *The Journal of Antibiotics* 43:1207–1222.

Tomoda, H., Igarashi, K., Cyong, J. C. and Ōmura, S. 1991. Evidence for an essential role of long chain acyl-CoA synthetase in animal cell proliferation000 Inhibition of long chain acyl-CoA synthetase by triacsins caused inhibition of Raji cell proliferation. *Journal of Biological Chemistry* 266:4214–4219.

Tsujita, Y., Kuroda, M., Shimada, Y., Tanzawa, K., Arai, M., Kaneko, I., Tanaka, M., Masuda, H., Tarumi, C., Watanabe, Y. and Fujii, S. 1986. CS-514, a competitive inhibitor of 3-hydroxy-3- methylglutaryl coenzyme A reductase. Tissue-selective inhibition of sterol synthesis and hypolipidemic effect on various animal species. *Biochimica et Biophysica Acta* 877:50–60.

Part 4
Pharmacologically Active Substances

9

Immunomodulators

Haruki Yamada

9.1 Introduction

A variety of information has accumulated concerning most immune defense mechanisms and the immune response system. Immunological defense reflects a complicated interplay between nonspecific and specific cellular and humoral immune responses, stimulation and suppression of immunocompetent cells, and the influence of endocrine and other mechanisms on the immune system. It is now widely recognized that a functioning host immune defense system consists of a variety of lymphoid cells, including effector and affector cells consisting of both T and B types. Mononuclear phagocytic cells such as macrophages are also involved in these processes. These cells are involved not only in the clearance of particulate matter, including bacteria, viruses, and fungi, but also presumably in host defense against neoplasia.

Microbial infections are associated with immunoregulation in that enhanced antibody or cell-mediated immune responses may occur. Much enhancement often may be associated with autoimmune phenomena as well as polyclonal activation of a variety of lymphoid cells. The immunomodulating substances that attack specific target sites of cells are thought to be potent tools for the study of cellular and biochemical events of the immune responses and also provide useful drugs for immunotherapy. For example, it has been long recognized that many of the "adjuvants" such as *Mycobacterium bovis* BCG that enhance the immune response in nonspecific ways are derived from microorganisms.

Human immunity decreases with old age. The decrease in immunity also can result from many diseases such as cancer, diabetes, arthritis, nephrosis, and infection. For the treatment of a decrease in immunity and resultant immunodeficiency disease, such as acquired immune deficiency syndrome (AIDS), immunopotentiators may be helpful. On the other hand, immunosuppressors may also be useful for reducing rejection of organ transplantation and for the treatment of autoimmune diseases such as rheumatoid arthritis, systemic lupus erythematosus (SLE), dermatomycosis, membrane glomerulonephritis, inflammatory bowel disease, and autoimmune hemolytic anemia.

The Fujisawa group and others have tried to search for immunomodulating substances among microbial metabolites using immunological assay systems, and some new, immunoactive, low molecular weight substances such as FK-506 have been discovered. Establishing assay systems for the immunomodulating substances with limited target spectra could be valuable, because the immune system involves multiple interactions between lymphocytes belonging to different subsets by direct contact as well as via soluble factors such as cytokines. Several immunoregulatory targets such as complement activation, interferon production, antibody formation, cytokine production, and activation of the reticuloendotherial system could be useful for the screening of immunomodulators.

This chapter summarizes the screening methods used to detect immunomodulators from microbial cultures.

9.2 General Screening Methods

The screening of immunomodulators has been carried out by using several immunological assays that are able to yield data on direct or indirect effects by the potentiation or suppression of cellular immunity (Table 9.1).

General immunological assay systems frequently used for the screening of immunomodulators from microbial cultures are described as follows.

In vitro methods

Plaque-forming cells Plasma cells secrete antibody, which can be used to quantify the number of antibody-producing cells in an organ. The plaque assay was developed by Jerne and Nordin (1963) to detect cells producing antibody against erythrocyte antigens. Spleen cells from immune mice are incubated in an agar gel with the immunizing erythrocytes. The specific antibody produced by some of the lymphoid cells is released and diffuses from the central cells; antibody is trapped by antigen in the areas immediately surrounding the plaque-forming cells (PFC). In the presence of complement, plaques appear as a consequence of lysis of sensitized red cells.

Only direct plaques [mainly immunoglobulin M (IgM) antibody] are detected by this method because of the high, hemolytic efficiency of this class of antibody. To be detected IgG plaques must be developed with an antiserum against mouse IgG (indirect plaques). Cunningham (1965) also proposed a modification of the hemolytic plaque technique in which where lymphoid cells and indicator sheep erythrocytes (SRBC) are incubated without supporting gels as monolayers in a chamber. When immunopotentiators are administered starting from the day of immunization, it is found that the plaque-forming cells increase more than in the drug-free assays. When an immunosuppressor is administered starting from the day of immunization, it is

Table 9.1 Screening methods of immunomodulators

Screening methods	Immuno potentiator	Microbial products	Immuno suppressor	Microbial products
In vitro methods				
Mitogen-induced spleen cell proliferation ([³H]thymidine uptake)	Enhance	Swainsonine (Hino et al., 1985) FR-900494 (Kifunensin) (Iwami et al., 1987) FR-900483 (Shibata et al., 1988)	Reduce	Concanamycin (Kinashi et al., 1984) Prodigiosin 25C (Harashima et al., 1967) Depsidomycin (Isshiki et al., 1990)
Mixed-lymphocyte reaction ([³H]-thymidine uptake)	Enhance		Reduce	FK-506 (Kino et al., 1987a) FR-900520 (Hatanaka et al., 1988b) FR-900523 (Hatanaka et al., 1988b) FR-65814 (Hatanaka et al., 1988a)
IL-2 production	Enhance	Lentinan (Okuda et al., 1972)	Suppress	Cyclosporin A¹
Total complement hemolysis (anti complementary activity)	Possible to enhance		Possible to enhance	K-76 (Miyazaki et al., 1980) Complestatin (Kaneko et al., 1989)
In vivo methods				
Experimental infections in mice	Protect	FK-156 (Gotoh et al., 1982) Aladapsin (Shiraishi et al., 1990)	No protect	
Antitumor activity against allogenic tumor cells	Yes	Lentinan (Chihara et al., 1969)		

¹ First screened as not being immunomodulators.

shown that the plaque-forming cells decrease more than in the drug-free assays.

Mitogen induced mouse spleen cell proliferation Many plant lectins induce blast cell transformation and mitosis in a manner similar to that of antigens. The mitogen binds to specific cell-surface receptors, as does the antigen, and the signal thus generated causes derepression of the nucleus and entry of the lymphocyte into the cell's cycle. Unlike antigens, mitogens stimulate a large proportion of the lymphocytes. As for antigen stimulation of lymphocytes in vitro, it has been shown that an approximate correlation exists between the in vitro response to mitogens and the immune status of individual. The degree of lymphocyte stimulation may therefore be assayed in the presence of a mitogen, either by determining the percentage of blast cells in the culture or by measuring the amount of radioactive DNA analogue (generally [^{3}H]thymidine) incorporated into newly synthesized DNA. When the lymphocytes are incubated in the presence of mitogen and immunopotentiator, the incorporation of [^{3}H]thymidine into lymphocytes is enhanced more than in the free potentiator. When the lymphocytes are incubated in the presence of mitogen and immunosuppressor, the incorporation is more reduced than with the free suppressor alone.

Sugawara et al. (1984) employed a colorimetric 3-(4,5-dimethylthiazole-2-yl)-2,5-diphenyl tetrazolium bromide (MTT) assay to measure lymphocyte blast transformation as an application of the method of Mosmann (1983). Because this method does not involve radioisotopes, it is convenient and can be used to screen immunomodulators. Murine B lymphocytes blastoid transformation is well correlated with alkaline phosphatase activity (Mosbach-Ozman and Loor, 1981; Mosbach-Ozman et al., 1986; Ohno et al., 1986). Therefore, measurement of the alkaline phosphatase activity also would be a useful method to assay murine blastoid B lymphocytes. However, these colorimetric assays must be used carefully because of the interference from the color of culture broth.

The typical mitogens of specific T lymphocytes are two plant lectins, concanavalin A (ConA) and phytohemagglutinin (PHA). Both lectins activate T lymphocytes and cause them to secrete various lymphokines. However, the population of their target cells and their dependency on accessory cells for expression of mitogenic activity is different. One of the most popular murine B lymphocyte mitogens is bacterial lipopolysaccharide (LPS); when LPS activates B lymphocytes directly, they differentiate to immunoglobulin-secreting cells. Employing these three different mitogens, Nakamura et al. (1986) established the method for the screening of immunomodulators. The activity of samples in the assay system was evaluated by the differential effects on [^{3}H]thymidine incorporation into spleen cells that were stimulated by the mitogens.

All three mitogens, ConA, PHA, and LPS, afford their maximal effect on DNA synthesis of lymphocytes 3 days after initiating the culture. The optimal concentration for induction of mitogenic response is 2–3 μg ml^{-1} for ConA or PHA. In LPS the stimulation is nearly constant at concentrations greater than 4 μg ml^{-1}. In this screening system, a suboptimal dose of the mitogen is adopted to detect both augmentation and inhibition of mitogenic response by the added samples.

Cytotoxicity as well as mitogenic activity of the samples is also assayed in the cultures incubated in the absence of a mitogen. Cyclosporin A is a potent immunosuppressive agent and has been used to prevent organ allograft rejection (White et al., 1982). Cyclosporin A suppressed response to ConA and PHA more extensively than to LPS. The results are in good agreement with the notion that cyclosporin A affects T-cell function greatly while leaving B-cell functions relatively intact. This further indicates that this screening system (Nakamura et al., 1986) may be used rationally to detect immunosuppressors of the cyclosporin A type. Antimycin A_3 suppressed, as well as funiculosin-suppressed, cell proliferation is induced by ConA and LPS more significantly than proliferation induced by PHA (Nakamura et al. 1986).

Assay method for mitogen-induced mouse spleen cell proliferation (Nakamura et al.,1986) Normal mice are sacrificed and the spleen teased with a plastic syringe in a Petri dish containing 5 ml of Eagle's minimal essential medium (MEM). The product is gently strained through a 100 mesh screen to remove clumps and yields a single cell suspension. These cells are washed three times with MEM and resuspended in the complete tissue culture medium [designated Roswell Park Memorial Institute (RPMI)-1640]. All media used contain 100 U ml^{-1} of benzylpenicillin and 100 μg ml^{-1} of streptomycin sulfate and 5–10% fetal calf serum.

The cells are suspended in the tissue culture medium and diluted to contain 5×10^5 cells ml^{-1} with or without ConA, PHA, or LPS, or a test sample, in flat-bottom plastic microtrays (A/S NUNC, Rockilde, Denmark). Most of the test samples are applied as methanol solutions. Because 0.5% methanol solution in the culture medium shows no effect on the mitogenic responses, the final concentration of methanol is adjusted to 0.5%. The cells are cultured for 72 h at 37°C in a humidified atmosphere of 5% CO_2 and air. Then, [^{3}H]thymidine (113 Ci mmol^{-1}, NEN, Boston, MA) (0.2 μCi well^{-1}) is added to the cells in each well, and culturing is continued for 5 h followed by harvesting with a cell harvester. The amount of [^{3}H]thymidine incorporated into the cells is determined by a liquid scintillation counter.

Concanamycins A, B, and C, prodigiosin 25C, and depsidomycin also were screened as immunosuppressors by this method (Harashima et al., 1967; Kinashi et al., 1984; Isshiki, et al., 1990).

Depsidomycin was isolated as a new depsipeptide antibiotic from the culture broth of *Streptomyces levenclofoliae* MI951–62F2. It is primarily active against gram-positive microorganisms and has immunosuppressive activity. Depsidomycin inhibited LPS-induced blastogenesis at 25 μg ml^{-1} but not

ConA-induced blastogenesis, even at 100 μg ml^{-1}. Depsidomycin inhibited the mixed lymphocyte reaction in rat cells; the IC$_{50}$ was 1.6 μg ml^{-1}.

Mixed lymphocyte reactions   A mitotic response is also obtained when cells taken from two inbred strains or from two outbred individuals of any species are mixed in an in vitro culture. This is called a mixed lymphocyte reaction (MLR). The proliferation of target cells in either cell type may be blocked with mitomycin C or by x-ray irradiation. The degree of the responder cell stimulation may be assayed by measuring the amount of [^{3}H]thymidine incorporation into the target cells. With immunopotentiators it is possible to enhance this reaction whereas with immunosuppressors it is possible to reduce it. A recently discovered potent immunosuppressor, FK-506, was first screened by this method (Goto et al., 1987; Kino et al., 1987a).

Assay method for MLR (Hino et al., 1985)   Female C57 BL/6 mice and female BALB/c mice (both at 8 weeks old) are sacrificed. Their spleens are aseptically removed and single-cell suspensions are prepared as sources of stimulating and responding cells, respectively. Before culture, the stimulator cells are treated with 200 μg ml^{-1} of mitomycin C for 45 min at 37°C, followed by three washes with phosphate-buffered saline (PBS, 0.15 M sodium chloride and 0.01 M phosphate buffer, pH 7.4). The cells are used at a final concentration of 0.2 × 10^6 cells 0.2 ml^{-1} of reaction mixture and cultured in a microtitration plate (Falcon, No. 3040) using RPMI 1640 medium supplemented with 10% fetal bovine serum and 5 × 10^{-5} M 2-mercaptoethanol. Then, 0.1 ml of the cell suspension and 0.1 ml of the test sample are poured into each hole in the plate. The culture is continued for 120 h and is pulse-labeled with [^{3}H]thymidine for 24 h before incubation is terminated. The cells are then collected and the radioactivity is counted.

FK-506, FR-900520, FR-900523, and FR-65814 were screened as immunosuppressors by this method (Kino et al., 1987a; Hatanaka et al., 1988a, 1988b).

In vivo methods

Bloodstream clearance   Microorganisms, or their experimental equivalent of carbon particles, are readily engulfed by circulating and tissue-fixed phagocytes. Some immunopotentiators enhance this clearance of carbon particles (Stossel and Cohn, 1976).

Delayed-type hypersensitivity   One type of cell-mediated immunity is known as delayed-type hypersensitivity (DTH). It is caused by the injection of an antigen into the skin of an individual previously immunized by the same antigen. The reaction is characterized by a reddening of the skin and a localized inflammation that reaches its height at 24–48 h. An antigen, ovalbumin, was given to guinea pigs together with the immunopotentiator, FK-

156, and Freund's incomplete adjuvant as the emulsion. The skin reaction (DTH) was performed 16 days after the immunization, and the DTH reaction was strongly enhanced at 48 h after the antigen challenge in comparison with that FK-156-free conditions (Gotoh et al., 1982).

Biochemical methods During the course of studies on enzyme inhibitors, Umezawa's group established quantitatively exact screening methods to find small molecular weight immunomodifiers produced by microorganisms (Umezawa, 1981). This group found that the antitumor metabolite coriolin and a derivative (diketocoriolin B) increased the number of antibody-forming cells in mouse spleen (Ishizuka et al., 1972), and that diketocoliorin B inhibited a Na^+, K^+-ATPase that is located on the cell-surface membrane (Kunimoto and Umezawa, 1973). The action of diketocoliorin B in increasing the number of antibody-forming cells may be caused by its binding to a Na^+, K^+-ATPase in the membrane of B lymphocytes. Therefore, Umezawa et al. assumed that even a small molecular weight compound that is bound to cells involved in the immune response may enhance or suppress the effects. It also appeared that inhibitors of enzymes on the cell surfaces can bind to immune cells, and the investigators decided to search for inhibitors of such enzymes. Several immunomodulators such as bestatin, forphenicine, esterastin, amastatin, and coriolin have been found by this biochemical screening method (Table 9.2). A detailed screening method for bestatin is shown in Chapter 7. Immunomodulating activities of these enzyme inhibitors are summarized in Table 9.3.

Table 9.2 Immunomodulators discovered by the screening of enzyme inhibitor

Microbial products	Target of the enzyme	Reference
Immunopotentiator		
Bestatin	Aminopeptidase Leucine aminopeptidase	(Umezawa et al., 1976)
Forphenicine	Alkaline phosphatase (chicken intestine)	(Aoyagi et al., 1978a)
Forphenicinol (chemically modified from Forphenicine; effective by oral administration)	Alkaline phosphatase (chicken intestine)	(Ishizuka et al., 1982)
Amastatin	Aminopeptidase A	(Aoyagi et al., 1978b)
Coriolin	Na^+-K^+-ATPase	(Takeuchi et al., 1969)
Diketocoriolin	Na^+-K^+-ATPase	(Takeuchi et al., 1971)
Nucleoticidin	5'-Nucleotidase	(Ogawara et al., 1985a; Uchino et al., 1985a; Uchino et al., 1986)
Melanocidin A, B	5'-nucleotidase	(Ogawara et al., 1985b; Uchino et al., 1985b; Uchino et al., 1986)
Immunosuppressor		
Esterastin	Hog pancreas lipase	(Umezawa et al., 1978)

Table 9.3 Immunomodulating activities of the enzyme inhibitors

Inhibitors	Immunomodulating activities
Bestatin (Umezawa, 1981)	1. Enhance delayed-type hypersensitivity (DTH) (p.o. and i.p) 2. Increase the number of antibody-forming cells at the high dose 3. Enhance lymphocyte proliferation (The action of bestatin on macrophages induces the blastogenesis of T cells) 4. Recover the immunosuppression by cyclophosphamide 5. Recover the immune response in tumor-bearing mice 6. Enhance IL-1 induction 7. Enhance IL-2 induction 8. Enhance differentiation of bone marrow cells 9. Inhibit the decrease of leucocytes 10. Inhibit the growth of IMC carcinoma 11. Enhance the natural killer cell activity 12. Prevent the opportunistic infection
Forphenicine (Umezawa, 1981)	1. Increase the number of antibody-forming cells by the action on macrophages 2. Enhance the DTH 3. Enhance lymphocyte proliferation 4. Enhance phagocytosis 5. Inhibit the growth of IMC carcinoma
Amastatin (Umezawa, 1981)	1. Increase the number of antibody-forming cells
Esterastin (Umezawa, 1981)	1. Suppress the DTH 2. Suppress antibody formation
Nucleoticidin (Uchino et al., 1986)	1. Antitumor activity against sarcoma-180 2. Enhance cytotoxicity of peritoneal exudate macrophages 3. Enhance superoxide-generating activity 4. Enhance IL-1 secretion 5. Secrete tumor necrosis factor (TNF) moderately
Melanocidin A and B (Uchino et al., 1986)	1. Anti-tumor activity against melanoma B-16 2. Enhance cytotoxicity of peritoneal exudate macrophages 3. Enhance O_2^- generating activity 4. Enhance IL-1 secretion 5. Secrete tumor necrosis factor (TNF) moderately

9.3 Characteristic Screening Methods for Immunomodulators

Immunopotentiators

Competitive action against immunosuppressive factors Some investigators have reported that tumor extracts and sera of tumor-bearing hosts contain immunosuppressive factors (Stuart, 1978; Kumar et al., 1981; Oh and Moolten, 1981; Tamura et al., 1981; Hino et al., 1985) partially purified an immunosuppressive factor from the serum of tumor-bearing mice. This factor also suppressed both ConA-induced lymphocyte proliferation and mixed lymphocyte reactions in vitro.

Methods

Preparation of an immunosuppressive factor from serum of tumor-bearing mice (Hino et al. 1985) Mice were inoculated subcutaneously with 0.2 ml of Sarcoma 180 (S-180) cell suspensions (5×10^6 cells ml^{-1}). S-180-bearing mice were bled from the heart under light ether anesthesia between 7 and 9 days after transplantation, and serum was collected. The immunosuppressive factor is partially purified from the S-180 tumor-bearing mouse serum according to the method described by Oh and Moolten (1981).

Assay of competitive action against the immunosuppressive factor Competitive action against the immunosuppressive factor can be detected by assaying mitogen-induced mouse spleen cell proliferation (see this chapter, p. 175) in the presence of immunosuppressive factor or a test sample.

Swainsonine Hino et al. (1985) detected a substance that was competitive against immunosuppressive factors present in the serum of tumor-bearing mice and thus found swainsonine (**1**) (Figure 9.1), which was isolated from the fungus *Metarhizium* sp.

Swainsonine has the capacity to reverse the depression of mitogenic responses of mouse spleen cells caused by immunosuppressive factors (Hino et al., 1985). The administration of swainsonine restored the capacities of the immunodeficient mice to produce antibody against SRBC (Kino et al., 1985). Swainsonine completely inhibited the growth of S-180 ascite tumor in mice and also reduced lung metastasis of B16 melanoma in mice. These findings suggest that swainsonine has a potential as an immunomodulator for the treatmant of immunocompromised hosts.

Swainsonine had first been isolated from *Swainsona* sp. (Colegate et al.,

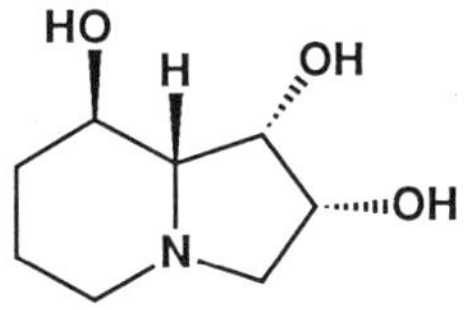

Figure 9.1 Structure of swainsonine (**1**).

1979) and locoweed plants (Molyneux et al., 1982) that induce livestock to a condition resembling the lysosomal storage disease, mannosidosis. Swainsonine is a potent inhibitor of α-mannosidase, and blocks the processing of N-linked oligosaccharides in glycoproteins (Colegate et al., 1979; Elbein et al., 1983; Tulsiani and Touster, 1983). Hino et al. (1985) also suggested that swainsonine accelerated the appearance of new ConA receptors, which may result from an increase in high mannose (or hybrid) types of oligosaccharides of receptors and a decrease in the complex types. Other new immunopotentiators, kifunensine (FR-900494) (2) (Figure 9.2) and FR-900483 have also been found in the fermentation broth of *Kitasatosporia kifunense* sp. No. 9482 and *Nectria lucida* F-4490, respectively, by the same screening method (Iwami et al., 1987; Shibata et al., 1988). In contrast, kifunensin had a potent α-mannosidase (from jack bean) inhibitory activity (Kayakiri et al., 1991). FR-900483 inhibits α-glucosidase activity from yeast, the gastrointestinal tract of rat, or the porcine small intestine (Shibata et al., 1988). The structure of kifunesin is a representative of a new class of 1, 5-iminopyranoses (Kayakiri et al., 1989, 1991).

Protective efficacy in experimental infections in mice   Gotoh et al. (1982) have established a screening system for the isolation and evaluation of new low molecular weight immunostimulating agents. This system is based on the protective efficacy of preadministered culture filtrates on experimental infections in mice (from the passage of whole cultures through molecular filters to remove high molecular weight substances).

Assay for protective efficacy in experimental infections in mice (Gotoh et al., 1982)   Cells of *Escherichia coli* No. 22 are cultured overnight at 30 C in Difco nutrient broth. The fully grown culture is diluted in fresh nutrient broth (× 1/10) and incubated at 30°C for a further 2 or 3 h. When a density of 1 × 10^7 cells ml^{-1} is obtained, the culture is used to infect experimental animals. The culture (0.2 ml) is injected intraperitoneally (i.p.) into mice. This challenge dose kills the untreated mice within 48 h of infection. Test samples prepared from a culture broth are administered subcutaneously to mice 1, 4, 5, and 6

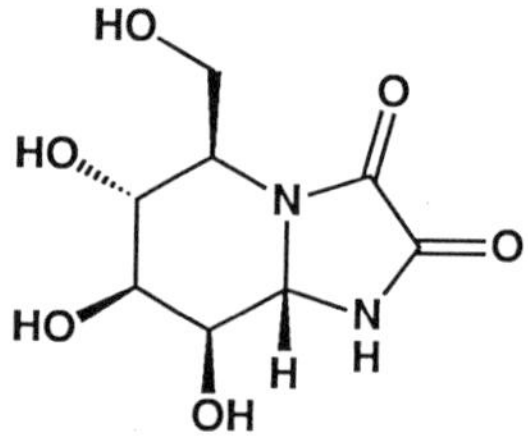

Figure 9.2   Structure of kifunesine (2).

days before the infection. The numbers of the dead and surviving mice are recorded for 2 days after the infection.

FK-156 An immunoactive peptide, FK-156 (**3**) (Figure 9.3), has been found in the fermentation broths of *Streptomyces olivaceogriseus* sp. nov. and *S. violaceus* by this screening method (Gotoh et al., 1982).

FK-156 afford marked protection against rapidly lethal infection in mice. It significantly enhances the proliferation of mouse spleen cells and tumor transplants and stimulates humoral antibody production and the delayed-type hypersensitivity reaction when administered with antigens (Gotoh et al., 1982). It also enhances the phagocytic function of the reticuloendothelial system of animals as measured by the carbon clearance method (Gotoh et al., 1982). The therapeutic effect of ticarcillin or gentamicin against *Pseudomonas* infection in immunosuppressed mice was enhanced markedly by combined use with FK-156 (Yokota et al., 1983). The killing activity of macrophages and polymorphonuclear leukocytes of immunosuppressed mice was also markedly enhanced by dosing with FK-156 (Mine et al., 1983a, 1983b).

Aladapcin was also screened from a culture broth of *Nocardia* sp. SANK60484 by the protection of mice against infection with *Escherichia coli* SANK73175 (Shiraishi, et al., 1990). Before screening, all the culture filtrates were passed through an Amicon Centriflo CF 25, and were used for this experiment to remove the effect of the culture medium (Shiraishi et al., 1990). Aladapcin enhanced the superoxide anion-releasing activity of polymorphonuclear leucocytes (PMN) after stimulation with ConA and cytochalacin D, suggesting that aladapcin, like other microbial immunomodulators, enhances host resistance by stimulating the activities of PMN (Shiraishi et al., 1990).

Immunosuppressors

Inhibition of IL-2 production Cyclosporin, an effective immunosuppressant, inhibits the production of cell-derived soluble mediators such as interleukin 2 (IL-2), interleukin 3 (IL-3) and gamma interferon (IFN-γ) induced

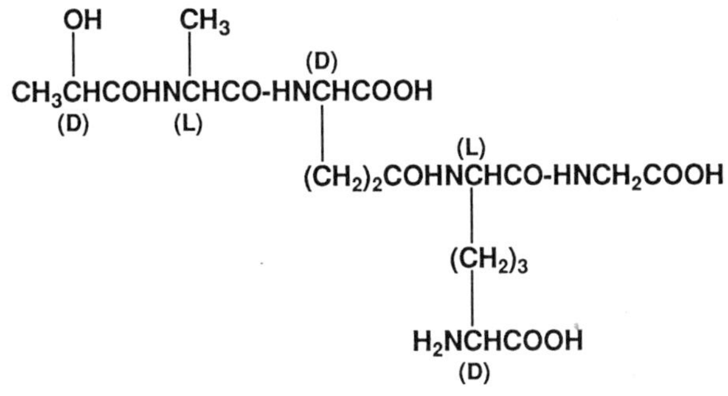

```
    OH        CH₃
    |         |        (D)
CH₃CHCOHNCHCO-HNCHCOOH
   (D)       (L)       |
                       |           (L)
                   (CH₂)₂COHNCHCO-HNCH₂COOH
                              |
                              |
                           (CH₂)₃
                              |
                              |
                          H₂NCHCOOH
                              (D)
```

3

Figure 9.3 Structure of FK-156 (**3**).

by antigens and lectins (Kino et al., 1987b). Therefore, the assay for IL-2 production can be used to screen for immunosuppressors. The immunosuppressive agents detected that can attack specific target sites of cells are likely to provide a useful prototype of drugs for clinically useful immunotherapy.

Methods

Preparation of supernatant solution containing IL-2 from mixed-lymphocyte cultures (Kino et al., 1987b) The bulk-mixed mouse lymphocyte reaction is performed in 24-well tissue culture plates (Corning Glass Works, Corning, NY); each well contains 5×10^6 C57BL/6 spleen cells (responder cells, H-2^b), 5×10^6 mitomycin C-treated (25 µg ml^{-1} mitomycin C at 37°C for 30 min and washed three times with RPMI 1640 medium) BALB/C spleen cells (stimulator cells, H-2^d) in 2 ml of RPMI 1640 complete medium. The cells are incubated at 37°C in a humidified atmosphere of 5% CO$_2$:95% air. The supernatant solution from the culture is used as the IL-2 fraction.

IL-2 assay (Kino et al., 1987b) IL-2 activity is measured according to the method described by Gillis and Smith (1977). The IL-2-dependent murine T-cell line, CTIL-2, is used to quantify IL-2 activity. CTIL-2 cells, 4×10^3, are cultured in flat-bottom microtiter plates at 37°C for 24 h with various dilutions of the supernatant solution containing IL-2 from the bulk-mixed lymphocyte reaction (0, 24, 48, 72, and 96 h) in 0.1 ml of Dulbecco's modified Eagle's complete medium (DMEM). To assay IL-2 production, [^{3}H]thymidine uptake is measured by pulsing cultures with 0.5 µCi of [^{3}H]thymidine for 6 h. The unit value is calculated by dilution analysis of the test sample and is compared with a laboratory standard IL-2 preparation in which 100 U is equivalent to the amount of IL-2 necessary to achieve 50% proliferation of CTIL-2 cells. The laboratory standard preparation is prepared from the supernatant solution from cultured AOFS cells (IL-2-producing hybridoma). AOFS cells (5×10^5 ml^{-1}) are incubated with ConA (10 µg ml^{-1}) in tissue culture flasks (F75, Corning) in a volume of 30 ml DMEM complete medium for 24 h; the supernatant solutions are then harvested, centrifuged to remove the cells, and filtered through a membrane filter (0.22 µm).

FK-506 Mixed-lymphocyte reactions (MLR) have been regarded as an in vitro correlate of allograft rejection and as an IL-2-dependent reaction (Morgan et al, 1976). Fujisawa's group screened the substance that inhibits IL-2-dependent T cell proliferation, MLR, but does not inhibit IL-2 independent proliferation of EL-4 cells, to find T- cell specific immunosuppressors from a wide range of fermentation broths (Goto et al., 1987; Kino *et al.*, 1987a, 1987b). As a result, a strain of *Streptomyces tsukubaensis* No. 9993 was found to produce a potent neutral immunosuppressive macrolide, designated by the code number FK-506 (**4**) (Figure 9.4). FK-506 shows antifungal activity against *Aspergillus fumigatus* and *Fusarium oxysporum*; it has no inhibitory effect on

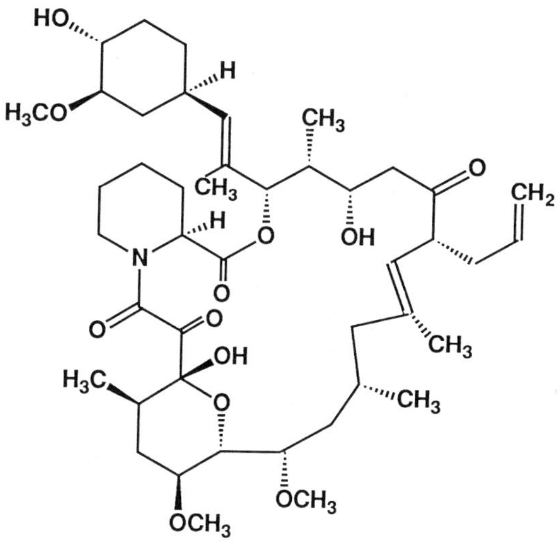

Figure 9.4 Structure of FK-506 (**4**).

bacteria or yeast. The in vitro studies on a mouse MLR correlate the results of in vivo allograft responses. The IC_{50} values of FK-506 and cyclosporin for mouse MLR were 0.32 and 27 nmol liter^{-1}, respectively.

FK-506 shows much higher activity than cyclosporin in suppression of the mouse MLR (Goto et al., 1987). The MLR can result in the generation of alloreactive cytotoxic T lymphocytes (CTL), which specifically recognize and kill the appropriate target antigen.

FK-506 significantly decreased the induction of CTL in a dose-dependent manner similar to cyclosporin A (Goto et al., 1987). Dose-dependent inhibitory effects of FK-506 were also demonstrated on the production of various T-cell-derived soluble mediators such as IL-2, IL-3, and IFN-γ, and on the expression of alloantigen-induced human lymphocyte IL-2 receptors and transferrin receptors (Thomson, 1989). However, FK-506 does not inhibit the secondary proliferation of activated human T cells stimulated by IL-2 (Thomson, 1989). FK-506 also does not directly affect B cell activation by LPS stimulation, macrophage activation, or colony formation for bone marrow cells (Thomson, 1989).

Oral administration of FK-506 strongly suppressed plaque-forming cell (PFC) response in mice (Goto et al., 1987). Consequently, FK-506 suppressed humoral immunity, but this effect would not be directly cytotoxic against lymphocytes (Goto et al., 1987). Oral treatment with FK-506 suppressed delayed type hypersensitivity (DTH) responses; therefore, FK-506 also strongly suppressed cellular immunity (Goto et al., 1987). FK-506 has been shown to prevent graft rejection after organ transplantation in animals and in man (Thomson, 1989). In several systems, FK-506 inhibits immune responses with a potency 10- to 100-fold greater than that of cyclosporin A (Thomson, 1989). The mechanism of suppression of FK-506 differes from cyclosporin A (Thomson, 1989).

FK-506 is not only a unique but is also an attractive way to study the lymphocyte activation cascade. It binds with high affinity to a specific protein (FKBP) distinct from cyclosporin A binding protein, cyclophilin (Siekierka et al., 1989). FKBP and cyclophilin are in the same family of proteins, known as immunophilins. Both FKBP and cyclophilin possess activity of peptidyl-prolyl *cis–trans* isomerase, which catalyzes the folding of prolin-containing proteins and peptides by the interconversion of *cis* and *trans* rotamers of the peptidyl-prolyl amide bond, but with different substrate specificity (Fischer et al., 1989; Harding et al., 1989; Siekierka et al., 1989; Takahashi et al., 1989). Cyclosporin A and FK-506 inhibited this enzyme activity from each binding protein (Siekierka et al., 1989). These observations indicate that the inhibitory activity of peptidyl-prolyl *cis–trans* isomerase and the antagonistic effect for the binding to FKBP or cyclophilin will be used as a potential target for the screening of immunosuppressors.

Anticomplementary substance   The complement system consists of nine kinds of complement components: C1, C2, C3, C4, C5, C6, C7, C8, and C9; and their regulators. It is known that the complement system is activated via classical or alternative pathways, and that by the classical pathway the complement system is activated from Cl in the presence of the immune complex, whereas by the alternative pathway it is activated nonspecifically from C3 by some activators such as lipopolysaccharide (Figure 9.5). When normal human serum is incubated with some complement regulators, and the remaining complement titer is measured by the use of sensitized sheep erythrocytes as an antigen–antibody complex, the inhibition of hemolysis is anticomplement activity, and the inhibitory substance for hemolysis is an anticomplement substance.

Therefore, anticomplementary activity involves both activation and inhibition of the complement system. The activators have the possibility of enhancing the immune system because the complement system performs the following functions: activation of the immune system via activation of macrophages, cytolysis of target cells, and opsonization, where complement facilitates the phagocytosis of antigens such as bacteria. Several antitumor polysaccharides also have anticomplement activity (Okuda et al., 1972). One antitumor polysaccharide, lentinan, has the ability to split C3 into C3a and C3b, and it markedly activates the alternative pathway of the complement system, which produces nonspecific activation of macrophages (Okuda et al., 1972; Hamuro et al., 1978). Complement activation clearly has been shown

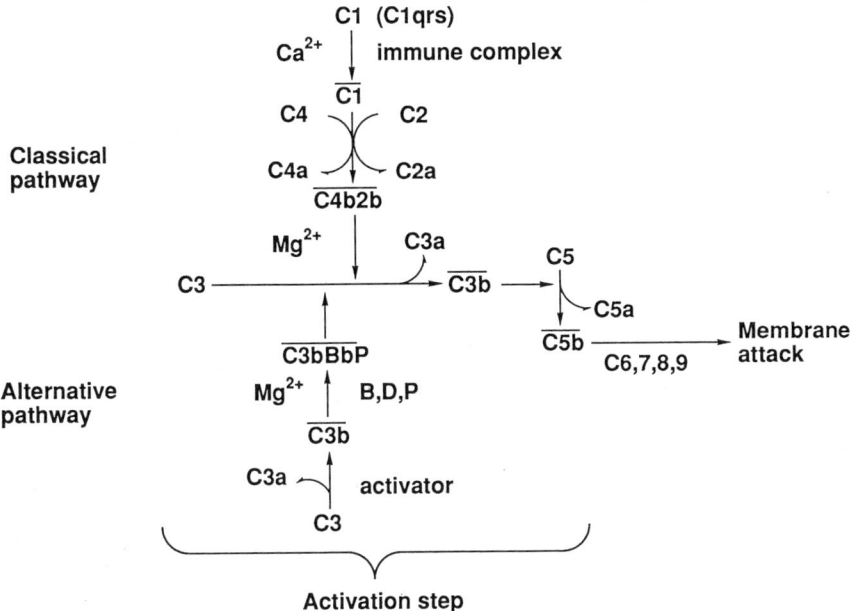

Figure 9.5 Complement system.

to occur in rheumatoid arthritis, lupus erythematosus, and glomerulonephritis, and is believed to be an important effector mechanism in these diseases. It is generally accepted that treatment of certain diseases may be effected through control of complement consumption, by interrupting the generation of, or the action of, cellular stimuli accompanying inflammatory processes. Such stimuli include chemoattractants for polymorphonuclear leukocytes, induction of noncytotoxic enzyme secretion, histamine-releasing agents, and permeability factors.

To a large degree these collective activities are expressed by the pharmacological action of C5a and C3a, cleavage products of the fifth (C5) and the third (C3) components of complement. Therefore, complement inhibitors are expected to have utility for the treatment of these diseases. In recent years, several synthetic inhibitors of complement have been developed and used for a number of specific purposes (Patrick and Johnson, 1980; Asghar, 1984). The inhibitors have been used to study the possibility of immunosuppression at the level of complement. A general assay method for anticomplement activity has been used with the functional assay being total complement-dependent hemolysis.

Assay for anticomplement activity (Miyazaki et al., 1980) Gelatin veronal buffered saline (pH 7.3) containing 0.15 mM $CaCl_2$ and 1.0 mM $MgCl_2$ (GVB^{++}) is prepared as described by Mayer (1961). Isotonic phosphate-buffered saline containing 2 mM $MgCl_2$ and 10 mM ethyleneglycol-*bis*(aminoethyl)-tetraacetate (EGTA), pH 7.2 (Mg–EGTA–PBS) is prepared as described by Martin et al. (1976). Guinea pig serum is obtained and centrifuged for 30 min at 30,000 r.p.m. to remove lipids (Inoue et al., 1977). Normal human serum (NHS) is obtained from a healthy adult. Serial dilutions of the samples are made in GVB^{++} using microtiter plates. The solutions are mixed with sensitized sheep erythrocytes (EA) at a final concentration of 2×10^7 cells ml^{-1} and 1:600 guinea pig serum (complement) in GVB^{++} and incubated at 37°C on vibrators. Inhibition is scored as the dilution of each supernatant giving 50% hemolysis. To detect the agents blocking the alternative pathway, the hemolytic agarose plate method of Martin et al. (1976) is used; the supernatants are put into wells punched in 1.2% agarose containing 5% normal human serum, as complement and 0.75% guinea pig erythrocytes in Mg–EGTA–PBS and incubated for 60 min at 37°C. The diameters of the inhibition zones are then measured.

K-76 Miyazaki et al. (1980) have screened anticomplementary agents produced by various *Actinomycetes* and fungi, and have isolated K-76 (5) (Figure 9.6), a potent inhibitor of complement activation, from a species of Fungi Imperfecti, *Stachybotrys complementi*, nov. sp. K-76.

K-76 is a sesquiterpene compound and can be oxidized to a monocarboxylic acid derivative, K-76-COOH (6) (see Figure 9.6), the sodium salt of which is very soluble and much less toxic than K-76. The LD_{50} of K-76 is 40 mg kg^{-1} in mice, whereas that of K-76-COOH is 500 mg kg^{-1} body weight. K-76 and K-76-COOH both inhibit complement activation by either the classical or alternative pathway. They inhibit generation of the chemotactic factor to human polymorphonuclear leukocytes from human serum by aggregated immunoglobulin. K-76-COOH mainly blocks the C5 intermediate step in the complement activation cascade. K-76-COOH did not inhibit any proteases or esterases tested, except when tested at high concentration. Therefore, K76-COOH should be useful in the prevention or treatment of various allergic, inflammatory, or immune complex diseases. Complestatin (7) (see Figure 9.6) was also screened by the same methods that were used in K-76 from the culture broth of *Streptomyces lavendulae* SANK 60477 (Kaneko et al., 1989).

9.4 Immunomodulators, But First Screened as Antibiotics

Cyclosporin A A mixture of metabolites containing cyclosporin A was first isolated as an antimicrobial agent from the crude extract of *Tolypocladium inflatum* Gams (Borel, 1982). The extract had a narrow spectrum of antifungal activity in vitro and in vivo. This activity was, moreover, coupled with an unusually low toxicity, and for this reason the mixture of metabolites was

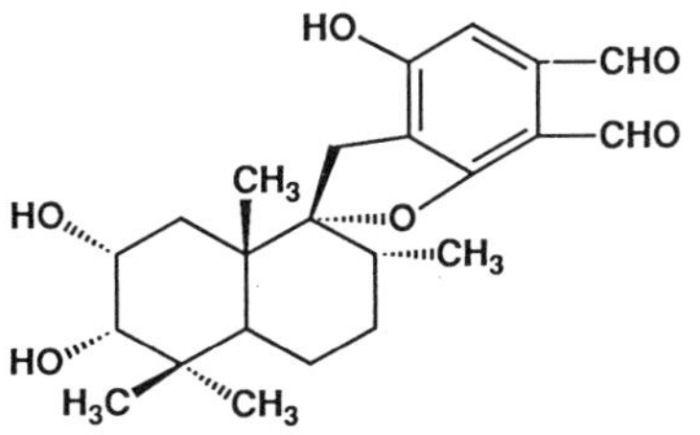

Figure 9.6 Structure of K-76 (**5**), K-76COOH (**6**) and complestatin (**7**).

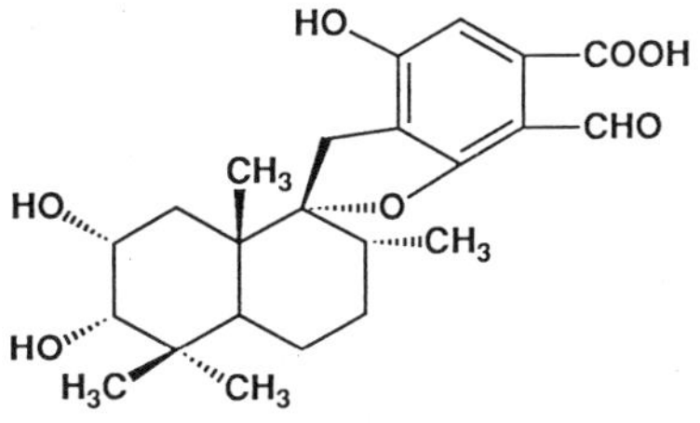

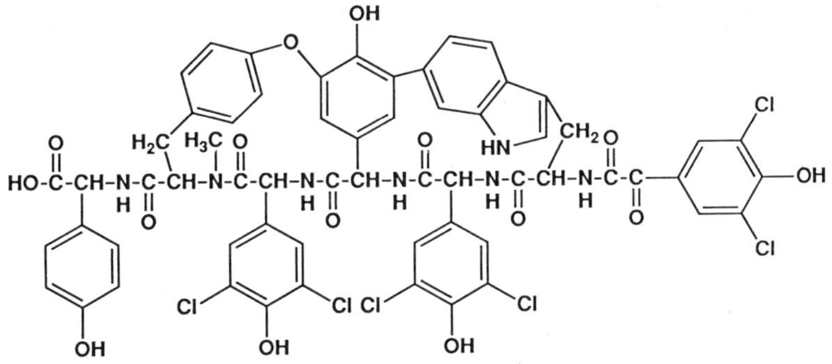

tested in a pharmacological screening program. In 1972, the immunosuppresive action of the metabolite mixture was discovered using a mouse model involving inhibition of hemagglutination, and this activity was associated with cyclosporin A (**8**) (Figure 9.7). This immunosuppression was not linked to a general cytostatic activity. Cyclosporin A-binding protein, cyclophilin, was isolated, and it was found that this protein has the activity of peptidyl prolyl *cis–trans* isomerase, which may be involved in T-cell activation that is suppressed by cyclosporin A (see p. , Takahashi et al., 1989; Fischer et al., 1989). Cyclosporin A has been in clinical use as an immunosuppressive drug for a number of years.

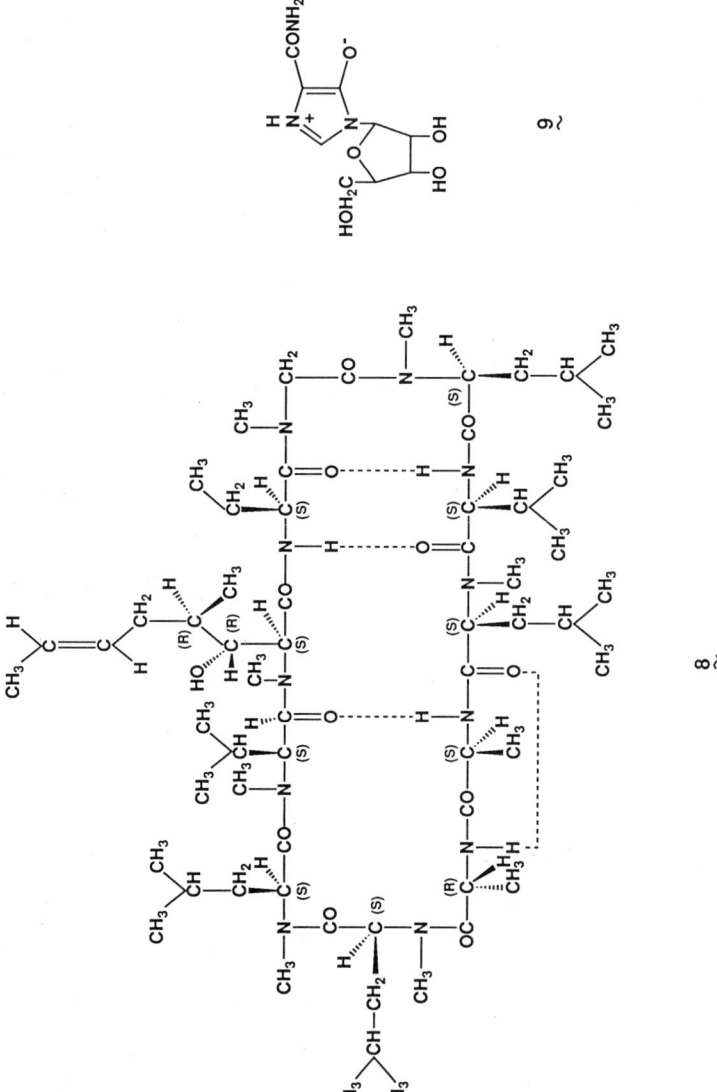

Figure 9.7 Structure of cyclosporin A (**8**) and bredinin (**9**).

Bredinin   During screening for new antibiotics, Mizuno et al. (1974) found that a fungal culture of *Eupenicillium brefeldianum* M-2166 produced an antibiotic, bredinin (**9**) (see Figure 9.7), which is an imidazole nucleoside partially active against *Candida albicans*.

Although bredinin was primarily isolated as an antibiotic, it showed no efficacy in treating experimental candidiasis of mice. The failure of bredinin to protect mice against systemic candidiasis probably depends not only on its weak activity but on its immunosuppressive activity; as a good example, azathioprine has been used as a secondary immunosuppressive agent against infection in experimental mycosis (Linquist et al., 1973). By extensive pharmacological studies, bredinin was found to have potent immunosuppressive activity, which was assayed by the plaque-forming cell procedure.

Bredinin was more active than azathioprine and showed potent activity on oral administration.

Rapamycin   Rapamycin (**10**) (Figure 9.8) was first isolated as an antifungal actibiotic from a culture of *Streptomyces hygroscopicus* (Seghal, S. N. et al., 1975). Rapamycin was also found to inhibit immunological responses (Martel et al., 1977) and to prevent graft rejection followed by prolongation of the survival of transplanted grafts (Morris et al., 1990). Rapamycin and FK-506 are structurally related macrolide antibiotics, and they have no structural relationship to cyclosporin A. Like FK-506, rapamycin appears to be 10- to

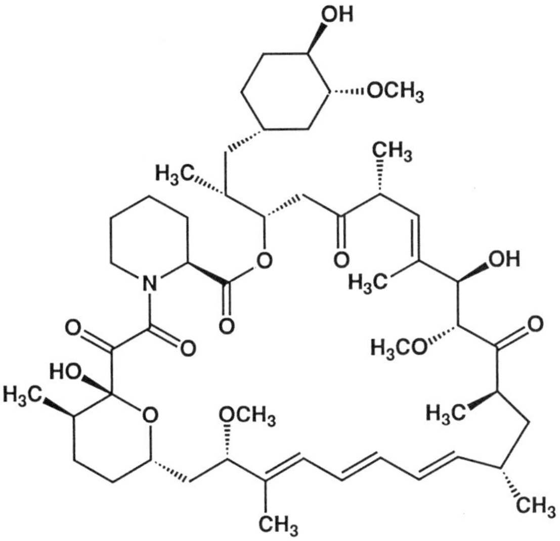

Figure 9.8   Structure of rapamycin (**10**).

100 fold more active than cyclosporin A in various models (Morris, 1991). Because rapamycin and FK-506 would suppress the immune system by identical mechanisms, both competitively inhibit the peptidyl-prolyl isomerase activity of the same immunophilin previously characterized as the major cytosolic FK-506-binding protein (FKBP) (Morris, 1991). Although there is a triene segment in the 31-membered ring of FK-506, both molecules contain the same unusual hemiketal masked α, β-diketopipecolic acid amide substructure (Morris, 1991).

Although there is no direct proof, it has suggested that the ketone carbonyl next to the homoprolyl amide bond in both FK-506 and rapamycin may inhibit peptidyl-prolyl isomerese by mimicking a twisted X-prolyl amide bond in peptides (Morris, 1991). Rapamycin blocks T-lymphocyte proliferation at a much later stage than does FK-506. It also inhibits human, porcine, and murine T- and B-lymphocyte activation by all pathways tested, including pathways that are insensitive to FK-506, such as IL-2-mediated proliferation of IL-2-dependent T cells and activation of murine B lymphocytes by LPS (Kay et al., 1991). These two macrolides, which bind competitively to the same major intracellular receptor protein, immunophilin, inhibit T- and B-lymphocyte activation by quite different mechanisms (Kay et al., 1991). Rapamycin, like FK-506 and cyclosporin-A, inhibits both the human and mouse mixed lymphocyte reactions (Henderson et al., 1991). Rapamycin was a poor inhibitor of IL-2 production, although it inhibited cellular responses to IL-2 (Henderson et al., 1991). FK-506 and cyclosporin-A affect IL-2 gene transcription by inhibiting of the appearance of the transcription factor but rapamycin does not (Henderson et al., 1991).

Deoxyspergualin   Deoxyspergualin (11) (Figure 9.9), an analog of spergualin (which was found from the culture broth of *Streptomyces lecterosporus*; Takeuchi et al., 1981), has strong antitumor (Iwasawa et al., 1982) and immunosuppressive activity (Nemoto et al., 1987). Deoxyspergualin administration to animal cells inhibits cellular immune responses including allograft rejection, xenograft rejection, graft-versus-host diseases, and delayed type hypersensitivity. It also has therapeutic effects on experimental animal models for autoimmune diseases that are responsible for humoral immunity. It has

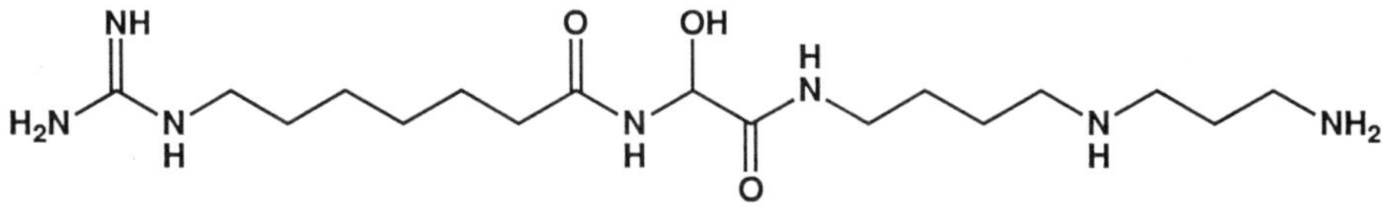

Figure 9.9   Structure of 15-deoxyspergualin (11).

been suggested that deoxyspergualin affects the proliferative stage of B lymphocytes in such a way as to inhibit their growth and antibody production (Fujii et al., 1990). Although dexyspergualin has less activity in vitro, it was found that this is caused by decomposition of deoxyspergualin in the culture medium in vitro (Fujii et al., 1989).

MRL/lpr mice develop an autoimmune disease, systemic lupus erythematosus-(SLE-)like lesions, and this mouse is a good model to evaluate or screen the therapeutic activities of immunosuppressive agents because some studies have demonstrated the efficacy of immunosuppressive agents in retarding these SLE lesions. Treatment with dexyspergualin in the early phase of the disease at doses of 1.5 and 3 mg kg^{-1} strongly suppressed the development of SLE-like lesions developed in MRL/lpr mice (Nemoto et al., 1990).

9.5 Concluding Remarks

Low molecular weight immunomodulators such as bestatin and cyclosporin A have already been used clinically. Bestatin has been used to treat of host-mediated antitumor effects, and cyclosporin A has been used to prevent rejection of transplanted organs as a potent immunosuppressor. FK-506 is also under development for clinical use. However, when immunomodulators are screened from microbial cultures by the several assay methods described here that concern the immune response, several problems occur frequently during screening. General screening methods concerning the immune response are very sensitive; therefore these activities are frequently affected by constituents in the culture medium used to grow microorganisms, especially metal ions such as Ca^{2+} and Mg^{2+} and immunoactive polysaccharides from yeast extract.

Some high molecular weight substances such as proteins and polysaccharides have immunomodulating activity. Many immunomodulators have also been discovered among the primary metabolites of microorganisms, such as cell wall components. Therefore, pretreatment of the test sample in the microbial culture may be required before screening for immunomodulators. Although biochemical screening methods may have a problem in that screened samples occasionally have no activity in vivo, this method is still very useful. It avoids problems concerning the immunological method, and if the biochemical mode of action of the known immunomodulators becomes clarified, more specific biochemical screening methods will be available. A few inhibitors of glycoprotein processing (glycosylation steps) have been observed to have immunomodulating activity (Bowlin and Sunkara, 1988). This is very interesting because a few differentiation inducers are also known to have inhibitory activity on glycosylation (Morin and Sartorelli, 1984).

Cyclosporin A has not only immunosuppressive activity but also side effects such as nephrotoxicity, hepatotoxicity, and antimicrobial activity. Because the cyclosporin A receptor protein cyclophilin is present in a wide variety of living cells such as yeast, insects and bacteria as well as in several internal organs of mammals, it is possible that these activities, including the side effects of cyclosporin A, also may be caused by the action of cyclophilin. If this is true, the antagonist or enzyme inhibitor for the peptidyl prolyl *cis-trans* isomerase activity of the cells involved only in the immune response should be screened to find the immunomodulator with low side effects.

If an immunophilin specific to the cells involved in the immune response, such as helper T cells were discovered, this specific immunophilin also will be an available target for the screening of specific immunosuppressors. FK-506 inhibits the activity of the IL-2 gene transcriptional activator (nuclear factor of activated T cells) but rapamycin does not; therefore, the screening of the inhibitory activity of these nuclear factors also may be useful in finding specific immunosuppressors.

Many cytokine receptors, such as the IL-1 receptor, IL-2 receptor, and transferin receptor, have recently been discovered. Therefore, the screening of an antagonist for these receptors may be available for the screening of immunomodulators, and the use of the binding method in this screening will be to screen many microbial culture broths simultaneously.

The clarification of the molecular mechanism of action of immunomodulators such as FK-506, cyclosporin A, and rapamycin has progressed greatly, and this advance gave new ideas for the screening of new immunomodulators. The discovery of a new type of immunomodulator will also contribute to the clarification of the immune system and related diseases at the molecular level.

References

Aoyagi, T., Yamamoto, T., Kojiri, K., F., Hamada, M., Takeuchi, T. and Umezawa, H. 1978a, Forphenicine, an inhibitor of alkaline phosphatase produced by actinomycetes. *Journal of Antibiotics* (Tokyo) 31:244–246.

Aoyagi, T., Tobe, H., Kojima, F., Hamada, M., Tekeuchi, T. and Umezawa, H. 1978b, Amastatin an inhibitor of aminopeptidase A produced by actinomycetes. *Journal of Antibiotics* (Tokyo) 31:636–638.

Asghar, S. S. 1984. Pharmacological manipulation of complement system. *Pharmacological Reviews* 36:223–244.

Borel, T. F. 1982. The history of cyclosporin A and its significance. In: D. J. G. White (editor), *Cyclosporin*, p.5. Elsevier, Amsterdam.

Bowlin, T. L. and Sunkara, P. S. 1988. Swainsonine, an inhibitor of glycoprotein processing, enhances mitogen induced interleukin 2 production and receptor expression in human lymphocytes. *Biochemical Biophysical Research Communnications* 151:859–864.

Chihara, G., Maeda, Y., Hamuro, J., Sasaki, T. and Fukuoka, F. 1969. Inhibition of mouse sarcoma 180 by the polysaccharides from *Lentinus edodes* (Berk.) Sing. *Nature* (London) 222:687–688.

Colegate, S. M., Dorling, P. R. and Huxtable, C. R. 1979. A spectroscopic investigation of swansonine: An α-mannosidase inhibitor isolated from *Swainsona canescens*. *Australian Journal of Chemistry*: 2257–2264.

Cunniningham, A. J. 1965. A method of increased sensitivity for detecting single antibody-forming Cells. *Nature* (London) 207:1106–1107.

Elbein, A. D., Pan, Y. T., Soef, R. and Vosbeck, K. 1983. Effect of swainsonine, an inhibitor of glycoprotein processing on cultured mammalian cells. *Journal of Cell physiology* 115:265–275.

Fischer, G., Wittmann-Liebold, B., Lang, K., Kiefhabel, T. and Schmid, F. X. 1989. Cyclophilin and peptidyl-prolyl cis–trans isomerase are probably identical proteins. *Nature* (London) 337:476–478.

Fujii, H., Takada, T., Nemoto, K., Abe, F. and Takeuchi, T. 1989. Stability and immunosuppressive activity of deoxyspergualin in comparison with deoxymethylspergualin. *Transplantation Proceedings* 21:3471–3473.

Fujii, H., Takeda, T., Nemoto, K., Yamashita, T., Abe, F., Fujii, A. and Takeuchi, T. 1990. Deoxyspergualin directly suppresses antibody formation *in vivo* and *in vitro*. *Journal of Antibiotics* (Tokyo) 43:213–219.

Gillis, S. and Smith, K. A. 1977. Long term culture of tumour specific cytotoxic T cells. *Nature* (London) 268:154–156.

Goto, T., Kino, T., Hatanaka, H., Nishiyama, M., Okuhara, M., Kohsaka, M., Aoki, H. and Imanaka, H. 1987. Discovery of FK-506, a novel immunosuppressant isolated from *Streptomyces tsukubaensis*. *Transplantation Proceedings* 19 (suppl. 6):4–8.

Gotoh, T., Nakamura, K., Nishiura, T., Hashimoto, M., Kino, T., Kuroda, Y., Okuhara, M., Kohsaka, M., Aoki, H. and Imanaka, H. 1982. Studies on a new immunoactive peptide, FK-156 II. Fermentation, extraction and chemical and biological characterization. *Journal of Antibiotics* (Tokyo) 35:1286–1292.

Hamuro, J., Hadding, U. and Bitter-Seuermann, D. 1978. Solid phase activation of alternative pathway of complement by β-1,3- glucans and its possible role for tumor regressing activity. *Immunology* 34:695–705.

Harashima, K., Tsuchida, N., Tanaka, T. and Nagatsu, J. 1967. Prodigiosin-25C. Isolation and the chemical structure *Agricultural and Biological Chemistry* 31:481–489.

Hatanaka, H., Kino, T., Hashimoto, M., Tsurumi, Y., Kuroda, A., Tanaka, H., Goto, T. and Okuhara, M. 1988a. FR65814, A novel immunosuppressant isolated from a *Penicillium* strain. Taxonomy, fermentation, isolation, physico-chemical and biological characteristics and structure assignment. *Journal of Antibiotics* (Tokyo) 41:999–1008.

Hatanaka, H., Kino, T., Miyata, S., Inamura, N., Kuroda, A., Goto, T., Tanaka, H. and Okuhara, M. 1988b. FR-900520 and FR- 900523, novel immunosuppressants isolated from a *Streptomyces*. II. Fermentation, isolation and physico-chemical and biological characteristics. *Journal of Antibiotics* (Tokyo) 41:1592–1601.

Harding, M., Galat, A., Uehling, D. E. and Schreiber, S. A. 1989. A receptor for the immunosuppressant FK-506 is a cis–trans peptidyl-prolyl isomerase. *Nature* (London) 341:758–760.

Henderson, D. J., Naya, I., Bundick, R. V., Smith, G. M. and Schmidt, J. A. 1991. Comparison of the effects of FK-506, cyclosporin A and rapamycin on IL-2 production. *Immunology* 73:316–321.

Hino, M., Nakamura, O., Tsurumi, Y., Adachi, K., Shibata, T., Terano, H., Kohsaka, M., Aoki, H. and Imanaka, H. 1985. Studies of an immunomodulator, swainsonine I. Enhancement of immune response by swainsonine *in vitro*. *Journal of Antibiotics* (Tokyo) 38:926–935.

Ishizuka, M., Iinuma, H., Takeuchi, T. and Umezawa, H., 1972. Effect of diketocoriolin B on antibody-formation. *Journal of Antibiotics* (Tokyo) 25:320–321.

Ishizuka, M., Ishizeki, S., Matsuda, T., Momose, A., Aoyagi, T., Takeuchi, T. and Umezawa, H. 1982. Studies on effects of forphenicinol on immune responses. *Journal of Antibiotics* (Tokyo) 35:1042–1048.

Isshiki, K., Sawa, T., Naganawa, H., Koizumi, Y., Matsuda, N., Hamada, M., Takeuchi, T., Iijima, M., Osono, M., Masuda, T. and Ishizuka, M. 1990. Depsidomycin, a new immunomodulating antibiotic. *Journal of Antibiotics* (Tokyo) 43:1195–1198.

Iwami, M., Nakayama, O., Terano, H., Kohsaka, M., Aoki, H. and Imanaka, H. 1987.

A new immunomodulator, FR-900494: Taxonomy, fermentation, isolation, and physico-chemical and biological characteristics. *Journal of Antibiotics* (Tokyo) 40:612–622.

Iwasawa, H., Kondo, S., Ikeda, D., Takeuchi, T. and Umezawa, H. 1982. Synthesis of (-)-15-deoxyspergualin and (-)-spergualin- 15-phosphate. *Journal of Antibiotics* (Tokyo) 35:1665–1669.

Jerne, N. K. and Nordin, A. A. 1963. Plaque formation in agar by single antibody-producing cells. *Science* 140:405.

Kay, J. E., Kromuel, L., Doe, S. E. A. and Denyer, M. 1991. Inhibition of T and B lymphocyte proliferation by rapamycin. *Immumology* 72:544–549.

Kaneko, I., Kamoshida, K. and Takahashi, S. 1989. Complestatin, a potent anti-complement substance produced by *Streptomyces lavendulae*. I. Fermentation, isolation and biological characterization. *Journal of Antibiotics* (Tokyo) 42:236–241.

Kayakiri, H., Takase, S., Shibata, T., Okamoto, M., Terano, H. and Hashimoto, M. 1989. Structure of kifunensine, a new immunomodulator isolated from an Actinomycete. *Journal of Organic Chemistry* 54:4015–4016.

Kayakiri, H., Takase, S., Shibata, T., Hashimoto, M., Tada, T. and Koda, S. 1991. Structure of Kifunensine, a new immunomodulator isolated from an actinomycete. *Chemical and Pharmaceutical Bulletin* (Tokyo) 39:1378–1381.

Kinashi, H., Someno, K. and Sakaguchi, K. 1984. Isolation and charecterization of concanamycins A, B and C. *Journal of Antibiotics* (Tokyo) 37:1333–1343.

Kino, T., Inamura, N., Nakahara, K., Kiyoto, S., Goto, T., Terano, H. and Kohsaka, M., Aoki, H. and Imanaka, H. 1985. Studies of an immunomodulator, swainsonine II. Effect of swainsonine on mouse immunodeficient system and experimental murine tumor. *Journal of Antibiotics* (Tokyo) 38:936–940.

Kino, T., Hatanaka, H., Hashimoto, M., Nishiyama, M., Goto, T., Okuhara, M., Kohsaka, M., Aoki, H. and Imanaka, H., 1987a. FK- 506, a novel immunosuppressant isolated from a *Streptomyces*. I. Fermentation, isolation, and physico-chemical and biological characteristics. *Journal of Antibiotics* (Tokyo) 40:1249–1255.

Kino, T., Hatanaka, H., Miyata, S., Inamura, N., Nishiyama, M., Yajima, T., Goto, T., Okuhara, M., Kohsaka, M., Aoki, H. and Ochiai, T. 1987b. FK-506, a novel immunosuppressant isolated from a *Streptomyces*. II. Immunosuppressive effect of FK-506 *in vitro*. *Journal of Antibiotics* (Tokyo) 40:1256–1265.

Kumar, P. K., Lykke, A. W. J. and Penny, R. 1981. Immunosuppression of a plasma suppressor factor in tumor-bearing mice. *Journal of the National Cancer Institute* 67:1277–1282.

Kunimoto, T. and Umezawa, H. 1973. Kinetic studies on the inhibition of ($Na^+ + K^+$)-ATPase by diketocoriolin B on antibody formation. *Biochimica et Biophysica Acta* 318:78–90.

Linquist, J. A., Rabinovich, S. and Smith, I. M. 1973. 5- Fluorocytosine in the treatment of experimental candidiasis in immunosuppresive mice. *Antimicrobial Agents and Chemotheraphy* 4:58–61.

Martel, R. R., Klicius, J. and Galet, S. 1977. Inhibition of the immune response by rapamycin, a new anti-fungal antibiotic. *Canadian Journal of Physiology and Pharmacology* 55:48–51.

Martin, A., Lachmann, P. J., Halbwachs, L. and Hobart, M. J. 1976. Hemolytic diffusion plate assay for factor B and D of the alternative pathway of complement activation. *Immunochemistry* 13:317–324.

Mayer, M. M. 1961. Complement and complement fixation, In: Kabat, E. A. (editor), *Kabat and Mayer's Experimental Immunochemistry*, 2nd Ed., pp. 133–240 Charles C. Thomas, Springfield.

Mine, Y., Yokota, Y., Wakai, Y., Fukada, S. and Nishida, M. 1983a. Immunoactive peptides, FK-156 and FK-565. I. Enhancement of host resistance to microbial infection in mice. *Journal of Antibiotics* (Tokyo) 36:1045–1050.

Mine, Y., Watanabe, Y., Tawara, S., Yokota, Y., Nishida, M., Goto, S. and Kuwahara, S. 1983b. Immunoactive peptides FK-156 and FK-506. III Enhancement of host defence mechanisms against infection. *Journal of Antibiotics* (Tokyo) 36:1059–1066.

Miyazaki, Y., Tamaoka, H., Shinohara, M., Kaise, H., Izawa, T., Nakano, Y., Kinoshita, T., Hong, K. and Inoue, K. 1980. A complement inhibitor produced by *Stachybotrys complementi*, nov. sp. K-76, a new species of Fungi Imperfecti. *Microbiology and Immunology* 24:1091–1108.

Mizuno, K., Tsujino, M., Takada, M., Hayashi, M., Atsumi, K., Asano, K. and Matuda, T. 1974. Studies on bredinin. I. Isolation, cheracterization and biological properties. *Journal of Antibiotics* (Tokyo) 27:775–782.

Molyneux, R. J. and James, L. F., 1982. Loco intoxication: Indolidine alkaloids of spotted locoweed (*Astragalus lentiginous*). *Science* 216:190–191.

Morgan, D. A., Ruscetti, F. W. and Gallo, R. 1976. Selective *in vitro* growth of T lymphocytes from normal human bone marrows. *Science* 193:1007–1008.

Morin, M. J. and Sartorelli, A. C. 1984. Inhibition of glycoprotein biosynthesis by the inducers of HL-60 cell differentiation, aclacinomycin A and marcellomycin. *Cancer Research* 44:2807–2812.

Morris, R. E., Wu, T. and Shorthouse, R. 1990. A study of the contrasting effects of cyclosporine, FK-506 and rapamycin on the suppression of allograft rejection. *Transplantation Proceedings* 22:1638–1641.

Morris, R. E. 1991. Rapamycin; FK506's fraternal twin or distant cousin? *Immunology Today* 12:137–140.

Mosbach-Ozman, L. and Loor, F. 1981. Basal and lipopolysacchacide-inducible membrane alkaline phosphatase of lymphoid cells from mice with immune system dysfunctions. *Annales de l'Institut Pasteur Immunology* 138:549–560.

Mosbach-Ozman, L., Lehuen-Renard, A., Gaveriaux, C. and Loor, F. 1986. Membrane alkaline phosphatase activity: An enzymatic marker of B-cell activation. *Annales de l'Institut Pasteur Immunology* 137D:109–125.

Mosmann, T. 1983. Rapid colorimetric assay for cellular growth and survival: application to proliferation and cytotoxic assays. *Journal of Immunological Methods* 65:55–63.

Nakamura, A., Nagai., K., Suzuki, S., Ando, K. and Tamura, G., 1986. A novel method of screening for immunomodulating substances, establishment of an assay system and its application to culture broths of microorganisms. *Journal of Antibiotics* (Tokyo) 39:1148–1154.

Nemoto, K., Hayashi, M., Abe, F., Nakamura, T., Ishizuka, M. and Umezawa, H. 1987. Immunosuppressive activities of 15-deoxyspergualin in animals. Journal of Antibiotics (Tokyo) 40:1590–1596.

Nemoto, K., Mae, T., Sugawara, Y., Hayashi, M., Abe, F. and Takeuchi, T. 1990. Deoxyspergualin therapy in autoimmune MRL/lpr mice suffering advanced lupus-like disease. *Journal of Antibiotics* (Tokyo) 43:1590–1596.

Ogawara, H., Uchino, K., Akiyama, T. and Watanabe, S. 1985a. A new 5'-nucleotidase inhibitor, nucleoticidin I. Taxonomy, fermentation, isolation and biological properties. *Journal of Antibiotics* (Tokyo) 38:153–156.

Ogawara, H., Uchino, K., Akiyama, T. and Watanabe, S. 1985b. New 5'-nucleotidase inhibitors, melanocidin A and melanocidin B. I. Taxonomy, fermentation, isolation and biological properties. *Journal of Antibiotics* (Tokyo) 38:587–591.

Oh, S. K. and Moolten, F. L. 1981. Nonspecific immunosuppressive factors in malignant ascites: Further characterization and possible relationship to erythrocyte receptors of human peripheral cells. *Journal of Immunology* 127:2300–2307.

Ohno, N., Arai, Y., Suzuki, I. and Yadomae, T. 1986. Induction of alkaline phosphatase activity in murine spleen cells treated with various mitogens. *Journal of Pharmacobio-Dynamics* 9:593–599.

Okuda, T., Yoshioka., Y., Ikekawa, T., Chihara, G. and Nishioka, K. 1972. Anti complementary activity of antitumor polysaccharides. *Nature New Biology* 238:59–60.
Patrick, R. A. and Johnson, R. E. 1980. Complement Inhibitors, In: Hess, H-J. (Editor), *Annual Reports in Medicinal Chemistry*, Vol. 15, pp. 193–201 Academic Press New York.
Inoue, K., Kinoshita, T., Okada, M. and Akiyama, Y. 1977. Release of phospholipids from complement-mediated lesions of the surface structure of *Escherichia coli*. *Journal of Immunology* 119:65–72.
Seghal, S. N., Baker, H. and Vezina, C. 1975: Rapamycin (AY22,989), a new anti-fungal antibiotic. II. Fermentation, isolation and characterization. *Journal of Antibiotics* (Tokyo) 28:727–732.
Shibata, T., Nakayama, O., Tsurumi, Y., Okuhara, M., Terano. H. and Kohsaka, M. 1988. A new immunomodulator, FR-900483. *Journal of Antibiotics* (Tokyo) 41:296–301.
Shiraishi, A., Nakajima, M., Katayama, T., Matsuda, T., Niwa, T., Okazaki, T., Takamatsu, Y., Nagaki, H., Kinoshita, T., Takatsu, T. and Haneishi, T. 1990. Aladapsin, a new microbial metabolite that enhances host resistance against bacterial infection. Production, isolation, physico-chemical properties and biological activities. *Journal of Antibiotics*. (Tokyo) 43:634–638.
Siekierka, J. J., Hung, S. H. Y., Poe, M., Lin, C. S. and Sigal, N. H. 1989. A cytosolic binding protein for the immunosuppressant FK-506 has peptidyl-prolyl isomerase activity but is distinct from cyclophilin. *Nature* (London) 341:755–757.
Stossel, T. P. and Cohn, Z. A. 1976. Phagocytosis, In: Williams, C. A. and Chase, M. W. (editors), *Methods in Immunology and Immunochemistry*, Vol. V, pp. 261–301. Academic Press, New York.
Stuart, A. E. 1978. The pathogenesis of Hodgkin's disease. *Journal of Pathology* 126:239–254.
Sugawara, I., Ishizaka, S., Tsujii, T. and Nishiyama, T. 1984. MTT assay-rapid colorimetric assay applicable to cellular proliferation and cytotoxicity assays. *Igaku No Ayumi* 128:733–735.
Takahashi, N., Hayano, T. and Suzuki, M. 1989. Peptidyl-prolyl *cis–trans* isomerase is the cyclosporin A-binding protein cyclophilin. *Nature* (London) 337:473–475.
Takeuchi, T., Iinuma, H., Iwanaga, J., Takahashi, S., Takita, T. and Umezawa, H. 1969. Coriolin, a new basidiomycetes antibiotic. *Journal of Antibiotics* (Tokyo) 22:215–217.
Takeuchi, T., Takahashi, S., Iinuma, H. and Umezawa, H. 1971. Diketocoriolin B, an active derivative of coriolin B produced by *Coriolus consors*. *Journal of Antibiotics* (Tokyo) 24:631–635.
Takeuchi, T., Iinuma, H., Kunimoto, S., Matsuda, T., Ishizuka, M., Takeuchi, M., Hamada, M., Naganawa, H., Kondo, S. and Umezawa, H. 1981. A new antitumor antibiotic, spergualin: Isolation and antitumor activity. *Journal of Antibiotics* (Tokyo) 34:1619–1621.
Tamura, K., Shibata, Y., Matsuda, Y. and Ishida, N. 1981. Isolation and characterizetion of an immunosuppressive acidic protein from ascitic fluids of cancer patients. *Cancer Research* 41:3244–3252.
Thomson, A. W. 1989. FK-506-How much potential? *Immunology Today* 10:6–9.
Tulsiani, D. R. P. and Touster, O. 1983. Swainsonine causes the production of hybrid glycoproteins by human skin fibloblasts and rat liver golgi preparations. *Journal of Biological Chemistry* 258:7578–7585.
Uchino, K., Ogawara, H., Akiyama, T., Watanabe, S., Fukuchi, A., 1986. Effect of 5'-nucleotidase inhibitors on mouse immune system and experimental murine tumors. *Journal of Antibiotics* (Tokyo) 39:682–687.
Uchino, K., Ogawara, H., Akiyama, T. Fukuchi, A., Shibata, S., Takahashi, K. and Narui, T. 1985a. A new 5'-nucleotidase inhibitor, nucleoticidin II. Physico-chemical properties and structure elucidation. *Journal of Antibiotics* (Tokyo) 38:157–160.

Uchino, K., Ogawara, H., Akiyama, T., Fukuchi, A., Shibata, S., Takahashi, K. and Narui, T. 1985b. New 5'-nucleotidase inhibitors, melanocidin A and melanociden B. II. Physicochemical properties and structure elucidation. *Journal of Antibiotics* (Tokyo) 38:592–598.

Umezawa, H. 1981. Small molecular immunomodifiers of microbial origin, In: Umezawa, H. (editor), *Fundamental and Clinical Studies of Bestatin*, pp. 1–17, Japan Scientific Societies Press, Tokyo.

Umezawa, H., Aoyagi, T., Hazato., T., Uotani, K., Kojima, F., Hamada, M. and Takeuchi, T. 1978. Esterastin, an inhibitor of esterase, produced by actinomycetes. *Journal of Antibiotics* (Tokyo) 31:639–641.

Umezawa, H., Aoyagi, T., Suda, H., Hamada, M. and Takeuchi, T. 1976. Bestatin, an inhibitor of aminopeptidase B, produced by actinomycetes. *Journal of Antibiotics* (Tokyo) 29:97–99.

White, D. J. G. and Calne, R. Y. 1982. The use of cyclosporin A immunosuppression in organ grafting. *Immunological Reviews* 65:115–131.

Yokota, Y., Mine, Y., Wakai, Y., Watanabe, Y., Nishida, M., Goto, S. and Kuwahara, S. 1983. Immunoactive peptides, FK-156 and FK- 565. II. Restoration of host resistance to microbial infection in immunosuppressed mice. *Journal of Antibiotics* (Tokyo) 36:1051–1058.

10

Vasoactive Substances

Akira Nakagawa

10.1 Introduction

There is a larger percentage of old people in our population and, because antibacterial drugs have proliferated, the active development of other types of pharmaceutical products is to be expected, especially the vasoactive drugs effective against circulatory blood diseases. Although there are many reports of searches and of in vitro and in vivo evaluation of synthetic vasoactive drugs, there are few such reports concerning natural products. Recently, many attempts to find vasoactive substances among microbial metabolites have been carried out using the screening programs customary for pharmacologically active compounds.

In searching for vasoactive compounds from microorganisms, it is often important, in contrast to finding substances effective against bacteria and fungi, to use animals and animal tissues in assays. During the screening program there are some special problems, especially the strong influence of noise on the bioassay. Additionally, it is necessary to assay many samples within a short period of time. Vasoactive substances from microorganisms can be classified into four categories according to the screening method employed:

1. Inhibition of platelet aggregation
2. Superfusion method
3. Euglobulin clot lysis time
4. Arterial blood pressure and heart rate measurements.

In this chapter, the author describes recent developments with vasoactive substances from microorganisms. He focuses on the screening methods used and the pharmacological activities of representative substances discovered.

10.2 In Vitro and In Vivo Screening Methods for Platelet Aggregation Inhibitors

The formation of a thrombus arising from apoplexy or heart attack involves platelet aggregation in the blood. Phospholipase A_2 on the membrane of a platelet is activated by stimulation of aggregation inducers such as thrombin, adenosine diphosphate (ADP), arachidonic acid, collagen, and platelet-activating factor (PAF), and it plays a major role in the formation of arachidonic acid from phospholipids in the membrane. As a result, prostaglandin biosynthesis is accelerated and platelet aggregation results from the thromboxane A_2 formed. Therefore, studies on the inhibitors of platelet aggregation are concentrated on a search for a preventive and therapeutic agent against thrombus formation. The most convenient method for finding inhibitors of platelet aggregation among microbial metabolites is to examine the inhibitory effects of collagen-, ADP-, arachidonic acid-, and PAF-induced platelet aggregation in vitro. In vivo administration of arachidonate is also associated with platelet aggregation. The screening programs measuring inhibition of platelet aggregation in vitro and in vivo are described, and pharmacological activities of new bioactive compounds discovered in the programs are introduced.

Inhibition of platelet aggregation in vitro A general method (Umehara et al., 1984a) for in vitro screening of platelet aggregation inhibitors is as follows.

Blood collected from the central ear artery of male Japanese white rabbits is used for in vitro screening for platelet aggregation inhibitors. Sodium citrate to prevent coagulation is added to the blood. Platelet-rich plasma (PRP) is prepared by centrifugation of the blood at 230–270 $\times$ g for 10 min at room temperature. The PRP is diluted with platelet-poor plasma obtained by further centrifugation of the blood at 400 $\times$ g for 10 min. PRP (platelet counts, 3-4 $\times$ 10^5 platelets μl^{-1}) is kept at room temperature and used within 4 h of collection.

Aggregation studies are conducted with an aggregometer. The mixture of PRP-saline including drugs or cultured broth is incubated in an aggregometer with stirring (1000 rpm) at 37°C for 2–5 min, and the aggregating agent is added (final concentrations: collagens, 2–20 μg ml^{-1} for PRP; arachidonic acid, 100–150 μM; ADP, 1–5 μM; thrombin, 0.3 U ml^{-1}). Platelet aggregation is measured turbidimetrically by recording change in the light transmission of PRP during aggregation. Activities of inhibitors are expressed as IC$_{50}$ values, i.e., concentrations required to inhibit the platelet aggregation responses by 50%.

Arachidonate-induced platelet aggregation in vivo The procedure for the arachidonate-induced platelet aggregation in vivo system is as follows (Kohler et al., 1976). Fasted mice (or rabbits) are divided into control (starch-fed) and experimental (drug-fed) groups. Drugs in starch suspension are given by stomach lavage 2 h before an arachidonate challenge. A solution of sodium arachidonate is administered by slow intravenous injection into the tail vein of male mice (the marginal ear vein in the rabbit) at doses of 1 $-$ 100 mg kg^{-1}. The mice are observed for signs of respiratory distress (cyanosis, interrupted

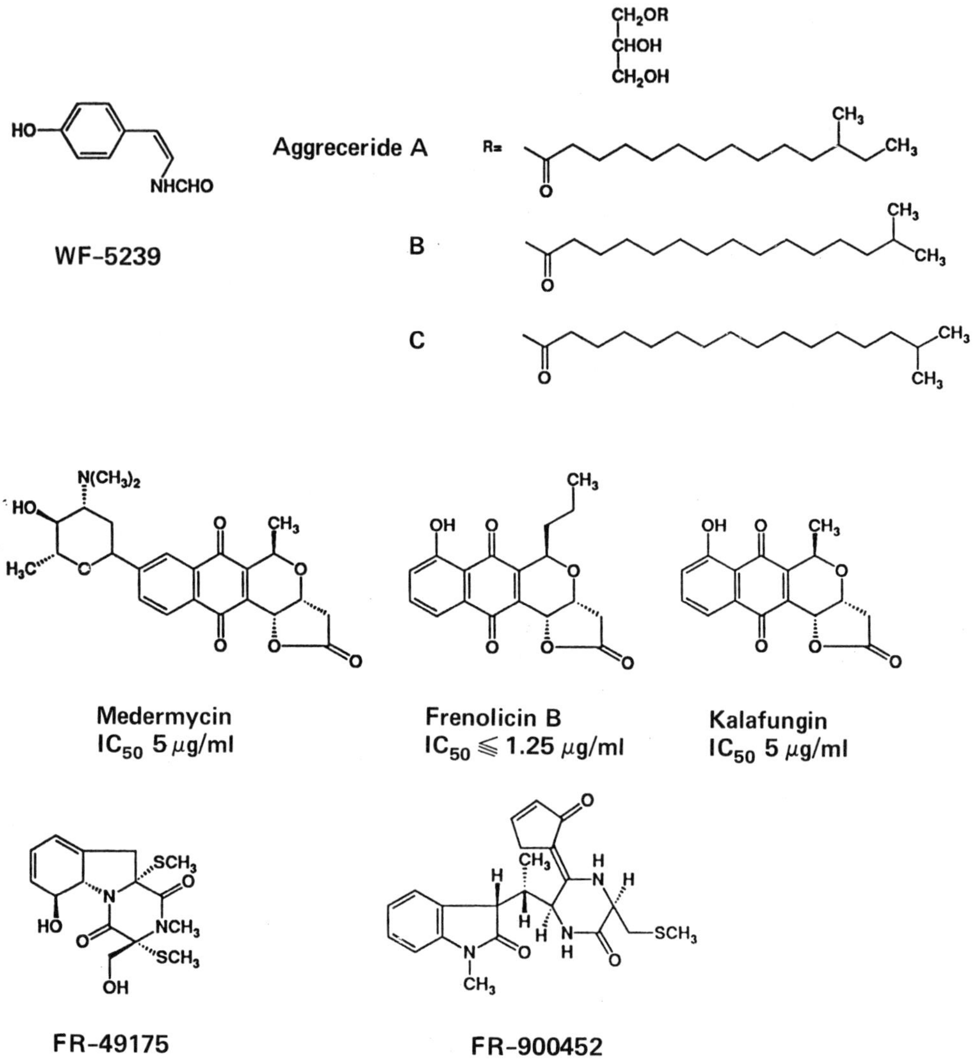

Figure 10.1 Structures of new microbial platelet aggregation inhibitors.

breathing rate, gasping), and the duration of the effect as well as recovery (regaining of exploratory behavior or righting reflex) are noted.

Inhibitors of platelet aggregation from microbial metabolites

WF-5239 In the course of the screening program just described for antithrombotic drugs, Umehara et al. (1984a) found a new platelet aggregation inhibitor, WF-5239 (Figure 10.1), in the cultured broth of *Aspergillus fumigatus*

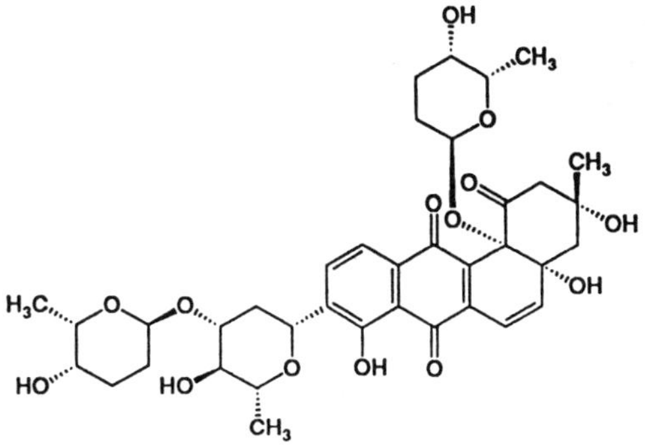

Aggreticin (OM-4842)

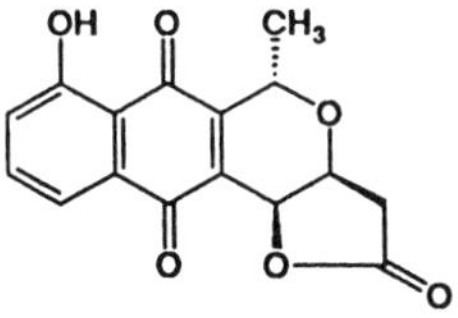

Nanaomycin D
$IC_{50} \leqslant 1.25$ μg/ml

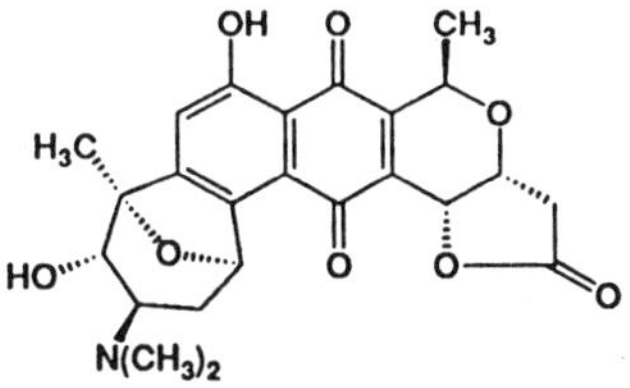

Sch 38519

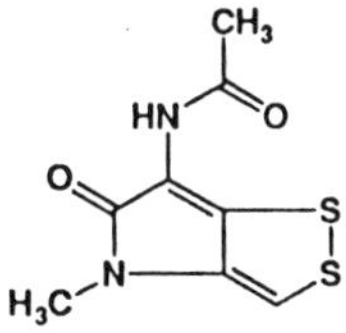

Thiolutin

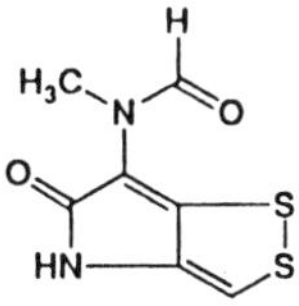

Thioaurin

Figure 10.1 *(cont.)*

Fresenius. The structure of WF-5239 is assigned as N-[2-*cis*(4-hydroxyphenyl)ethenyl] formamide. The compound exhibits inhibitory activity against rabbit platelet aggregation induced by arachidonic acid and collagen, with IC_{50} values of 1.25 and 5.0 μg ml^{-1}, respectively. The ability of the compound to block arachidonic acid-induced pulmonary thrombosis in mice was examined for 30 min after intraperitoneal administration to mice. WF-5239 appeared to prevent arachidonic acid-induced sudden death in at least 80% of the mice at 30 mg kg^{-1}.

Aggreceride New platelet aggregation inhibitors, named aggrecerides (Ōmura et al., 1986a), were found using a screening program involving incubation of washed platelets from rabbit blood with fermentation broth in a 24-well plate using thrombin or ADP as an inducer. Aggrecerides A, B, and C possess monoglyceride structures consisting of C_{15}, C_{16}, and C_{17} branched fatty acids, respectively.

The inhibitory effect of aggreceride A on thrombin-induced aggregation of washed platelets (2 U ml^{-1}, 50 µl) was 92% and 81% at concentrations of 50 µg ml^{-1} and 25 µg ml^{-1} of the substance, respectively. Three branched C_{15}, C_{16}, and C_{17} fatty acids produced together with aggrecerides in the fermentation broth also showed weak inhibitory activity on in vitro aggregation. The aggrecerides seem to act at the level of arachidonic acid metabolism or earlier, based on their inhibition of malondialdehyde (MDA) formation at a concentration of 50 µg ml^{-1} of component A. The new findings that fatty acids such as palmitic acid, stearic acid, and oleic acid, in addition to the constituent fatty acids in aggreceride, exhibit inhibitory activity against platelet aggregation are interesting because fatty acids may play a role in platelet aggregation in vivo.

Staurosporine Oka et al. (1986) found a potent platelet aggregation inhibitor, staurosporine, in the cultured broth of *Streptomyces* sp. strain M-193. Staurosporine had been isolated as a microbial alkaloid by chemical screening using Dragendorff's reagent by Ōmura et al. (1977). The compound strongly inhibits platelet aggregation induced by collagen, ADP, and arachidonic acid. Notably, the inhibitory activity against collagen-induced aggregation is more than 3000 times that of aspirin.

Tamaoki et al. (1986) found out that staurosporine markedly inhibits phospholipid/Ca^{2+}-dependent protein kinase (protein kinase C) from rat brain. Staurosporine is commercially available as a biochemical reagent.

Aggreticin (OM-4842), placetins, and benzoisochromanequinone antibiotics
A new anthraquinone, aggreticin, was isolated as a platelet aggregation inhibitor from *Streptomyces* sp. OM-4842 by Ōmura et al. (1988). The minimum inhibitory concentrations of aggreticin on aggregation induced by ADP (5 µM), arachidonic acid (100 M) and PAF (5 × 10 µM) were 12.5, 5.0, and 25.0 µg ml^{-1}, respectively. The compound also exhibits growth inhibition against adriamycin-resistant cells of P 388 mouse leukemia at a dose of 1.5 µg ml^{-1}.

Ozasa et al. (1990) have isolated the new aggreticin-related compounds placetins A, A_1, B, and B_1, which are active against platelet aggregation induced with PAF, ADP, and collagen. The IC_{50} values of A component for PAF and ADP were 6.4 and 9.7 µg ml^{-1}, respectively. Placetins A and B also exhibited strong cytotoxicities against P 388, L 1210, and HeLa cells. Lauer et al. (1991) isolated 2-methoxy-5-methyl-1,4-benzoquinone from basidiomycetes in the screening for antithrombotic compounds.

On the other hand, a benzoisochromanequinone antibiotic, medermycin, was isolated as an ADP-induced platelet aggregation inhibitor from *Streptomyces* sp. YKM-11749. Following this, the inhibitory activity on ADP-induced platelet aggregation for various isochromanequinone antibiotics, such as frenolicins, nanaomycins, and kalafungin, was examined by Nakagawa et al. (1987). Frenolicin B in the mouse model in vivo using an intravenous injection of sodium arachidonate significantly prolonged survival time of the mice. Further, nanaomycin D showed a hypotensive effect following intraperitoneal injection in rats.

Patel et al. (1989) isolated a novel isochromanequinone, Sch 38519, from the fermentation filtrate of *Thermomonospora* sp. SCC 1793. The compound inhibits aggregation of thrombin-stimulated human platelets (IC$_{50}$, 68 µg ml^{-1}). Sch 38519 also inhibits secretion of [^{3}H]serotonin over the same concentration range as observed for its inhibition of platelet aggregation.

FR-49175 [bis-dethiobis(methylthio)gliotoxin] and FR-900452   It has been suggested that PAF (1-O-alky1-2-acetyl-*sn*-glycerol-3-phosphocholine) plays an important role in various allergic and inflammatory reactions. This compound is an extremely potent inducer of platelet aggregation as well as causing hypotension and bronchoconstriction (Vargaftig et al., 1980).

During a screening program for PAF-induced rabbit platelet aggregation inhibitors, Okamoto et al. (1986a, 1986b, 1986c) have found FR-49175 [bis-dethiobis(methylthio)gliotoxin] and FR-900452 from the cultured broths of *Penicillium terlikowskii* No. 5348 and *S. phaeofaciens* No. 7739, respectively. FR-49175 is identical with a substance isolated from cultures of the wood fungus *Gliocladium delguescens* by Kirby et al. (1980).

The IC$_{50}$ values of FR-49175 and FR-900452 for PAF-induced rabbit platelet aggregation were 8.4 µM and 0.37 µM, respectively. Further, FR-49175 inhibited PAF-induced bronchoconstriction in an in vivo model using guinea pigs. In comparison, FR-900452 significantly inhibited PAF-induced bronchoconstriction in the guinea pigs and hypotension in the rat. When they found that FR-900452 prevented hypotension in rats induced by immunoglobulin(IgE)-mediated anaphylaxis, Okamoto et al. (1986c) presumed that PAF may play a role in the pathogenesis of IgE-medicated hypotension in rats.

Pyrrothine antibiotic   Ninomiya et al. (1980a) examined four bioassay systems as in vitro antiinflammatory screening probes applicable to microbial metabolites: heat hemolysis of rat erythrocytes, inhibition of protein denaturation (Mizushima and Sakai, 1969), and collagen- and ADP-induced platelet aggregation assays. Heat hemolysis of rat erythrocytes is tested by incubation of a reaction mixture of rat erythrocytes and drug (or fermentation broth) at 50°–55°C for 20–30 min. The activity of the drug is indicated as the concentration causing 50% inhibition of the heat hemolysis determined from A$_{540}$. By comparison of the effect of various known antiinflammatory compounds and antibiotics in the assay systems described here, Ninomiya et al. (1980a) concluded that these probes are simple and useful for routine screening. The effects of actinomycete metabolites on these four assay systems were examined with several hundred broths.

As a result, Ninomiya et al. (1980b) found three pigments from the cultured broth of *Streptoverticillium* sp. AR-10-AV-10 HB that possess a stabilizing activity on rat erythrocyte membranes (inhibitory activity on rat erythrocyte hemolysis). These compounds were identified as pyrrothine antibiotics consisting of thiolutin, aureothricin, and isobutyropyrrothine; these compounds had been isolated as antibacterial substances in 1952. Thiolutin is three times as active as indomethacin in inhibiting heat hemolysis of rat erythrocytes and ADP- and collagen-induced platelet aggregation. From the mode of action of pyrrothines, the antibiotics exhibit membrane-stabilizing activity on rat erythrocytes toward heat hemolysis. Thioaurin, being one of the pyrrothine antibiotics that has been isolated as an antiplatelet compound from *Streptomyces* OM-5123 by Ōmura et al. (unpublished data), seems to possess the same mode of action as thiolutin.

10.3 Superfusion Method

The superfusion method (Gaddum, 1953; Vane, 1964), which is frequently used in research on prostaglandin metabolism, has been applied to screening for new vasodilator drugs from microbial metabolites. This method is a sensitive and simple procedure used to measure contraction of the rabbit (or rat) mesenteric artery caused by thromboxane A_2, which is produced via reaction of rat platelets and arachidonic acid. In the following subsection, a typical superfusion method for screening of vasodilator-active substances from microorganisms is described. The new vasodilators, WS-1228 A, B, WS-30581 A, B, and amauromine found in this screening are also described.

Superfusion method Male Sprague-Dawley rats (8–10 weeks of age) are killed by a blow on the head and the thoracic aorta is quickly removed. After removing fatty tissues, spiral strips (2 mm in width and 50 mm in length) are cut from the aorta and are suspended under a resting tension of 1 g in a 30-ml organ bath containing warm (37°C), oxygenated (95% O_2:5% CO_2) tyrode solution of the following composition: NaCl (137 mM), KCl (2.7 mM), $CaCl_2 \cdot 2H_2O$ (1.8 mM), $MgCl_2 \cdot 6H_2O$ (1.02 mM), $NaHCO_3$ (11.9 mM), $NaH_2PO_4 \cdot 2H_2O$ (0.42 mM) and glucose (5.5 mM). The tissues are equilibrated for 90 min and then superfused with tyrode solution containing a low concentration of noradrenaline (0.03 µg ml^{-1}) or KCl (30 mM), which increases the tension of the tissues by 55 mg. Changes of tension of the tissues are measured isometrically by means of force displacement tranducers coupled to a polygraph.

Vasodilators from microbial metabolite

WS-1228 A, B Yoshida et al. (1982) and Tanaka et al. (1982) found that *S. aureofaciens* produces hypotensive vasodilator substances designated as WS-1228 A and B. These compounds possess a unique *N*-hydroxytriazene skeleton (Figure 10.2). The antiplatelet activity of WS-1228 B is markedly greater than that of the A component. The comparative activity of WS-1228 B with the calcium antagonist, nifedipine, indicated that WS-1228 B showed a slight reverse order of activities on the two preparations (KCl-treated and noradrenaline-treated aorta). Ōmura et al. (1986b) has isolated these compounds as acyl-CoA synthetase inhibitors from *Streptomyces.*

WS-30581 A, B Umehara et al. (1984b) found the new antiplatelet compounds WS-30581 A and B from *Streptoverticillium waksmanni* using an index of the inhibitory effects on platelet aggregation and thromboxane A_2 synthesis. Thromboxane A_2 synthesis via reaction between rat platelets and arachidonic acid is bioassayed by contraction of the rabbit mesenteric artery (superfusion method). The bioactive compounds were identified as pimprinine (5,3′-indolylmethyloxazole) analogs. Pimprinine is known to possess antiepileptic and monoamine oxidase inhibitory effects, but its inhibitory effects on platelet aggregation and thromboxane A_2 synthesis were unknown. The inhibitory activity (IC_{50}, 3–6 $\mu g\ ml^{-1}$) of WS-30581 on arachidonate-induced platelet

WS-1228A

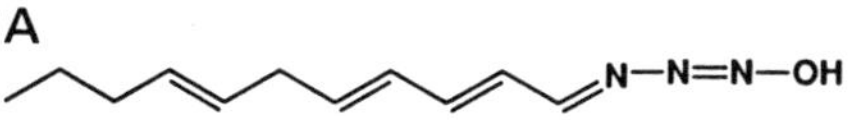

WS-1228B

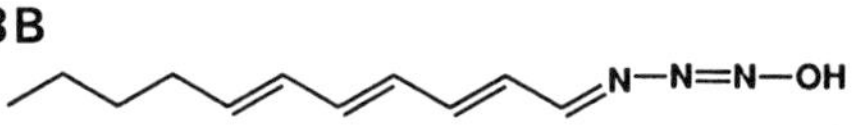

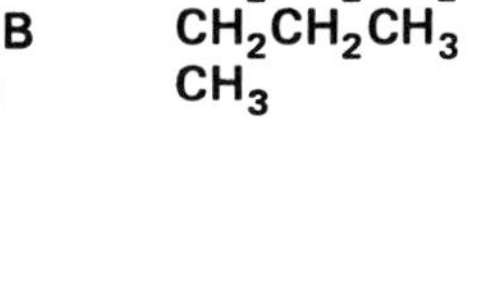

WS-30581 A R = $CH_2CH_2CH_2CH_3$
WS-30581 B $CH_2CH_2CH_3$
Pimprinine CH_3

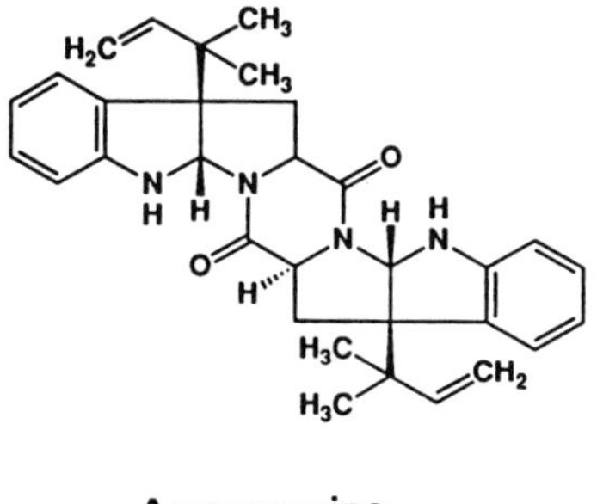

Amauromine

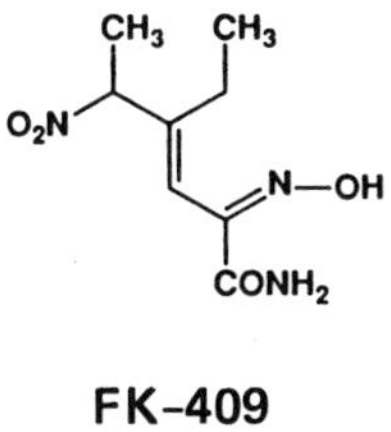

FK-409

Figure 10.2 Structures of new microbial vasodilators.

aggregation in vitro is about 10 fold higher than that of aspirin. Further, the inhibition of thromboxane A_2 synthesis by WS-50381 is 40 to 80 fold stronger than that of imidazole, which is known as an inhibitor of thromboxane A_2 synthesis.

Amauromine   Amauromine, a new vasodilator, was isolated from the culture of *Amauroascus* sp. No. 6237 by Takase et al. (1984a) by screening using the superfusion technique. Amauromine possesses a novel dimeric structure including a L-tryptophane moiety (Takase et al., 1984b). The vasodilating property of amauromine is characterized by comparing its activity on rat aortic strips contracted by KCl and norepinephrine. The concentration that produced a 50% relaxation of a rat aortic strip contracted with KCl (50 mM) was 1.15×10^{-8} µg ml^{-1}, but the relaxation activity on noradrenaline-treated strips was extremely low.

FK-409   Hino et al. (1989a, 1989b) found a novel vasodilator FK-409 with antiplatelet aggregation activity from the acid-treated fermentation broth of *Streptomyces griseosporeus* using a superfusion technique. FK-409 exhibited a marked relaxation activity in doses of 0.01 and 0.1 µg on noradrenaline-induced contraction of rat aorta. In addition to the vasodilating activity, the compound showed a marked hypotensive effect and a potent inhibitory activity on the aggregation induced by PAF, collagen, thrombin, ADP, and arachidonic acid.

10.4   Euglobulin Clot Lysis Test

Fibrinolytic therapy is important in treating thromboembolic disease. Fibrinolytic agents such as urokinase, streptokinase, and tissue plasminogen activator with high molecular weights have been used for thrombolytic therapy. Some efforts have been made to develop low molecular weight fibrinolytic drugs. Several tests such as euglobulin clot lysis, diluted blood clot lysis time, the fibrin plate method, and fibrin degradation have been reported for measuring the fibrinolytic activator activity in blood. In this section, the euglobulin clot lysis test by Nilsson et al. (1978) is described as an example. This is a suitable screening test for routine use and clinical practice.

The euglobulin clot lysis time test   Fibrinolytic agents present in a fermentation broth are detected by their stimulating activity on the euglobulin clot lysis time with rabbit plasma. The blood from the carotid artery of a male Japanese white rabbit is treated with sodium citrate (130 mM). Citrated rabbit plasma (450 µl), which is prepared by centrifugation (400 $\times$ g, 20 min, 4°C) of the blood is mixed with cultured broth (50 µl) and incubated at 37°C for 15 min and then diluted with 0.25% acetic acid solution (4 ml). The solution is kept on ice for 60 min and then centrifuged (400 $\times$ g for 10 min at 4°C) to obtain a precipitate of the euglobulin fraction. Bovine thrombin (250 μl, 5 U ml⁻¹) is added to the euglobulin solution in vernal buffer:saline to get a fibrin clot. The fibrin clot is incubated at 37°C and the spontaneous lysis time of the clot is determined visually. The activity of an activator is expressed as the ED_{50} value, i.e., the concentration required to stimulate the clot lysis time by 50%.

Fibrinolytic agents from microbial metabolites

WB-3559 A, B, C, and D Yoshida et al. (1985) found the new fibrinolytic Compounds WB-3559 A, B, C, and D from *Flavobacterium* sp. No. 3559 (Figure 10.3). The active compound present in the fermentation broth was detected by its stimulating activity on euglobulin clot lysis using rabbit plasma as described. The structures of WB-3559 consist of an alkyl chain containing a fatty acid and two amino acids, serine and glycine (Uchida et al., 1985). The ED_{50} values of WB-3559 A, B, C, and D on the euglobulin clot lysis time of rabbit plasma were 100, 52, 32, and 40 μg ml⁻¹, respectively.

10.5 Arterial Blood Pressure and Heart Rate Measurement

The measurements of arterial blood pressure and heart rate are useful for the evaluation of both antihypotensive and vasodilating activities. The method

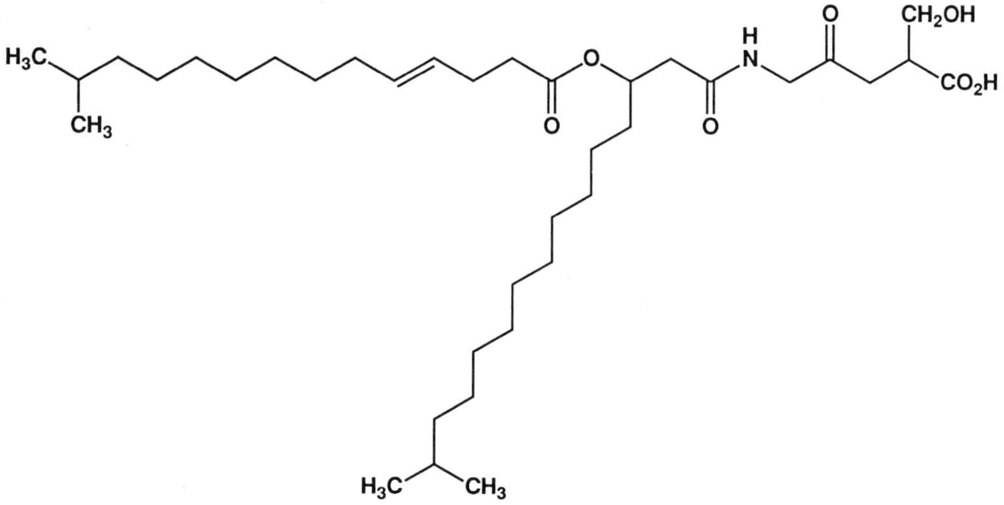

WB-3559 A

Figure 10.3 Structure of new fibrinolytic substance, WB-3559 A.

is useful in that one is able to detect directly the pharmacological activities in an in vivo system. Ando et al. (1988a, 1988b) and Yano et al. (1986a, 1986b) have applied this method to screen new bioactive compounds from microbial metabolites to find vinigrol and the actinopyrones, respectively.

The arterial blood pressure and heart rate method Arterial blood pressure in anesthetized male Sprague–Dawley rats (350–400 g) is measured with a pressure tranducer via a polyethylene catheter (PE_{50}) that is inserted into the femoral artery. Blood pressure is recorded on a biophysiograph system 180. The pulse pressure signal is used to trigger a tachometer for a measurement of heart rate. Active compounds present in a fermentation broth or in a preparation therefrom were detected by their hypertensive effect.

Antihypertensive compounds and vasodilators as microbial metabolites

Vinigrol Ando et al. (1988a, 1988b) found the new antihypertensive compound vinigrol (Figure 10.4) from a fungal strain identified as *Virgaria nigra* by testing culture broths against anesthetized normotensive rats. The compound possesses a novel diterpenoid structure (Uchida et al., 1987). In anesthetized Sprague–Dawley rats, vinigrol caused a reduction of the systolic arterial blood pressure at 10–200 μg kg^{-1}. Further, the antihypertensive activity of orally given vinigrol in conscious spontaneously hypertensive rat resulted in a decrease of blood pressure of the rat by approximately 15% at 2 mg kg^{-1}, and the duration of the effect was 6 h or more. From the experiments of the effect on vinigrol on the tension of smooth muscle and on a radioreceptor binding assay, Ando et al. (1988b) speculated that vinigrol may act as a Ca^{2+} agonist, showing inhibitory effect on Ca^{2+} move at lower concentrations.

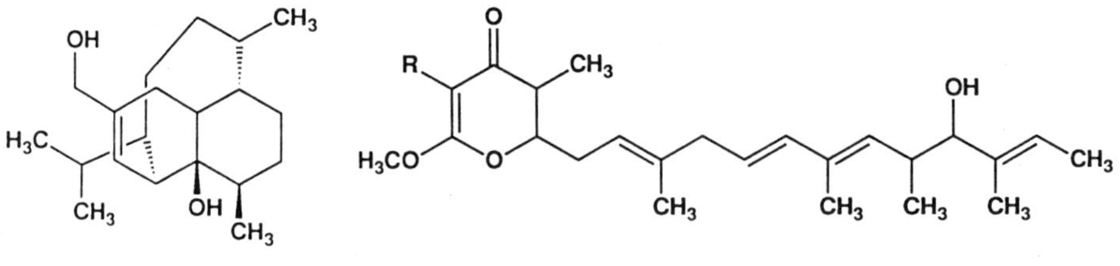

Vinigrol

Actinopyrone A R = CH_3
            B   H
            C CH_2CH_3

Figure 10.4 Structures of new microbial antihypertensive substance (vinigrol) and vasodilators (actinopyrones).

Actinopyrones A, B, and C Yano et al. (1986a, 1986b) found new compounds with vasodilating activity, the actinopyrones, from the cultured broth of *Streptomyces pactum*. The coronary vasodilating activities of actinopyrones A, B, and C (see Figure 10.4) were estimated by their action in circumflex coronary artery of the heart, exposed from a male mongrel dog, to measure the blood flow by the electromagnetic flow meter. Systemic blood pressure is measured directly from the right femoral artery through a pressure tranducer, and the heart rate is also measured by the cardiotachometer from RR intervals of the electrocardiogram. Actinopyrones A and B increase coronary blood flow and the relative activity of both compounds are approximately 100 fold greater than that of papaverine, a standard vasodilating drug.

10.6 Concluding Remarks

The foregoing description is mainly concerned with the various screening methods of vasoactive substances from the microbial sources and the interesting metabolites thus discovered. Although none of the compounds cited has been put to practical use, of particular interest are the vasodilating activity of amauromine and the hypotensive activity of vinigrol, which should be further investigated as to their medical usefulness.

Most of the in vivo pharmacological tests have been applied to a variety of synthetic compounds, the structures of which have been designed on the basis of the lead compound. Some of the in vivo tests have recently been improved significantly to be applicable for screening the active microbial metabolites.

It is hoped that the in vivo system utilizing animals or living organs will be of great help to scientists in pharmacology, biochemistry, and microbiology for discovering and developing new pharmacoactive microbial metabolites, by overcoming some of the difficulties that most of the active compounds resulting from in vitro screening programs suffer in the process of development. Consequently, the combination of in vitro and in vivo systems is desired.

References

Ando, T., Tsurumi, Y., Ohata, N., Uchida, I., Yoshida, K. and Okuhara, M. 1988a. Vinigrol, a novel antihypertensive and platelet aggregation inhibitory agent produced by a fungus, *Virgaria nigra*. I. Taxonomy, fermentation, isolation, physicochemical and biological properties. *Journal of Antibiotics* (Tokyo) 41:25–30.

Ando, T., Yoshida, K. and Okuhara, M. 1988b. Vinigrol, a novel antihypertensive and platelet aggregation inhibitory agent produced by a fungus, *Virgaria nigra*. II. Pharmacological characteristics. *Journal of Antibiotics* (Tokyo) 41:31–35.

Gaddum, J. H. 1953. The technique of superfusion. *British Journal of Pharmacology* 8:321–326.

Hino, M., Iwami, M., Okamoto, M., Yoshida, K., Haruta, H., Okuhara, M., Hosoda, J., Kohsaka, M., Aoki, H. and Imanaka, H. 1989a. FK409, a novel vasodilator isolated from the acid-treated fermentation broth of *Streptomyces griseosporeus*. I. Taxonomy, fermentation, isolation, and physico-chemical and biological characteristics. *Journal of Antibiotics* (Tokyo) 42:1578-1583.

Hino, M., Ando, T., Takase, S., Itoh, Y., Okamoto, M., Kohsaka, M., Aoki, H. and Imanaka H. 1989b. FK409, a novel vasodilator isolated from the acid-treated fermentation broth of *Streptomyces griseosporeus*. II. Structure of FK409 and its precursor FR- 900411. *Journal of Antibiotics* (Tokyo) 42:1584-1588.

Kirby, G. W., Robins, D. J., Sefton, M. A. and Talekar, R. R. 1980. Biosynthesis of bisdethiobis(methylthio) gliotoxin, a new metabolite of *Gliocladium deliquescens*. *Journal of the Chemical Society Perkin Transactions* 1:119-121.

Kohler, C., Wooding, W. and Ellenbogen, L. 1976. Intravenous arachidonate in the mouse: A model for the evaluation of antithromobotic drugs. *Thrombosis Research* 9:67-80.

Lauer, U., Anke, T. and Hansske, F. 1991. Antibiotics from basidiomycetes XXXVIII. 2-Methoxy-5-methyl-1,4-benzoquinone, a thromboxane A_2 receptor antagonist from *Lentinus adhaerens*. *Journal of Antibiotics* (Tokyo) 44:59-65.

Mizushima, Y. and Sakai, S. 1969. Stabilization of erythrocyte membrane by nonsteroid anti-inflammatory drugs. *Journal of Pharmacy and Pharmacology* 21:327-328.

Nakagawa, A., Fukamachi, N., Yamaki, K., Hayashi, M., Oh-ishi, S., Kobayashi, B. and Ōmura, S. 1987. Inhibition of platelet aggregation by medermycin and its related isochromanequinone antibiotics. *Journal of Antibiotics* (Tokyo) 40:1075-1076.

Nilsson, I. M., Hedner, U. and Pandolfi, M. 1978. The measurement of fibrinolytic activities. In: *Handbook of Experimental Pharmacology*, Vol.46, pp. 107-134. Springer-Verlag, Berlin.

Ninomiya, Y. T., Yamada, Y., Onitsuka, M., Tanaka, Y., Maeda, T. and Maruyama, H. B. 1980a. Biochemically active substances from microorganisms. IV. Establishment of several in vitro antiinflammatory probes applicable to microbial broths and effects of non-steroidal antiinflammatory drugs and antibiotics on them. *Chemical and Pharmaceutical Bulletin* 28:2553-2564.

Ninomiya, Y. T., Yamada, Y., Shirai, H., Onitsuka, M., Suhara, Y. and Maruyama, H. B. 1980b. Biochemically active substances from microorganisms. V. Pyrrothines, potent platelet aggregation inhibitors of microbial origin. *Chemical and Pharmaceutical Bulletin* (Tokyo) 28:3157-3162.

Oka, S., Kodama, M., Takeda, H., Tomizuka, N. and Suzuki, H. 1986. Staurosporine, a potent platelet aggregation inhibitor from a *Streptomyces* species. *Agricultural and Biological Chemistry* 50:2723-2727.

Okamoto, M., Yoshida, K., Uchida, I., Nishikawa, M., Kohsaka, M. and Aoki, H. 1986a. Studies of platelet activating factor (PAF) antagonists from microbial products. I. Bis-dethiobis (methylthio)gliotoxin and its derivatives. *Chemical and Pharmaceutical Bulletin* (Tokyo) 34:340-344.

Okamoto, M., Yoshida, K., Uchida, I., Kohsaka, M. and Aoki, M. 1986b. Studies of platelet activating factor (PAF) antagonists from microbial products. II. Pharmacological studies of FR-49175 in animal models. *Chemical and Pharmaceutical Bulletin* (Tokyo) 34:345-348.

Okamoto, M., Yoshida, K., Nishikawa, M., Hayashi, K., Uchida, I., Kohsaka, M. and Aoki, H. 1986c. Studies of platelet activating factor (PAF) antagonists from microbial products. III. Pharmacological studies of FR-900452 in animal models. *Chemical and Pharmaceutical Bulletin* (Tokyo) 34:3005-3010.

Ōmura, S., Nakagawa, A., Fukamachi, N., Otoguro, K. and Kobayashi, B. 1986a. Aggreceride, a new platelet aggregation inhibitor from *Streptomyces*. *Journal of Antibiotics* (Tokyo) 39:1180-1181.

Ōmura, S., Tomoda, H., Xu, Q. M., Takahashi, Y. and Iwai, Y. 1986b. Triacsins, new inhibitors of acyl-CoA synthetase produced by Streptomyces. *Journal of Antibiotics* (Tokyo) 39:1211–1218.

Ōmura, S., Nakagawa, A., Fukamachi, N., Miura, S., Takahashi, Y., Komiyama, K. and Kobayashi, B. 1988. OM-4842, a new platelet aggregation inhibitor from Streptomyces. *Journal of Antibiotics* (Tokyo) 41:812–813.

Ōmura, S., Iwai, Y., Hirano, A., Nakagawa, A., Awaya, I., Tsuchiya, H., Takahashi, Y. and Masuma, R. 1977. A new alkaloid AM-2282 of Streptomyces origin taxonomy, fermentation, isolation and preliminary characterization. *Journal of Antibiotics* (Tokyo) 30:275–282.

Ozasa, T., Suzuki, K., Yamada, T., Suzaki, K., Nohara, C., Kobori, M. and Saito, T. 1990. Placetins, platelet aggregation inhibitors from *Streptomyces* sp. Q-1043. *Journal of Antibiotics* (Tokyo) 43:331–335.

Patel, M., Hegde, V., Horan, A., Barrett, T., Bishop, R., King, A., Marquez, J., Hare, R. and Gullo, V. 1989. Sch 38519, a novel platelet aggregation inhibitor produced by a *Thermomonospora* sp. Taxonomy, fermentation isolation, physico-chemical properties, structure and biological properties. *Journal of Antibiotics* (Tokyo) 42:1063–1069.

Takase, S., Iwami, M., Ando, T., Okamoto, M., Yoshida, K., Horiai, H., Kohsaka, M., Aoki, H. and Imanaka, H. 1984a. Amauromine, a new vasodilator, taxonomy, isolation and characterization. *Journal of Antibiotics* (Tokyo) 37:1320–1323.

Takase, S., Kawai, Y., Uchida, I., Tanaka, H. and Aoki, H. 1984b. Structure of amauromine, a new alkaloid with vasodilating activity produced by *Amauroascus* sp. *Tetrahedron Letters*: 4673–4676.

Tamaoki, T., Nomoto, H., Takahashi, I., Kato, Y., Morimoto, M. and Tomita, F. 1986. Staurosporine, a potent inhibitor of phospholipid/Ca^{++} dependent protein kinase. *Biochemical and Biophysical Research Communications* 135:397–402.

Tanaka, H., Yoshida, K., Itoh, Y. and Imanaka, H. 1982. Studies on new vasodilators, WS-1228 A and B. II. Structure and synthesis. *Journal of Antibiotics* (Tokyo) 35:157–163.

Uchida, I., Yoshida, K., Kawai, Y., Takase, S., Itoh, Y., Tanaka, H., Kohsaka, M. and Imanaka, H. 1985. Studies on WB-3559 A, B, C and D, new potent fibrinolytic agents. II. Structure elucidation and synthesis. *Journal of Antibiotics* (Tokyo) 35:1476–1468.

Uchida, I., Ando, T., Fukami, N., Yoshida, K., Hashimoto, M., Tada, T., Koda, S. and Morimoto, Y. 1987. The structure of vinigrol, a novel diterpenoid with antihypertensive and platelet aggregation-inhibitory activities. *Journal of Organic Chemistry* 52:5292–5293.

Umehara, K., Yoshida, K., Okamoto, M., Iwami, M., Tanaka, H., Kohsaka, M. and Imanaka, H. 1984a. Studies on WF-5239, a new potent platelet aggregating inhibitor. *Journal of Antibiotics* (Tokyo) 37:469–474.

Umehara, K., Yoshida, K., Okamoto, M., Iwami, M., Tanaka, H., Kohsaka, M. and Imanaka, H. 1984b. Studies on new antiplatelet agents WS-30581 A and B. *Journal of Antibiotics* (Tokyo) 37:1153–1160.

Vane, J. K. 1964. The use of isolated organs for detecting active substances in the circulating blood. *British Journal of Pharmacology* 23:360–373.

Vargaftig, B. B., Lefort, J., Chignard, M. and Benveniste, J. 1980. Platelet-activating factor induces a platelet-dependent bronchoconstriction unrelated to the formation of prostaglandin derivatives. *European Journal of Pharmacology* 65:185–192.

Yano, K., Yokoi, K., Sato, J., Oono, J., Kouda, T., Ogawa, Y. and Nakashima, T. 1986a. Actinopyrones A, B, and C, new physiologically active substances. I. Producing organism, fermentation, isolation and biological properties. *Journal Antibiotics* (Tokyo) 39:32–37.

Yano, K., Yokoi, K., Sato, J., Oono, J., Kouda, T., Ogawa, Y. and Nakashima, T. 1986b.

Actinopyrones A, B and C, new physiologically active substances. II. Physico-chemical properties and chemical structures. *Journal of Antibiotics* (Tokyo) 39:38–43.

Yoshida, K., Okamoto, M., Umehara, K., Iwami, M., Kohsaka, M., Aoki, H. and Imanaka, H. 1982. Studies on new vasodilators, WS- 1228 A and B. I. Discovery, taxonomy, isolation and characterization. *Journal of Antibiotics* (Tokyo) 35:151–156.

Yoshida, K., Iwami, M., Umehara, Y., Nishikawa, M., Uchida, I., Kohsaka, M., Aoki, H. and Imanaka, H. 1985. Studies on WB-3559 A, B, C and D, new potent fibrinolytic agents. I. Discovery, identification, isolation and characterization. *Journal of Antibiotics* (Tokyo) 38:1469–1475.

Part 5
Agrochemicals

11

Fungicides and Antibacterial Agents

Shigenobu Okuda and Yoshitake Tanaka

11.1 Introduction

Fungal and bacterial infections in agriculture are found worldwide on about 1200 kinds of crop plants, fruit trees, and vegetables. Decreases in crop yields caused by microbial diseases are estimated to reach 9%–10% each in rice, wheat, and corn production. With the additional losses because of weeds and insects, the total decreases in crop yields are estimated to be more than 40%, 20%, and 30%, for rice, wheat, and corn, respectively.

Because these crops and the potato are used as staple foods, their steady supply and appropriate storage are of global importance in view of the increasing world population, particularly in developing countries. For this, antimicrobial agents have played a pivotal role (Burg, 1986; Mirsa, 1986; Woodruff and Burg, 1986).

Efforts toward control of plant diseases have evolved two approaches: host plant and pathogenic microorganisms. The former is concerned with defense systems in plants and with attempts to strengthen it; the latter deals with antibacterial and antifungal agents of microbial and synthetic origins. Today the latter approach appears to play a predominant role in controlling fungal diseases. This chapter is concerned with screening methods for agricultural fungicides.

11.2 Plant Disease and Fungicides

Table 11.1 lists major plant diseases and pathogenic microorganisms found in agricultural fields. One plant, e.g., the rice plant, is at the risk of being infected by many different pathogenic microorganisms. The actual causative pathogen depends on the weathering conditions of the year, the district, and the country, and on the variety of rice plant. On the other hand, one species of fungi, e.g., *Botrytis cinerea*, can infect different kinds of fruit trees and

Table 11.1 Typical diseases of crops

Crop	Disease	Pathogen
Rice	Blast	*Pyricularia oryzae*
	Sheath blight	*Rhizoctonia solani*
	Bacterial leaf blight	*Xanthomonas campestris* pv. *oryzae*[1,2]
Barley	Stem rust	*Puccinia graminis*
	Powdery mildew	*Erysiphe graminis* f. sp. *hordei*
Wheat	Leaf rust	*Puccinia recondita*
	Powdery mildew	*Erysiphe graminis* f. sp. *tritici*
Maize	Leaf spot	*Cochliobolus heterostrophus*
Potato	Late blight	*Phytophthora infestans*
Citrus	Canker	*Xanthomonas campestris* pv. *citri*[1]
	Melanose	*Diaporthe citri*
Apple	Alternaria leaf spot	*Alternaria mali*
	Scab	*Venturia inaequalis*
	Powdery mildew	*Podosphaera leucotricha*
	Fire blight	*Erwinia amylovora*[1]
Grape	Gray mold	*Botrytis cinerea*
	Downy mildew	*Plasmopara viticola*
Vegetables	Gray mold	*Botrytis cinerea*
	Soft rot	*Erwinia carotovora*[1]
	Powdery mildew	*Sphaerotheca fuliginea* (cucumber)

[1] Bacterial pathogens.
[2] Abbreviations: pv., pathovar; f. sp., forma specialis.
According to recent subclassification of plant pathogenic microorganisms, pv. (pathovar) is a subspecies or lower subclass which includes individuals with identical plant pathogenicity, and f. sp. (forma specialis) those with identical special physiological properties such as specific parasitism. The former is discussed mainly for bacterial pathogens, and the latter for fungal pathogens.

vegetables to cause a disease called "grey mold disease." The relationship between disease and pathogenic microorganism is not always one to one. Powdery mildew, a disease of many vegetables and fruit trees with the common characterictic of leaf biodegradation, is caused by different species of microorganisms that are specifically infectious to individual plants.

11.3 Control of Fungal and Bacterial Diseases by Microbial Metabolites

In early times the control of fungal plant diseases was begun with Bordeaux mixture, dithiocarbamates, and also with organomercuric and arsenic fungicides. The high toxicity of the latter organo metallic fungicides, despite their potent activity, prompted Japanese researchers to undertake screening for safer compounds from microorganisms. Blasticidin S was thus discovered and introduced into agriculture for the control of rice blast caused by *Pyricularia oryzae*. The success of blasticidin S encouraged further screening, which eventually brought about polyoxin, kasugamycin, validamycin, and more recently mildiomycin. It is noteworthy that these compounds were discovered and developed as agricultural fungicides by Japanese research groups (Misato, 1982). For bacterial control, antibacterial antibiotics discovered in the 1940s and the 1950s were also used for agriculture. Table 11.2 lists fungicides and bactericides of current application in agriculture and their modes of action. Representative chemically synthesized fungicides are also included in Table 11.2. Biological characteristics of microbial metabolites are described next.

Blasticidin S Blasticidin S is a nucleoside derivative (Figure 11.1) discovered as a metabolite of *Streptomyces griseochromogenes* (Takeuchi et al., 1958). It is moderately active in vitro against plant pathogenic fungi and some bacteria. It also shows antiviral and anticancer activities, so the activity is not specific to plant pathogenic fungi. The efficacy in controlling rice blast was detected by the pot test.

An application of 30–40 kg ha^{-1} (formulated product basis) is usually recommended. Both preventive and curative treatments are applicable. Blasticidin S potently inhibits the development of *Pyricularia oryzae* mycelia, and, to a lesser extent, also spore germination and spore formation. It is as potent as organomercuric fungicides but is less toxic.

In *P. oryzae*, blasticidin S inhibits protein synthesis. In an *E. coli*, cell-free, protein-synthesizing system, it inhibits the release of nascent peptide chain from ribosomes, in contrast to the action of puromycin, which enhances

Table 11.2 Agricultural fungicides and bactericides and their mechanism of action

Target of inhibition	Microbial metabolite	Synthetic chemicals
Sterol biosynthesis		Azols, piperazines
Respiration		Cu, SH reagents, dithiocarbamates
Protein synthesis	Antibacterial	
	Streptomycin	
	Chloramphenicol	
	Oxytetracycline	
	Novobiocin	
	Antifungal	
	Kasugamycin	
	Blasticidin S	
	Cycloheximide	
	Mildiomycin	
Cell wall synthesis	Polyoxin	Organophosphates
Cell function	Griseofulvin	Benzimidazols
Pathogen–plant interaction	Validamycin	Probenazole

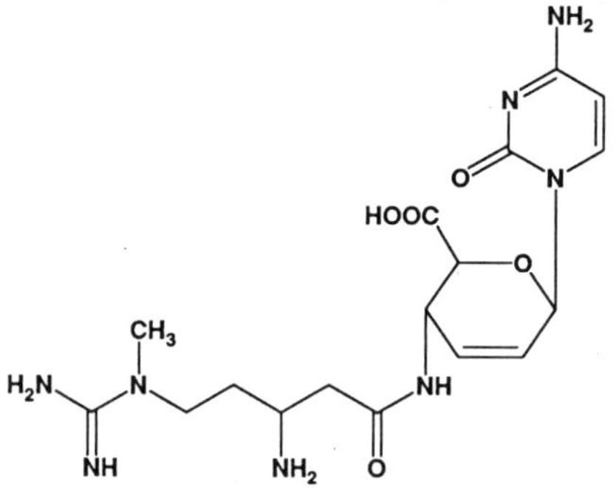

Blasticidin S

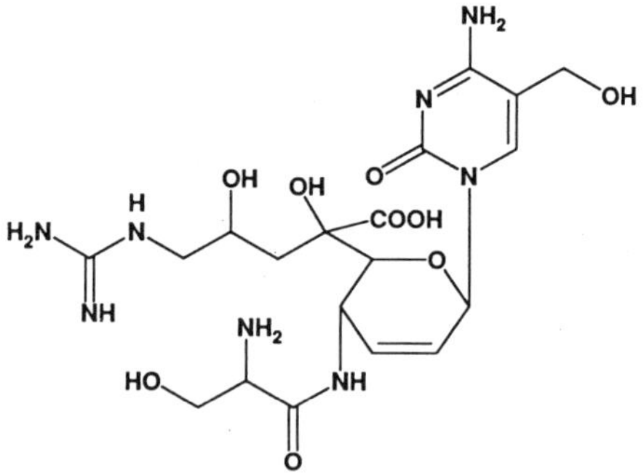

Mildiomycin

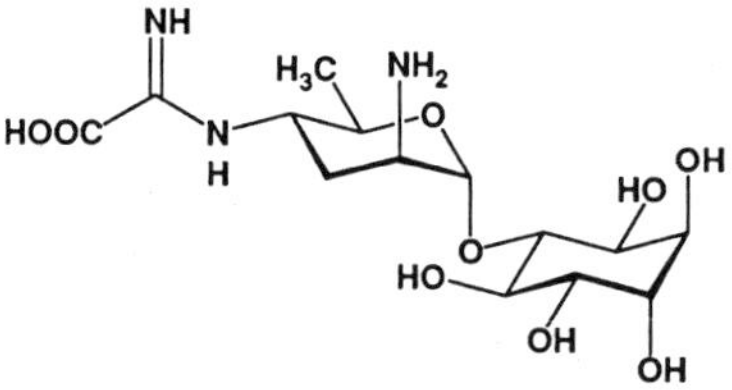

Kasugamycin

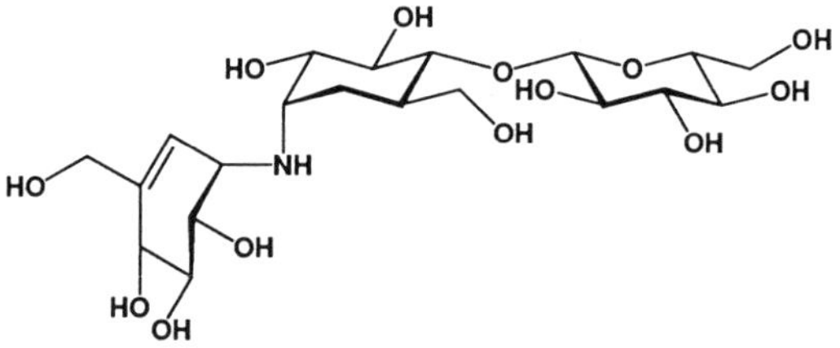

Validamycin A

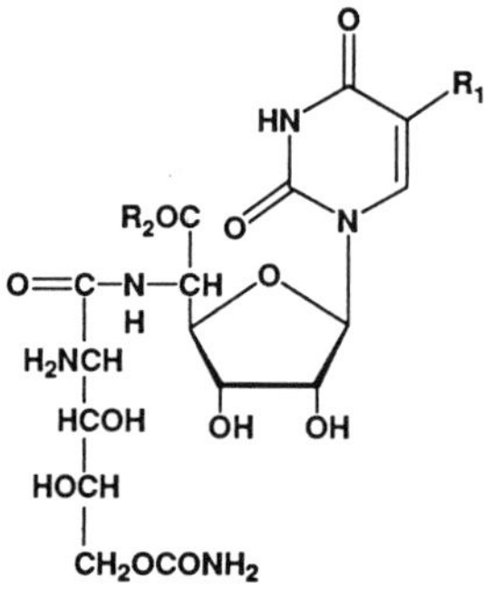

	R_1	R_2
Polyoxin A	CH_2OH	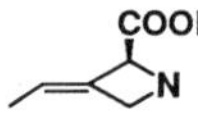
Polyoxin B	CH_2OH	OH
Polyoxin D	COOH	OH
Polyoxin L	H	OH

Figure 11.1 Agricultural fungicides of microbial origin.

the peptide release. Blasticidin S causes eye irritation; therefore, caution must be taken to protect farm workers from direct exposure to this fungicide. Because calcium ion and a microbial product (Yonehara et al., 1973) were later found to be an effective detoxicant for eye irritation, blasticidin S is formed with calcium acetate for commercial production.

Kasugamycin   Kasugamycin is an amino-sugar compound (see Figure 11.1) discovered in 1965 (Umezawa et al.) as a metabolite of *Streptomyces kasugaensis* and of *S. kasugaspinus*. It is active in vitro against yeast and filamentous fungi including *P. oryzae*, a plant pathogen, and some bacteria, e.g., *Pseudomonas*. The excellent plant-protecting efficacy was discovered by pot tests. On preventive and curative treatments, kasugamycin at 20 ppm suppresses, by 98%–100%, the development of *Pyricularia oryzae* mycelia on rice plant. It does not appear to inhibit spore germination. Like other amino-sugar antibiotics, kasugamycin inhibits protein synthesis. A characteristic property is that it does not induce miscoding, unlike other amino sugars such as streptomycin. It is stable in storage; after 3 years of storage at room temperature, more than 90% of the initial potency remained. It is nontoxic to animals, humans, and fish. The LD_{50} value is estimated to be more than 3 g kg^{-1} (i.p.) in mice and rats (Sato, 1983).

On repeated use of kasugamycin, kasugamycin-resistant strains of *P. oryzae* have emerged. However, the resistant strains disappeared within 2–3 years during which other fungicides were employed. It is suggested that kasugamycin resistance may occur but can be overcome by alternating the use of kasugamycin and other fungicides.

Polyoxin   Polyoxins are a family of nucleoside derivatives discovered in 1965 (Isono et al.) as metabolites of *S. cacaoi* var. *asoensis*. The name "polyoxin" represents "compound with many oxygen groups." They are active against many plant pathogenic fungi in vitro. The excellent in vivo efficacy was detected by pot tests early in the course of screening. Polyoxin D, after precipitation as a zinc complex, has been used for control of sheath blight of rice (*Rhizoctonia solani*); polyoxin B and L have been used for control of black spot of Japanese pear (*Alternaria kikuchiana*) and gray mold disease (*Botrytis cineria*) of grape and vegetables.

In *P. oryzae*, polyoxin D inhibits the synthesis of chitin, which is a cell wall polysaccharide. When exposed to polyoxin D, the mycelia of *P. oryzae* become swollen, changing to a large round shape. The effect of polyoxin on cell wall chitin biosynthesis is attributed to the structural similarity to uridine 5′-diphosphate-(UDP-) *N*-acetylglucosamine, a precursor and building unit for chitin biosynthesis (Isono and Suzuki, 1979; Isono, 1989).

Continual applications of polyoxin fungicides to a field often leads to the increase of the polyoxin-resistant fungus population in that field. In such cases, polyoxin usage is discontinued and other types of fungicides with a different mode of action are used for treatment.

Validamycin Validamycin A is a unique pseudo-oligosaccharide produced by *Streptomyces hygroscopicus* var. *limoneus* (Iwasa et al., 1970). It was discovered by an interesting in vitro screening (described later). Validamycin A is effective for the control of rice sheath blight caused by *Rhizoctonia solani* (= *Pellicularia sasakii*). Validamycin A at 20–30 ppm causes retardation of the growth of *R. solani* on rice plant, thereby inducing abnormal branching of *R. solani* hyphae. Of interest is the fact that validamycin A shows no in vitro antimicrobial activity in conventional nutrient-rich agar media against a variety of plant pathogenic fungi such as *R. solani, Pyricularia oryzae, B. cinerea,* and *A. kikuchiana,* or yeast and bacteria.

Studies on its mode of action have demonstrated that validamycin A is a potent inhibitor of trehalase from *R. solani* (Shigemoto et al., 1989). A hydrolysis product of validamycin A, validoxylamine, is the most potent inhibitor among the validamycin family. Trehalose is a storage material in *R. solani*. Under nutrient-deficient conditions encountered by *R. solani* on the leaf surface, trehalose metabolism via trehalase is essential. Therefore, the inhibition of trehalose metabolism by validamycin A and validoxylamine results in the retardation of fungal growth. It is suggested that validamycin A is hydrolyzed in fungal cells to generate validoxylamine after incorporation into *R. solani* mycelia.

Validamycin A shows no toxicity when given to mice at 10 g kg^{-1} (i.p.) or 15 g kg^{-1} (s.c.) or to fish at 100 ppm.

Mildiomycin Mildiomycin (see Figure 11.1) is an aminoacylated nucleoside produced by *Streptoverticillium rimofaciens*. It was discovered in 1978 by Harada and Kishi using an in vivo screening method established to assay the efficacy of agents in controlling powdery mildew, which is caused by many obligately parasitic fungi. Mildiomycin is effective at 25–100 ppm against a variety of causative fungi on more than 10 species of plants. It is also active against benomyl-resistant *Sphaerotheca fuliginea,* a causative parasite on cucumber. The fungicide inhibits germination of conidia, inducing abnormal morphology of hyphae (Iwasa, 1983).

In a cell-free system from *Escherichia coli*, mildimycin at 10 μg ml^{-1} inhibited poly U-directed protein synthesis by 80%–90% (Om et al., 1984).

11.4 Screening Methods for Agricultural Fungicides and Bactericides

The microbial products described here that are utilized today as agricultural chemicals were discovered by screening methods established specifically for agricultural compounds. To screen other potential agricultural fungicides and bactericidal compounds, unique screening methods have also been developed and used. These are described next.

Paper disk method   The paper disk method is one of the most popular and convenient methods for screening of agricultural fungicides as well as many other antimicrobial agents. The merits and demerits of this method are described in Chapter 2. However, none of the microbial metabolites discovered by this technique only have yet been introduced into agriculture. Many antifungal compounds, probably discovered by the paper disk method with or without other evaluating techniques, have been reported. Some of these seem interesting and promising:

Compounds active against *Pyricularia oryzae*:
   mitropeptin, miharamycin
Compounds active against *Botrytis cinerea*:
   irumamycin, tautomycin, neopeptin, neopolyoxin
Compounds active against *Rhizoctonia solani*:
   dapiramicin, notonesomycin
Compounds active against *Puccinia recondita*
   rustmicin, neorustmicin
Compounds active against *Valsa ceratosperma*:
   propanocin.

Pot test method   A pot test technique is one of the most reliable methods by which the efficacy of a chemical in field trials can be predicted. This method is employed in an early step of screening (e.g., blasticidin S and polyoxin) or in a later stage of screening (e.g., kasugamycin and validamycin). A typical process is as follows.

A host plant is grown in a small pot 9–12 cm in diameter under greenhouse conditions with dark-and-light control until true leaves emerge. Spores of *P. oryzae*, for example, that have been previously grown on an agar medium are suspended in saline with or without a supplemented surfactant like Tween 20 (0.1 mg ml^{-1}). The spore suspension is sprayed onto the leaves of the host plant, which is then incubated overnight in a humid room to facilitate microbial adhesion and infection. A methanol (or water) solution containing varying concentrations of a test compound is sprayed onto the infected plant, and the infection is allowed to develop. One to 2 weeks after drug application, the lesions from the microbial infection are counted or scored, and the curative (or preventive) effect of the test compound is evaluated in comparison with the scores of untreated control.

Modified in vitro systems   A large gap exists between in vitro potency and in vivo efficacy of many antifungal compounds. The gap is often magnified when in vitro potency is measured on nutrient-rich agar media, which permits speedy assay of antifungal activity. One potential solution of this problem is to employ pot tests at the earliest screening step. An alternative approach may be to use modified in vitro systems, which simulate in vivo environments yet allow a large number of samples to be tested. Attempts in this line are described next.

Rice plant–infusion agar method The rice plant–infusion agar method is a modified in vitro system that mimics the natural circumstances at the infected plant leaves. This method has been developed and utilized in the discovery of kasugamycin (Hamada et al., 1965).

Living rice plants (100 g) are cut into small pieces, boiled in water (1 liter) for 30 min, and gauze-filtered. To the infusion solution (500 ml) are added sucrose (1.5%) and agar (2.2%) and the pH is adjusted to 5.0. After autoclaving, the medium is mixed with an equal volume of sterile McIlvaine buffer (0.1 M, pH 5.0), and poured in 10-ml portions into petri dishes. After solidification, 4 ml of *P. oryzae* spores suspended in the same agar medium (2×10^5 spores ml^{-1}) are overlaid. In this rice plant infusion agar, *P. oryzae* is much more sensitive to many antifungal drugs. In an actual screening process, those microbial cultures which show good activity in this rice plant infusion agar method are reserved and evaluated further via the pot test.

Spot inoculation method A modified in vitro assay, called the spot inoculation method, was successfully employed in the discovery of validamycin (Iwasa et al., 1971; Iwasa, 1978). This method uses an in vitro model of infected plant surface. The assay plate consists of double layers. On the lower basal layer (10 ml of GLB medium, glycerol, 3%; peptone, 0.2%; meat extract, 0.2%; NaCl, 0.2%; agar, 1.5%; pH 7), four soil isolates to be tested are spot inoculated; these are allowed to grow and form colonies and to produce active metabolites at 28°C for 4 days. Overlaid on the colonies is V8 agar [5 ml, V8 vegetable juice (Suntory, Tokyo)], 20 vol/vol %; sucrose, 2%; agar 1.5%; pH 6). At the center of the top surface, an agar piece (10 mm in diameter) of GB medium (glucose, 1%; beef extract, 0.5%; peptone, 0.5%; NaCl, 0.5%; agar, 1.5%; pH 7) is placed for inoculation, on which *Rhizoctona solani* has previously been grown at 27°C for 2 days. The petri dishes are incubated for a further 2 days. Antifungal activity is estimated by the inhibition zone appearing around the colonies.

In a preliminary experiment, 5 media were tested as upper layers, and V8 agar and rice straw agar gave 6 positives of 200 cultures; the same 6 were also identified as effective in a parallel in vivo evaluation by Kohsaka's method, using the same 200 cultures. Thus, a good correlation was obtained between in vitro and in vivo results in testing by a plant extract-containing agar layer sandwiched by a fungal pathogen on the upper side and by active metabolites on the lower side. It is interesting to note that the primary feature of the afore-mentioned V8 agar layer is that it simulates, in vitro, the fungicide-treated plant leaf on which *R. solani* hyphae are proliferating.

A modification of the spot inoculation method using an animal tissue extract in place of a plant extract may extend the application of this method to the screening of antifungal and antibacterial agents of medical use.

Of about 39,000 cultures subjected to the spot inoculation assay, approximately 5,200 showed anti-*R. solani* activity, while 11,000 have exhibited both anti-*R. solani* and antibacterial activities. Of the 5,200, when detected by the cross-streak method 2,100 cultures represented antifungal activity distinct from those of known antifungals such as polyenes, cycloheximides, antimycins, blasticidin S, and polyoxins. These 2,100 cultures were evaluated in vivo by Kohsaka's method; 1 was later identified as a validamycin producer.

Leaf disk method   A modified in vitro method named the leaf disk method has been employed by Ko (1983) for screening of microbial metabolites active against gray mold disease caused by *Botrytis cinerea*, as follows:

Cucumber leaves are cut into small disks (20 mm in diameter) with a cork borer. The leaf disks are dipped into test solutions for a short period, then put on a glass slide in a humid chamber as in a petri dish. Onto the leaf are placed inoculation paper disks that are preimmersed in a solution containing *B. cinerea* conidia (about 10^5 spores ml^{-1}), glucose (5%), and yeast extract (1%). After the chamber is incubated for 1–2 days, the diameters of the lesions are measured.

This method may be applied to screening agents active against obligate parasites or viruses, if infectious conidia (or virus particles) and suitable host plants are available.

Brush inoculation method for systemic translaminar agents   The Brush inoculation method has been developed to evaluate the efficacy of chemicals exhibited after their translocation in plant tissues when the chemicals are applied on plant parts other than infected lesions. The method utilized in the discovery of mildiomycin is described here (Kusaka et al., 1979).

Cereals, for example, barley, are grown for a week to reach 7–9 cm in height. Roots are washed and wrapped with cotton or gauze, which is then moistened with 3 ml of test compound solution; foliars are inoculated with conidia of barley powdery mildew fungus, *Erysiphe graminis*, with a small brush.

The whole plant is incubated at 18°C for an additional week in a test tube to develop powdery mildew. The efficacy of the test compounds is evaluated by counting or scoring the lesions.

This method was applied to examine 1680 cultures of soil isolates containing potential antifungal activity. Among them, 26 cultures showed activity. Next, the 26 cultures were evaluated by the pot test method described here (wind-mediated infection test). One of the 26 cultures was reserved and was found later to be a mildiomycin producer.

The pot test, the wind-mediated infection test employed by Takeda's group seems noteworthy. Two pots are prepared. Barley in one pot is treated with mildiomycin-producing culture but is kept free from conidia; barley in the other pot is actively infected with powdery mildew fungus and is not treated with any drug. The two pots are placed in a chamber with an electronic fan. The mildiomycin-producing culture showed a very promising effect in this test.

Other methods Screening methods for enzyme inhibitors (Chapters 7 and 8) and fermentation techniques (Chapter 16) have been reported. These are useful in screening antifungal agents for agriculture.

11.5 Future Prospect

New and effective fungicides are needed in agriculture, because many plant diseases remain outside our control although a steady and plentiful supply of crops is essential to the modern world. In spite of the undesirable problems brought about by the use of fungicides, which include drug accumulation in soils, change in natural microbial flora, and pollution of agricultural fields, forests, rivers, etc., the use of fungicides in agriculture is expected to continue to increase in the near future, provided that safer fungicides become available. The automation in agriculture may also accelerate the increased use, mainly for economic reasons.

The discovery of new and safe fungicides for agriculture is highly possible if we rely on microorganisms and on the strategy and techniques that have been utilized in the screening for medicines, enzyme inhibitors, and other bioactive microbial metabolites. The requirements for the agricultural fungicides to be obtained are excellent potency against a variety of pathogenic microorganisms and safety, not only for humans, animals, and the host plant, but also for the ecosystem. Microbial metabolites are expected to meet the latter requirement, because they are biodegradable.

The mechanism of action and structural feature of today's fungicides fall into a narrow range (see Table 11.2). For example, sterol biosynthesis inhibitors must also be found from microorganisms. In addition to compounds with fungicidal action, those which interfere with plant-pathogen interaction are of current interest. Validamycin is one of the nonfungicide type of agents that was discovered by a unique approach. Additional examples of this type of action may be available. This is surely promoted by good cooperation between microbiologists and plant pathologists.

Recombinant DNA technology is expanding the ways of pesticide use, as is exemplified by the herbicides of microbial and synthetic origins. The genes conferring resistance to the herbicides phosphinothricin and glyphosate have been introduced into plants, and the plants have acquired tolerance to these herbicides (Kishore and Shah, 1988). Newer fungicides will also provide new tools useful for plant engineering.

References

Burg, R. W. 1986. Trends in the use of fermentation products in agriculture. In: Moats, W. A. (editor), *Agricultural Uses of Antibiotics*, ACS Symposium series 320, pp. 61–72. American Chemical Society, Washington, D. C.

Hamada, M., Hashimoto, T., Takahashi, T., Yokoyama, S., Miyake, M., Takeuchi, T., Okami, Y. and Umezawa, H. 1965. Antimicrobial activity of kasugamycin. *Journal of Antibiotics* (Tokyo) Series A 18:104–106.

Harada, S. and Kishi, T. 1978. Isolation and characterization of mildiomycin, a new nucleoside antibiotic. *Journal of Antibiotics* (Tokyo) 31:519–524.

Isono, K. 1989. New aspect in research and development of agricultural antibiotics. *Nippon Nogeikagaku Kaishi* 63:1351–1358.

Isono, K. and Suzuki, S. 1979. The polyoxins: Pyrimidine nucleoside peptide antibiotics inhibiting fungal cell wall biosynthesis. *Heterocycles* 14:333–351.

Isono, K., Nagatsu, J., Kobinata, K., Sasaki, K.and Suzuki, S. 1965. Studies on polyoxins, antifungal antibiotics. Part I. Isolation and characterization of polyoxins A and B. *Agricultural and Biological Chemistry* 29:848–854.

Iwasa, T. 1978. Antifungal characteristics and bioassay methods of validamycin, an antibiotic effective against the sheath blight of rice plants. *Journal of Takeda Research Laboratories* 37:307–352.

Iwasa, T. 1983. Mildiomycin, an effective eradicant for powdery mildew. In: Takahashi, N., et al. (editors), *Pesticide Chemistry, Human Welfare and the Environment, Vol. 2, Natural Products*, pp. 57–62. Pergamon Press, Oxford.

Iwasa, T., Higashide, E., Yamamoto, H. and Shibata, M. 1970. Studies on validamycins, new antibiotics. II. Production and biological properties of validamycins A and B. *Journal of Antibiotics* (Tokyo) 23:595–602.

Iwasa, T., Higashide, E. and Shibata, M. 1971. Studies on validamycins, new antibiotics. III. Bioassay methods for the determination of validamycin. *Journal of Antibiotics* (Tokyo) 24: 114–118.

Kishore, G. M. and Shah, D. M. 1988. Amino acid biosynthesis inhibitors as herbicides. *Annual Review of Biochemistry* 57:627–663.

Ko, K. 1983. How to discover new antibiotics for fungicidal use. In: Takahashi, N., et al. (editors), *Pesticide Chemistry, Human Welfare and the Environment, Vol. 2, Natural Products*, pp. 247–252. Pergamon Press, Oxford.

Kusaka, T., Suetomi. K. and Iwasa, T. 1979. Screening system for anti-mildew substance using the barley seedling. *Journal of Pesticide Science* 4:345–348.

Mirsa, A. K. 1986. Antibiotics as crop protectants. In: Moats, W. A. (editor), *Agricultural Uses of Antibiotics*, ACS Symposium Series 320, pp. 50–60. American Chemical Society, Washington, D. C.

Misato, T. 1982. Present status and future prospects of agricultural antibiotics. *Journal of Pesticide Science* 7:301–305.

Om, Y., Yamaguchi, I. and Misato, T. 1984. Inhibition of protein synthesis by mildiomycin, an anti-mildew substance. *Journal of Pesticide Science* 9:317–323.

Sato, K. 1983. Biological properties of kasugamycin. In: Takahashi, N, et al. (editors), *Pesticide Chemistry, Human Welfare and the Environment, Vol. 2, Natural Products*, pp. 293–299. Pergamon Press, Oxford.

Shigemoto, R., Okuno, T. and Matsuura, K. 1989. Effect of validamycin on the activity of trehalase of *Rhizoctonia solani* and several sclerotial fungi. *Annals of the Phytopathology Society of Japan* 55:238–241.

Takeuchi, S., Hirayama, K., Ueda, K., Sakai. H. and Yonehara, H. 1958. Blasticidin S, a new antibiotic. *Journal of Antibiotics* (Tokyo) Series A 11:1–5.

Umezawa, H., Okami. Y., Hashimoto, T., Suhara, Y., Hamada, M. and Takeuchi, T. 1965. A new antibiotic, Kasugamycin. *Journal of Antibiotics* (Tokyo) Series A 18:101–103.

Woodruff, H. B., Burg, R. W. 1986. The antibiotic explosion. In: Parnham, M. J. and Bruinvels, J. (editors), *Pharmacological Methods, Receptors and Chemotherapy*, pp. 303–351. Elsevier, Amsterdam.

Yonehara, H., Seto, H., Shimazu, A., Aizawa, S., Hidaka, T., Kakinuma, K. and Otake, N. 1973. Production, isolation and biological properties of detoxin complex. *Agricultural and Biological Chemistry* 37:2771–2776.

12

Herbicides

Shigenobu Okuda

12.1 Introduction

The farmland of the world includes about 150,000,000 hectares or 10% of terra firma, and there are more than 200 families (6000 species including the analogs) of weeds that damage agricultural production on our planet's earth. Weeding is often required for production of a crop.

Before chemical herbicides appeared early in the 1950s, weeding was done by hand and by machinery, requiring more than half the total hours of farming labor. Nowadays, the appropriate application of herbicides has brought about a great reduction of labor and an increase in crop yield to support the world population.

Herbicides for practical use today are mostly synthetic compounds that may be classified as follows, on the basis of their mode of action (Yoshida, 1989).

1. Herbicides inhibiting photosynthetic electron transfer
2. Diphenyl ether-type herbicides necessary to photoactivation
3. Herbicides affecting plant hormonal actions
4. Herbicides interfering with nutritive metabolism
5. Herbicides affecting cell division
6. Microherbicides.

Sooner or later, weeds acquire resistance to the existing herbicide and thus the continued development of new and potent drugs for controlling them is required. However, this development of new herbicides must always take into consideration the problem of environmental pollution.

As a solid foothold for designing unique herbicides, extensive investigations on the mode of action of herbicides have been carried out. Efforts have been directed to designing superior compounds, based on their mode of action. This has led to potent inhibitors of plant-specific enzymes and biosynthetic paths. The ideal herbicide should have potent activity against weeds, minimum toxicity against living things other than plants, high selectivity between crop plants and weeds, and cause no damage to the environment by residual material.

Microorganisms produce a large number of metabolites that possess a broad variety of biological activities and structures, and these are potentially biodegradable. The search for new types of herbicides of microbial origin has a bright future and is making steady progress.

12.2 Herbicidal Microbial Metabolites

In 1976, selective and contact herbicidal activity of herbicidins A and B were reported following an investigation concerning discovery of a new herbicidal antibiotic. Until that time, the herbicidal activities of only a few antibiotics were known. Cycloheximide, a glutarimide-group antibiotic and a typical inhibitor of protein synthesis, showed a nonselective killing effect on mono- and dicotyledonous plants. Anisomycin and toyocamycin inhibited germination of various plant seeds.

Some of the typical herbicidal microbial metabolites that followed the discovery of herbicidin (Terahara et al., 1982) are summarized in Table 12.1 and Figure 12.1; among these, bialaphos has been widely used.

The characteristic feature of herbicidins A and B is their selectivity after spraying on the foliar part of the plant. A potent herbicidal activity against dicotyledonous plants and strong resistance of rice plants, among monocotyledonous plants, was observed. Herbicidin A had a more selective effect between rice plants and others than did herbicidin B (Arai et al., 1976).

Herbimycin A, isolated by Ōmura et al. (1979), showed a potent activity against most mono- and dicotyledonous plants, especially against *Cyperus microiria* Steud. However, rice plants showed a strong resistance to it. This compound was again isolated by Uehara et al. (1985) as an active substance that causes reversion of the morphology of Rous sarcoma virus-infected rat kidney cells.

Bialaphos, L-2-amino-4-[(hydroxy)(methyl)phosphinoyl)]butyryl-L-alanyl-L-alanine, was first isolated as an antibiotic against *Pellicularia sasakii*, a pathogen of the sheath bright disease of rice plants (Kondo et al., 1973). Thereafter, its very broad and potent weed-killing spectrum and rapid biodegradation after application to soil were demonstrated. This oligopeptide, containing phosphorus, has been used for weeding a field before planting, and in open spaces or orchard.

Bialaphos itself does not inhibit glutamine synthetase, but it is metabolized to L-amino-4-[(hydroxy)(methyl)phosphinoyl]butyric acid, phosphinothricin, which does inhibit glutamine synthetase. Phosphinothricin itself was first obtained as an acid hydrolysis product of phosphinothrycyl-alanyl-alanine from *Streptomyces viridochromogenes* Tu494 (Bayer et al., 1972). D,L-Phosphinothricin was patented as a herbicide by Hoechst (Rupp et al., 1977).

Table 12.1 Typical herbicidal microbial metabolites

Compound	Source	Activity	Reference
Herbicidins A and B	*Streptomyces saganonensis*	37.5 ppm; *Polygonum hydropiper*; damage, 95%–100%	Arai et al., 1976
	S. saganonensis	150 ppm; *P. hydropiper*; damage, 80%–95%	Ōmura et al., 1979
Herbimycins A and B	*S. hygroscopicus* AM-3672	12.5 ppm; *Cyperus microiria* Steud; damage, 90%–100%	Ōmura et al., 1979
Bialaphos (atibiotic SF-1239)	*S. hygroscopicus*	Field test; 9–16 g (32% liquid formulation) are^{-1}; application for common annual weeds	Tachibana and Kaneko, 1986
Phosalacine	*Kitasatosporia phosalacinea*	10 ppm; *Medicago sativa* L.; complete kill	Ōmura et al., 1984a
Oxetin	*Streptomyces* sp. OM-2317	125 ppm; *M. sastiva*; low activity	Ōmura et al., 1984b
Antibiotic SF-2494	*S. mirabilis*	*Digitaria ciliaris*; potent as bialaphos. *Polygonum lapathifolium*; more potent than bialaphos	Iwata et al., 1987
Antibiotic No. 6241-B	*Streptomyces* No.6241	500 ppm; *Punicum crus-galli*; remarkable activity	Kida, 1989
Homoalanosine	*S. galilaeus*	Plowed field test; 2.5 g are^{-1} *Polygonum persicaria*; complete kill	Fushimi et al., 1989
Phthoxazolin	*Streptomyces* sp. OM-5714	100 ppm; *Raphanus sativus*; complete inhibition, more potent than bialaphos	Ōmura et al., 1990
Hydantocidin	*S. hygroscopicus*		Nakajima et al., 1991

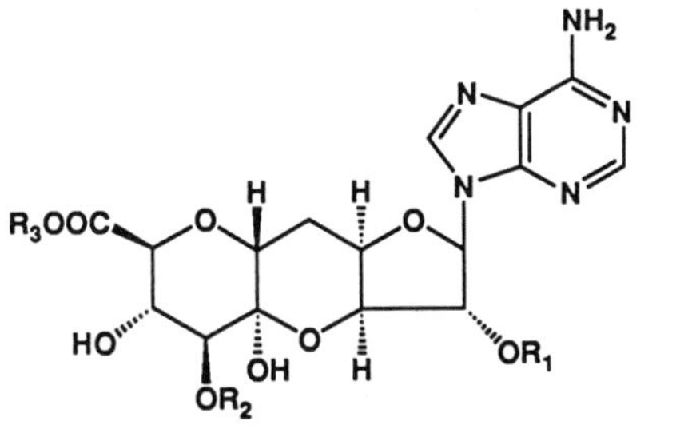

	R_1	R_2	R_3
Herbicidin A	CH_3	CH_3	$COC(CH_2OH)CH{=}CH_3$
Herbicidin B	CH_3	CH_3	H

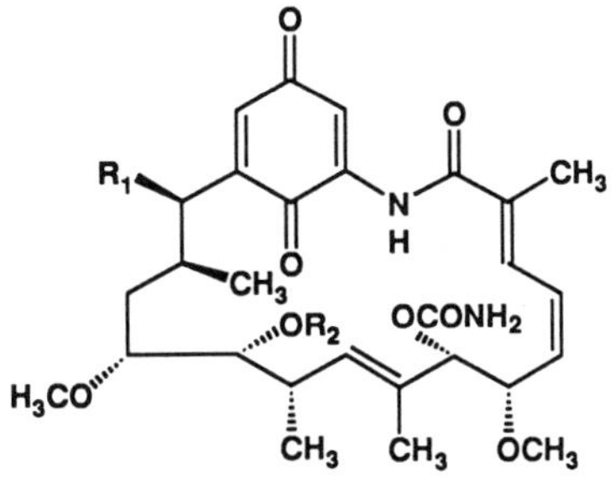

	R_1	R_2
Herbimycin A	OCH_3	CH_3
Herbimycin B	H	H

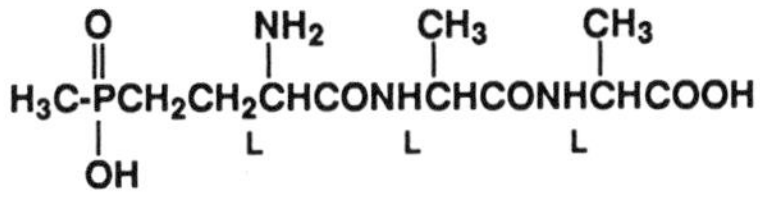

Bialaphos

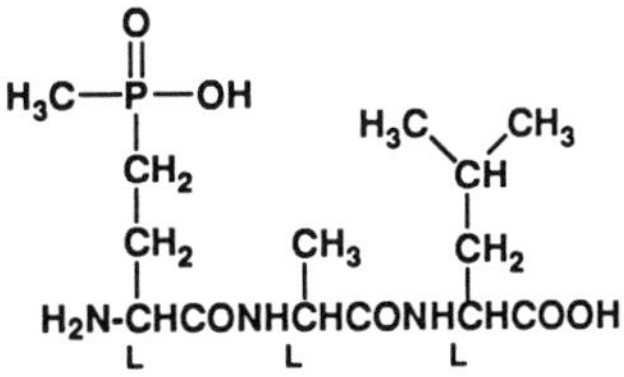

Phosalacine

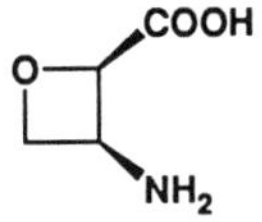

Oxetin

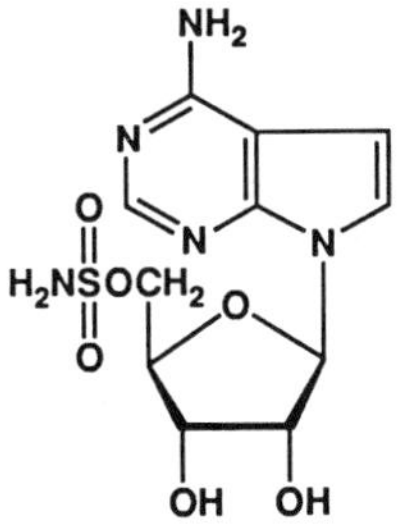

Antibiotic SF-2494

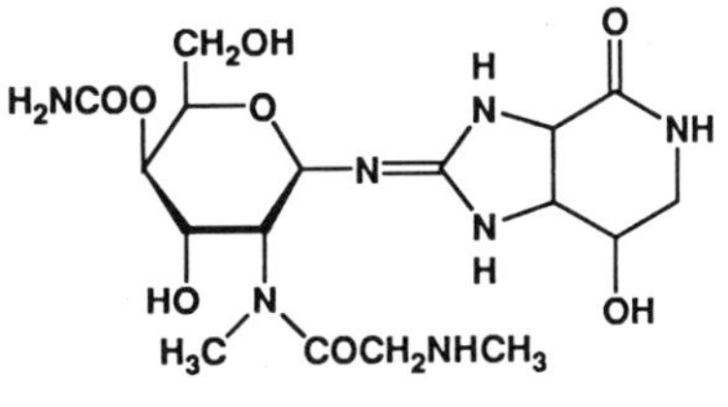

Antibiotic No 6241-B

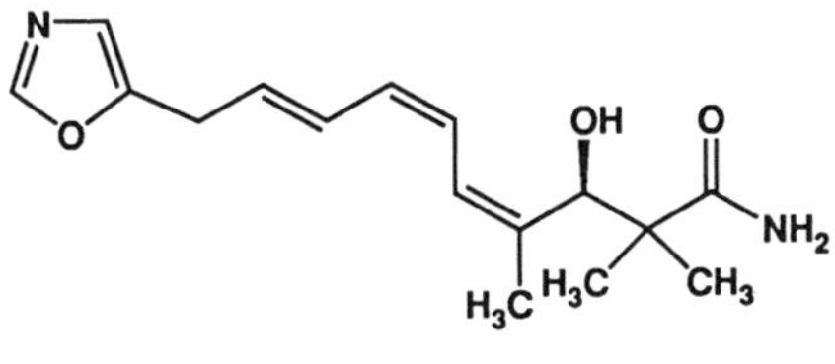

Phthoxazolin

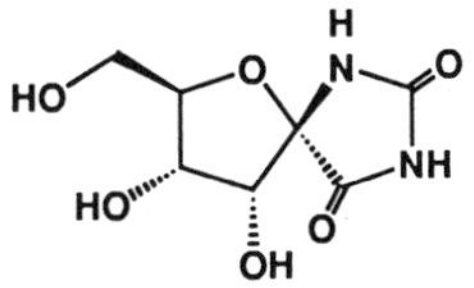

Hydantocidin

Figure 12.1 Structures of typical herbicidal microbial metabolites.

Tachibana et al. (1986) proposed that a primary factor in inducing herbicidal activity was not a decrease in the endogeneous glutamine content of plants by bialaphos but toxicity of the ammonia accumulated in treated plants. This was proposed because of the close correlation between ammonia content and herbicidal activity. However, Krieg et al. (1990) recently demonstrated that addition of as much as 40 times the standard level of ammonium nitrate to the culture media had no effect on alfalfa petiole-derived callus growth, although glutamine synthetase activity was inhibited by 50% and endogenous ammonia increased to 27-fold. They proposed that ammonia accumulation might not be the primary cause of cell death in alfalfa after exposure to phosphnothricin.

Phosalacine, containing the phosphinothricin moiety, was discovered while screening for glutamine antagonists (Ōmura et al., 1984a). This compound was metabolized in plants and then showed inhibition of glutamine synthetase activity as in the case of bialaphos.

Oxetin, the first natural product possessing an oxetane ring, was obtained by the same screening procedure as in the case of phosalacine (Ōmura et al. 1984b). This antibiotic itself inhibited (noncompetitively) glutamine synthetases from *Bacillus subtilis* and spinach leaves.

Antibiotic SF-2494, 5'-O-sulfamoyltubercidin, exhibited an activity as potent as that of bialaphos against crabgrass (*Digitaria ciliaris*) and more potent than that of bialaphos against pale smartweed (*Polygonum lapathifolium*) by foliar treatment (Iwata et al., 1987). The substitution of sulfamoyl group gave the herbicidal activity.

The streptothricin-like antibiotic, No 6241-B, inhibited de novo starch synthesis at 1.0 ppm but not the photosynthesis- dependent oxygen evolution of alga at 100 ppm (Kida, 1989). This antibiotic showed remarkable activity against barnyard millet (*Panicum crus-galli*) but only slight activity against rice plants at concentrations over 500 ppm. The selectivity may be caused by differing permeability with each plant.

Homoalanosine, L-2-amino-4-nitrosohydroxyaminobutyric acid, showed herbicidal activity especially against common cocklebur (*Xanthium strumarium*), velvetleaf (*Abutilon theophrasti*), and lady's thumb (*Polygonum persicaria*) in a plowed field (Fushimi et al. 1989). In a paddy field, this compound inhibited various weeds but not rice. The appearance of activity was very slow and the antibiotic was considered to be translocated in the plant symplastically, because this compound damaged buds and roots rather than the treated leaves. The racemate of this antibiotic was synthesized as a homolog of alanosine, and its insecticidal activity was also reported.

Phytophthora parasitica is a unique fungus containing cellulose as one constituent of its cell wall. Phthoxazolin was discovered by screening for selective antifungal activity against *P. parasitica* but not against common fungi lacking cellulose (Ōmura et al., 1990).

Hydantocidin, reported recently as a herbicidal antibiotic, showed more potent activity than bialaphos against mono- and dicotyledonous plants, both annual and perennial (Nakajima et al., 1991).

A number of phytotoxic metabolites from plant pathogenic microorganisms have been isolated and considered to be very attractive as lead compounds for herbicides of new types (Kenfield et al., 1988). However no practical developments along this line have yet been reported.

On the other hand, the spores of host-specific pathogens have successfully been used as microherbicides. Among these: DeVine (chlamydospores of *Phytophthora palmivora)* for control of strangler vine in Florida citrus groves and College (conidia, blastospore, and fission spore of *Colletotrichum gloeosporioides*) for control of northern joint-vetch *(Aeschynomene virginica)* in Arkansas rice and soybean fields (Stowell et al., 1989).

In its course to becoming a new herbicide on the market, a candidate selected by prescreening has to pass the following tests for practical utility: (1) selectivity for target and application processes; (2) a potency test both in greenhouse and field; (3) a hazard test against life other than the target weed; and (4) safety evaluation and quantitative analysis for the residual herbicide in crop or field after application.

12.3 Screening Methods

A survey of soil microorganisms for herbicidal activity (Heisey et al., 1985; Mishra et al., 1988) Bioassay with garden cress (*Lepidium sativum* L.) and barnyard grass (*Echinochloa crus-galli* L. Beauv.)

Methods Samples tested were applied to filter paper disks as aliquots of broths or in an organic solvent and air-dried. Twenty seeds each of cress and barnyard grass were placed on these filter paper disks in sterile petri dishes (1.5 × 6 cm), moistened with 1.5 ml of distilled water, and incubated at 25°C in the dark for 72 h to examine inhibition of germination and radicle growth compared to a control sample. Distilled water was used on control disks because noninoculated culture broth itself had an effect on radicle growth. Activity grades: "C," 100% inhibition of germination. "P," more than 80% inhibition of germination, seedlings dramatically smaller than controls, and chlorophyll development markedly reduced.

Secondary screening on weeds grown in flats (18 × 12 × 5 cm) in greenhouse Culture broths graded as "C" and "P" were subjected to secondary screening with barnyard grass, redroot pigweed, vervet leaf, purslane, large crabgrass, proso millet, and green foxtail. Broths were homogenized, diluted 1:1 or 1:9 with distilled water (vol/vol) and added with surfactant at 0.1% (vol/vol). The test solution of 7.5 ml was evenly sprayed on 10-day-old plants. The symptoms of plant injury were monitored up to 14 days.

Screening of microbial products affecting plant metabolism (Kida, 1989). The traditional pre- and postemergence spray test on plants is reliable but requires days. The compound tested must penetrate the plant and then be translocated to the site of action. There is some risk that an active substance may not be detected.

Kida (1989) developed a simple and sensitive assay system using blue green-algae and an excised leaf segment of barnyard millet. The active compounds isolated by these screening systems are pereniporin, which inhibits the root elongation of lettuce, from the culture of *Perenniporia medullaepanis* Aj 834; 7-deoxy-D-glycero-D-glucoheptose, a greening-inhibitor of dark-grown green algae, produced by *Streptomyces purpeofusclus* No 381; streptothricin-like antibiotics No. 6241-A and 6241-B, from the culture broth of *Streptomyces* sp. No. 6241 (antibiotic No. 6241-B was identified as SF-701).

Screening of inhibitors of greening of dark-grown *Scenedesmus obliquus* C-2A' *Scenedesmus* cells require light for formation of δ-aminolevulinic acid and chlorophyll.

Methods The culture medium was a basal medium, supplemented with 0.5% glucose and 0.25% yeast extract. 0.808 g KNO_3, 0.46 g NaCl, 0.358 g $Na_2HPO_4 \cdot 12H_2O$, 0.468 g $NaH_2PO_4 \cdot 2H_2O$, 0.015 g $CaCl_2 \cdot 2H_2O$, 0.246 g $MgSO_4 \cdot 7H_2O$, and 1 ml Arnon's microelement solution, per liter.

The algal cells were cultivated in this medium in the dark at 27°C for 6 days and washed twice with the same medium. A cell suspension, 3 mg of algae on a dry basis in 1.8 ml of medium, was poured into plastic petri dishes (1 cm in diameter) containing 0.2 ml of test solution. After illumination at an intensity of 14,000 lux for 18 h at 27°C, the newly synthesized pigments were extracted from algal cells with hot methanol and chlorophyll was assayed spectrophotometrically at 665 nm.

Detection of oxygen evolution The cells of *Scenedesmus obliquus*, dark grown as described, were washed twice with 50 mM HEPES (*N*-2-hydroxyethylpiperazine-*N*'-ethansulfonic acid) solution (pH 7.2). Into the reaction vessel, wrapped with aluminum foil, were poured 2 ml of cell suspension (about 5 mg of dry weight ml^{-1}) and 0.2 ml of 0.2 m*M* $NaHCO_3$, and then gently stirred (magnetic stirrer) at 27°C. Illumination with a projector lamp (150 W, light intensity about 40,000 lux) started photosynthesis. The rate of photosynthetic oxygen evolution was measured by a membrane-coated oxygen electrode. The photosynthetic ability of the cells used was 220 to 280 nmol oxygen evolved mg^{-1} $cell^{-1}$ h^{-1}. The rates of oxygen evolution was measured before and after adding 0.02 ml of each sample solution tested.

Oxygen evolution was also examined by the cotyledon disk (5 mm in diameter) of the pumpkin (*Cucurbita moschata* Duch. cv. Miyako) at 3000 lux illumination.

Simple bioassay by determining de novo starch synthesis in a leaf segment
Barnyard millet (*Panicum crus-galli*) and Italian ryegrass (*Lolium multiforum* Lam) were germinated and grown in a greenhouse at 25° to 30°C. The pots of plants at the age of a second leaf, on the tenth day after planting, were transferred to a dark chamber and kept at 25°C for 12 h. Part of the second leaf was cut into segments of about 5 mm, showing a negative starch reaction with iodine. Three segments were transferred to plastic petri dishes (1 cm diameter) containing 0.2 ml of test sample and 1.8 ml of reaction medium, 0.01 M potassium phosphate buffer (pH 6.5), and 0.01% (vol/vol) Tween 20 (wetting agent). After illumination at 14,000 lux at 25°C for 16 h, leaf segments were immersed in hot methanol for pigment extraction and then stained with a solution of 0.2% iodine and 2% KI.

A microtest system for serial assay of phytotoxic compounds using photoautotrophic cell suspension cultures of *Chenopodium rubrum* (Thiemann et al., 1989)

Sato et al. (1987) developed a bioassay for photosynthesis-inhibiting herbicides using a photoautotrophic plant cell culture (*Nicotiana tabacum* var. Samsum NN). However, with Sato's procedure, a serial assay, such as screening for herbicidal microbial metabolites, was difficult because of its requirement of time, labor, space, etc.

To increase the applicability of green cells for screening phytotoxic compounds, Thiemann et al. (1989) recently designed a microtest system using *Chenopodium rubrum* cell suspensions on a microtiter dish.

Methods

Cell culture Photoautotrophic *Chenopodium rubrum* cell suspension cultures, prepared according to Husemann and Barz (1977), were used at the early stage of stationary growth. Cells (100 mg of fresh weight in 1.5 ml) were suspended in fresh culture medium.

Loading of microtiter dishes Plastic microtiter dish (8.5 × 12.5 cm): this dish contained 24 wells (3 ml volume each) for cell culture and 15 additional compartments for the CO_2 generation solution.

The following preparations were made under aseptic condition: the cell culture suspensions were mixed with sterile-filtered methanol solutions of the samples tested. The organic solvent in the culture medium was set at 0.25% (vol/vol). The prepared cell suspensions (1.5 ml) were poured in each well and K_2CO_3:$KHCO_3$ buffer (0.75 ml), generating a 2% CO_2 partial pressure in the gas phase, was added in each compartment of the dish. Four microtiter dishes, thus prepared and tightly sealed with parafilm, were incubated on a shaker at 350 × rpm under continuous white light (80 E m^{-2} sec^{-1}) at 25°C.

Using the cell suspensions in the wells, which were removed from the microtiter dish under sterile condition at various time intervals, evaluation of cell growth was made by determining packed cell volume, cell number, chlorophyll content, and production of oxygen. Packed volume was determined by centrifugation (1400 × rpm for 10 min, Sigma 2KD-Centrifuge) of calibrated conical tubes (volume, 2 ml; Fa. Schlee, Witten, FRG). Cell number was determined by resuspending the cells from the packed volume assay, and a 200 µl aliquot was withdrawn. The cells were mixed with 4.8 ml chromic acid (10.5% vol/vol), incubated for 10 min, and counted in a Fuchs-Rosenthal chamber under a microscope. Chlorophyll content was measured photometrically using acetone extracts of the cells.

Production of oxygen by the cells was assayed with a Hansatech (Bachofer, Reutlingen, FRG) equipped with a Clark electrode. After a short preincubation period and determination of dark respiration, the net oxygen production was measured using cells from one well under light intensity of 1050 E m^{-2} sec^{-1}.

Bioassay using *Lemnaceae* plants (Ogawa and Kitamura 1988; Taguchi et al., 1988) Many species of duckweeds provide an excellent test system for studying growth and flowering because of their minute size, rapid growth, and easy manipulation under aseptic conditions. Various hormones, growth inhibitors including herbicidal compounds, and amino acids have been reported to be involved in floral induction in *Lemnaceae*.

Ogawa and Kitamura (1988) selected *Lemnaceae paucicostata* Hegelm (6746) from among 17 duckweeds as the plant most suitable for bioassay, based on its growth speed, uniformity of growth, and sensitivity to two herbicides, Paraquat and DTP. They carried out biological assays of plant-growth regulators, 20 herbicides, and 10 plant hormones or related compounds.

Methods

Stock cultures and experimental cultures As a typical experiment, the case of *Lemna paucicostata* 151 (Taguchi et al., 1988) is cited.

The stock cultures of *L. paucicostata* were cultivated in 100-ml Erlenmeyer flasks containing 40 ml of half-strength Hunter's medium supplemented with 1% sucrose under continuous illumination with daylight-type fluorescent lamps of about 1000 lux at 25°C. The medium was sterilized by autoclaving at 1.2 kg cm^{-2} for 10 min after adjusting to pH 4.7. Experimental cultures with test compounds were grown in 1/10 strength M medium supplemented with 15% sucrose, which was sterilized under the same conditions after adjusting to pH 4.6.

One three-frond colony of *L. paucicostata* 151 7 to 10 days old on the stock culture was transplanted to a 100-ml Erlenmeyer flask containing the experimental medium to which a test compound was added aseptically. The inoculated plants were grown for 7 days under the conditions described earlier.

Evaluation of herbicidal activity according to Ogawa's report (Ogawa and Kitamura 1988). The number of colonies counted by the naked eye was compared with that of a control. The symptoms of damage to the frond, such as bleaching, dwarfing, or crooking, were also scored.

Screening for enzyme inhibitors as a target for herbicides

Screening for glutamine antagonists Phosalacine (Ōmura et al., 1984a) was discovered by this screening method followed by use of the plant seeding assay.

Methods Bacillus subtilis PCI219 was used as the test organism. It was grown at 37 C overnight in Davis' minimal agar [(glucose, 0.2%, $(NH_4)_2SO_4$, 0.1%; KH_2PO_4, 0.2%; K_2HPO_4, 0.7%; sodium citrate, 0.05%; $MgSO_4 \cdot 7H_2O$, 0.01%; agar, 0.8–1%; pH 7.0). Cultures of soil isolates that showed anti-*Bacillus* activity but is antagonized by L-glutamine were reserved for further studies.

Screening for inhibitors of cellulose synthesis (Ōmura et al., 1990) Phthoxazolin was discovered by this screening approach followed by the seedling test.

Methods *Phytophthora parasitica* and *Candida albicans* were used as indicator microbial strains. *Phytophthora parasitica* was grown at 27°C with shaking for 2–3 days in glucose-V_8 medium (glucose, 0.9%; V_8 supernatant, 14 (vol/vol)%; agar, 0.2%; pH 6.5). The V_8 supernatant was prepared by centrifugation of V8 vegetable juice (Suntory, Tokyo) after addition of $CaCO_3$ (3 g) to V_8 vegetable juice (200 ml) with stirring and pH adjusted to 6.5 (Ko, 1978). After gauze-filtration of the culture, the filtrate was used to seed a glucose-V_8 agar medium and incubated at 27°C.

Candida albicans was grown at 27°C for 2 days in GY medium (glucose, 1%; yeast extract, 0.5%; pH 6.0). The culture was used to seed a GY agar medium. As the first step in screening for inhibitors of cellulose synthesis, the cultures of soil microorganisms were examined for their selective growth inhibitory activity against *P. parasitica* but with no activity against *C. albicans*.

The cultures showing the above-described pattern of selective antimicrobial activity were reserved and were then subjected to herbicidal screening. For example, five radish seeds were incubated at 27°C for 3 days under light on culture broth-containing wet cotton pieces in test tubes (2 × 10 cm) that were capped with aluminum.

12.4 Concluding Remarks

Future targets for herbicides suggested by Dodge (1987) are (1) enzyme inhibitors and (2) inhibitors of protective systems and detoxification. For example, herbicides may promote cell damage by activated oxygen species such as superoxide or hydroxy radicals, or act as inhibitors of scavenging enzymes against these toxic species; and (3) photosensitizers and allelochemicals; and (4) herbicide selectivity.

The inhibition of amino acid biosynthesis is well established as a target for herbicidally active compounds (Pillmoor, 1989). Powell (1989) summarized screening techniques for this type of inhibitor. Kishore and Shah (1988) also pointed out four targets: inhibition of biosynthesis of aromatic amino acids, branched-chain amino acids, glutamine, and histidine.

In a different study, the acetyl-CoA carboxylase of weeds was the target of a cyclohexane-1,3-dione type of herbicide. Compounds of this type show particular sensitivity to many members of the *Poaceae* family (Kobek and Lichtenthaler 1990).

The biosynthetic processes unique to a plant, such as those following cyclization of 2,3-oxydosqualene to cycloartenol, are also useful targets. Cell-free systems isolated from higher plants have been utilized for study of the detailed mechanism of inhibitor action (Taton et al., 1989). These enzyme systems may be applicable to new screening approaches. Thus, the knowledge on enzyme inhibition applicable to herbicide screening has been accumulating and the development of screening systems based on this knowledge is expected.

Finally, special emphasis is placed on finding phthoxazolin by the screening system in which a certain microorganism containing cellulose as a cell wall constituent is employed in place of the plant cellulose biosynthesis system. This invention for rapid prescreening is very useful for handing a large number of samples.

References

Arai, M., Naneishi, T., Kitahara, N., Enokita, R., Kawakubo, K. and Kondo, Y. 1976. Herbicidins A and B, two new antibiotics with herbicidal activity. *Journal of Antibiotics* (Tokyo) 30:863–869.

Bayer, E., Gugel, K. H., Hagele, K., Hagenmaier, H., Jessipow, S., Konig, W. A. and Zahner, H. 1972. Stoffwechselprodukt von Mikroorganismen 98. Mitteilung [1] Phosphinothricin und Phosphinothrithyl-Alanyl-Alanin. *Helvetica Chimica Acta* 55:224–239.

Dodge, A. D. 1987. Potential new targets for herbicides. *Journal of Pesticide Science* 20:301–313.

Fushimi, S., Nishikawa, S., Mito, N., Ikemoto, M., Sasaki, M. and Seto, H. 1989. Studies on a new herbicidal antibiotic, homoalanosine. *Journal of Antibiotics* (Tokyo) 42:1370–1378.

Heisey, R. M., DeFrank, J., Putnam, A. R. 1985. A survey of soil microorganisms for herbicidal activity. In: Thompson, A. C. (editor), *The chemistry of Allelopathy* ACS Symposium Series No. 268. pp.387–349. American Chemical Society, Washington D. C.

Iwata, M., Sasaki, T., Iwamatsu, H., Miyadoh, H., Tachibana, K., Matsumoto, K., Shomura, T., Sezaki, M. and Watanabe, T. 1987. A new herbicidal antibiotic, SF 2494

(5'-O-sulfamoyltubercidin) produced by *Streptomyces mirabilis*. *Scientific Reports of Meiji Seika Kaisha* 26:17–22.
Husemann, W. and Barz, W. 1977. Photoautotrophic growth and photosynthesis in cell suspension cultures of *Chenopodium ruburum*. *Physiologia Plantarum* 40:77–81.
Kenfield, D., Bunkers, G., Strobel, G. A.and Sugawara, F. 1988. Potential new herbicide—phytotoxin from plant pathogens. *Weed Technology* 2:519–524.
Kida, T. 1989. Screening of microbial products affecting plant metabolism. In: Demain, A. L., Somkuti, G. A., Hunter-Cevera, J. C. and Rossmoore, H. W. (editors), *Novel Microbial Products for Medicine and Agriculture* pp. 195–202 Elsevier, Amsterdam.
Kishore, G. M. and Shah, D. M. 1988. Amino acid biosynthesis inhibitors as herbicides. *Annual Review of Biochemistry* 57:627–663.
Ko, W. H. 1978. Heterothallic *Phytophthora*: evidence for hormonal regulation of sexual reproduction. *Journal of General Microbiology* 107:15–18.
Kobek, K. and Lichtenthaler, H. K. 1990. Effect of different cyclohexane-1,3-dione derivatives on the *de novo* fatty-acid biosynthesis in solated oat chloroplasts. *Zeitschrift fur Naturforshung* 45 Section C Biosciences: 84–88.
Kondo, Y., Shomura, T., Ogawa, Y., Tsuruoka, T., Watanabe, H., Totsukawa, K., Suzuki, T., Moriyama, C., Yoshida, J., Inouye, S. and Niida, T. 1973. Studies on a new antibiotic SF-1293. I. Isolation and physico-chemical and biological characterization of SF-1293 substance. *Scientific Reports of Meiji Seika Kaisha* 13:34–41.
Krieg, L. C., Walker, M. A., Senaratna, T. and McKersie, B. D. 1990. Growth, ammonia accumulation and glutamine synthetase activity alfalfa (*Medicago sativa* L.) shoots and cell cultures treated with phosphinothricin. *Plant Cell Reports* 9:80–83.
Mishra, S. K., Whitenack, C., Putnum, A. R. 1988. Herbicidal properties of metabolites from several genera of soil microorganisms. *Weed Science* 36:122–126.
Nakajima, M., Itoi, K., Takamatsu, Y., Kinoshita, T., Okazaki, T., Kawakubo, K., Shindo, M., Honma, T., Tohjigamori, M. and Haneishi, T. 1991 Hydantocidin: a new compound with herbicidal activity from *Streptomyces hygroscopicus*. *Journal of Antibiotics* (Tokyo) 44:293–300.
Ogawa, M. and Kitamura, T. 1988. Biological assay of plant growth-regulating compounds using *Lemnaceae* plants. *Sankyo Kenkyusho Nempo* 40:91–99.
Ōmura, S., Nakagawa, A. and Sadakane, N. 1979. Structure of herbimycin, a new ansamycin antibiotic. *Tetrahedron Letters* 44:4323–4326.
Ōmura, S., Murata, M., Hanaki, H., Hinotozawa, K., Oiwa, R. and Tanaka, H. 1984a. Phosalacine, a new herbicidal antibiotic containing phosphinothricin. *Journal of Antibiotics* (Tokyo) 37:829–835.
Ōmura, S., Murata, M., Imamura, N., Iwai, Y. and Tanaka, H. 1984b. Oxetin, a new antimetabolite from actinomycete. Fermentation, isolation, structure and biological activity. *Journal of Antibiotics* (Tokyo) 37:1234–1332.
Ōmura, S., Tanaka, Y., Kanaya, I., Shinose, M. and Takahashi, Y. 1990. Phthoxazolin, a specific inhibitor of cellulose biosynthesis, produced by a strain of *Streptomyces* sp. *Journal of Antibiotics* (Tokyo) 43:1034–1036.
Pillmoor, J. B. 1989. Amino acid biosynthesis—an Aladdin's cave of new pesticide targets. *Prospects for Amino Acid Biosynthesis Inhibitors in Crop Protection and Pharmaceutical Chemistry*, BCPC Monograph No. 42, pp. 23–30. British Crop Protection Council, London.
Powell, K. A. and Rees, B. 1989. Screening techniques for amino acid biosynthesis inhibitors. No. 42, *Prospects for Amino Acid Biosynthesis Inhibitors in Crop Protection and Pharmaceutical Chemistry*, BCPC Monograph NO. 42, pp.31–39.
Rupp, W., Finke, M., Bieringer, H. and Langelueddeke, P. 1977. Ger. Offen. 2,717,440. 1978. Herbicidal composition. *Chemical Abstracts* 88:70494e.
Sato, F., Takeda, S. and Yamada, Y. 1987. A comparison of effects of several herbicides on photoautotrophic, photomixotrophic and heterotrophic cultured tobacco cells and seedings. *Plant Cell Reports* 6:401–404.

Stowell, L. J., Nette, K., Heath, and Shutter, R. 1989. Fermentation alternatives for commercial production of a mycoberbicide. In: Demain, A. L., Somkuti, Hunter-Cevera, J. C., Shutter, R. and Rossmoore, H. W. (editors) *Novel Microbial Products for Medicine and Agriculture*, pp.219–227 Elsevier, Amsterdam.

Tachibana, K. and Kaneko, K. 1986. Development of a new herbicide, Bialaphos. *Journal of Pesticide Science* 11:297–304.

Tachibana, K., Watanabe, T., Sekizawa, Y. and Takematsu, T. 1986. Accumulation of ammonia in plant treated with bialaphos. *Journal of Pesticide Science* 11:33–37.

Taguchi, H., Kashimoto, A. L., Nishitani, H., Shimabayashi, Y. and Iwai, K. 1988. The effects of pyridine and pyrazine carboxylic acids derivatives on the growth of *Lemna paucicostata* 151. *Agricultural and Biological Chemistry* 52:85–89.

Taton, M., Benveniste, P. and Rahier, A. 1989. Microsomal $\Delta^{8,14}$-sterol Δ^{14}-reductase in higher plants. *European Journal of Biochemistry* 185:605–614.

Terahara, A., Haneishi, T., Arai, M., Hata. T., Kuwano, H., and Tamura, C. 1982. The revised structure of herbicidin. *Journal of Antibiotics* (Tokyo) 35:1711–1714.

Thiemann, J., Nieswandt, A. and Barz, B. 1989. A microtest system for the serial assay of phytotoxic compounds using photoautotrophic cell suspension cultures of *Chenopodium ruburum*. *Plant Cell Reports* 8:399–402.

Uehara, Y., Hori, M., Takeuchi, T. and Umezawa, H. 1985. Screening of agents which convert 'Transformed Morphology' of rous sarcoma virus infected rat kidney cells to 'Normal Morphology'. Identification of an active agent as hebimycin and its inhibition of intracellular srs kinase. *Japan Cancer Research* (Gann) 76:672–675.

13

Insecticides, Acaricides, and Anticoccidial Agents

Yoshitake Tanaka and Shigenobu Okuda

13.1 Introduction

For decades, pesticides have been utilized as crop protectants. They are useful for reduction of the harmful actions of field pests so as to increase crop yield and crop quality. The need for safe and excellent pesticides is more serious today than before because of the estimated shortage of food supply in the near future. The world population is estimated to reach 10 billion in the coming twenty-first century, while the area of available farmland is decreasing. Obviously, an increase of food production is one of the most serious problems urgently requiring solution.

The progress in plant engineering and related biotechnology is very rapid, offering hope for a marked improvement of food production. Actually, however, we will have to wait for 10 years or more until genetically engineered insect-tolerant plants are bred and successfully planted in widespread areas or until crops can be produced in systematic plant factories. Pesticides are expected to continue to play their important role as a significant help in crop production.

As for the causative agents, insects and mites are the representative two groups that may damage crop plants. Table 13.1 lists representative examples of crop-damaging insects and mites. A secondary effect of pests is viral, bacterial, and fungal infections, which are facilitated by the damage and injuries to crop plants caused by insects and mites. The soil nematode is an additional factor. Birds (e.g., swallow, crow), and small wild animals (field mice, etc.) may also be harmful but are not considered here.

Insects and mites produce damage and loss of crops in both quantity and quality in various ways. These damages occur during plant growth, postharvest storage, and transportation. The yield loss of crops by nibbling, injury, and growth retardation is estimated to be 20%–30%. An additional 10% is lost during postharvest storage and transportation. Decrease in crop quality causes further economic loss; for example, the sweetness, taste, color, shape,

Table 13.1 Representative crop-damaging insects and mites

Crops	Pest Common name	Scientific name	Spreading area
Rice	Green rice leafhopper	*Nephotettix cincticeps* Uhl.	Asia
	Rice stem borer	*Chilo oryzae*	Asia, Spain, Australia
	Brown (rice) planthopper	*Nilaparvata lugens* Stal	Asai, Australia
	Green vegetable bug	*Nezara viridula* L.	World (tropical)
Corn	Corn aphid	*Aphis maidis*	World
Maize	Asian maize borer	*Ostrinia furnacalis* Gn.	Asia
	Northern corn rootworm	*Diabrotica longicornis* Say	America
Potato	Potato aphid	*Macrosiphum euphorbiae* Ths.	World
	Peach-potato aphid	*Myzus persicae* Sulz.	World
Legumes	Green stink bug	*Acrosternum hilare* Say	Central/South America
	Bean aphid	*Aphis fabae* Scop.	World
	Soybean pod borer	*Leguminivora glycinivorella* Mats.	Asia, Soviet
Wheat	Oat aphid	*Phopalosiphum padi* L.	World
Barley	Green bug	*Schizaphis graminum* Rond.	World
Oats			
Sugar beet	Beet armyworm	*Spodoptera exigua* Hb.	Asia, America, Southern Europe
	Spinach leaf miner	*Pegomya hyoscyami* Panz.	Europe, America
Sugar cane	Armyworm	*Pseudaletia separeate* Wak.	Asia, Australia
	Pine borer	*Sesamia inferens* Wld.	Asia
	Cane aphid	*Longiunguis sacehari* Zhnt.	Africa, Caribbean Middle East

Table 13.1 (cont.)

Crops	Pest		Spreading area
	Common name	Scientific name	
Cotton	Tobacco budworm	Heliothis virescens F.	Central/South America
	Pale legume bug	Lygus elisus Van D.	America
	Armyworm	Spodoptera exigua Hb.	World
	Pink bollworm	Pectinophora gossypiella Saund.	World
	Cotton aphid	Aphis gossypii Glov.	World
Tobacco	Tobacco budworm	Helicoverpa armigera Hb.	World (except America)
	Peach-potato aphid	Myzus persicae Sulz.	World
Tea	Kanzawa spider mite	Tetranychus kanzawa Kishida	Asia
	Smaller tea tortrix	Adoxophyes spp.	
Vegetables	Small white butterfly	Pieris rapae L.	New Zealand, America, Europe, Asia
	Cabbage armyworm	Mamestra brassicae L.	Asia, Europe
	Cabbage moth	Plutella xylostella L.	World
	Peach-potato aphid	Myzus persicae Sulz.	World
Citrus	Citrus red mite	Panonychus citri McGregor	World
	Arrowhead scale	Unaspis yanonensis Kuwana	Japan
Apple	Summer fruit tortrix	Adoxophyes orana F. von Rosl.	World
	Peach fruit moth	Carposina nipponensis Wals.	Asia
	Apple leaf miner	Leucoptera scitella Zell.	Europe
	European red mite	Panonychus ulmi Koch	Europe
Pear	Oriental fruit moth	Grapholitha molesta Busck	Central/South America, Asia
	Two-spotted spider mite	Tetranychus urticae Koch	World
Peach	Peach leaf minor	Lyonetia clerkella L.	
Grape	Tea thrips	Scirtothrips dorsalis Hood	

nutritional, and water content are reduced as a consequence of the activities of live pests on crops.

Pesticides are useful in protecting crops from these harmful actions of insects and mites and from such economic losses. In fact, the cost effect of pesticides in rice production (net increase in rice production on the value base/cost of pesticides) was about 7–8 during 1980–1984 in Japan.

The world market of pesticides has increased to a value of 1.1 billion dollars in 1960, 3.6 billion in 1970, 13.3 billion in 1982, and 17.9 billion in 1986, according to the Wood Mackenzie report, a well-known annual report on the world market of agrochemical and veterinary products and on agro-industry information, reported by CNWM (County NatWest Securities Limited incorporating Wood Mackenzie & Co. Ltd.) Edinburgh, United Kingdom. The U.S. market is the largest today, accounting for about 25%, followed by about 15% (300–400 billion yen) in Japan. Of all pesticides, insecticide production (40%–45%) is largest in value although decreasing gradually; fungicides (25%–30%) and herbicides (20%–25%) follow.

Progress in developing of agricultural pesticides has not been without difficulties. The most serious problems have concerned toxicity to humans and animals, particularly in early times, and environmental side effects from the accumulation of pesticide chemicals in soils. Today it can be safely said that the efforts to solve these problems have largely succeeded, but not completely. This is exemplified by decreases in the amounts of pesticides applied in agricultural fields: 10 kg ha^{-1} for pentachlorophenol, an old herbicide, and 10–20 g ha^{-1} for chlorosulfuron, a newer herbicide. The selective toxicity index, that is, the LD$_{50}$ value (mg kg^{-1}) for animals/ED$_{50}$ value (mg kg^{-1}) for the target insect, increased correspondingly: 4 and 60 for parathion and DDT, both old insecticides, respectively, and greater than 1,700,000 for methoprene, a newer insect growth regulator. The continual efforts made in screening and chemical synthesis toward finding newer compounds have brought about this improvement.

This chapter presents methods of screening for insecticides, acaricides, and anticoccidial agents of microbial origin. Emphasis is placed on the concept that screening method is one of the key factors in the development of excellent agrochemicals.

13.2 Synthetic Insecticides

In the early 1920s, pest control was attempted using natural raw materials, for example, oils such as kerosene and whale oil against planthoppers, pyrethrum extract (containing pyrethroids, as was revealed later), delice extract (rotenone), tobacco leaf extract (nicotine), etc. Lead arsenate, copper sulfate (Bordeaux mixtures), and other minerals were also relied upon.

After discovery of the potent pesticidal activity of DDT [(1,1,1-trichloro-2,2-bis(p-chlorophenyl)ethane] in 1939, DDT and its derivatives became the major reagents for pest control. The effect was profound. For example, rice production in Japan increased to 3.9 tons ha^{-1} in 1955 after the introduction of DDT and γ-BHC (γ-benzene hexachloride), as compared to 3.2 tons ha^{-1} in 1950. An estimated loss of rice yield correspondingly decreased from 500 to 200 kg ha^{-1}. In spite of the excellent crop-protecting efficacy, the use of DDT and its derivatives (γ-BHC and others) was discontinued in the 1960s and by the early 1970s because of their accumulation in the environment and related toxicity to warm-blooded animals.

In the 1950s, synthetic organophosphorus type (parathion, diazinon, sumithion, etc.) and carbamate type pesticides (carbaryl, fenobucarb) and cartap were introduced into field applications. It is worth mentioning that the carbamates were derivatized with the natural products physostigmine and nereistoxin as models. All three types of synthetic pesticides show activity by inhibition of choline esterase, which results in the interference of signal transduction at the cholinergic synapse with choline as the neurotransmitter.

Early derivatives of the organophosphates and carbamates were associated with high toxicity. Those which were introduced into field trials were advanced derivatives with high activity and low toxicity. A problem has evolved with the emergence of resistant strains. Current efforts are therefore directed toward the synthesis and discovery of newer compounds active also against resistant strains (Georghiou, 1990).

13.3 Microbial Metabolites as Pesticides

Antibiotics had achieved a great success as chemotherapeutics early in the 1950s. This success encouraged attempts to evaluate many antibiotics as pesticides. Many of them were found to show potent pesticidal activities under laboratory conditions and in preliminary field trials. Table 13.2 lists these antibiotics.

Despite their potent pesticidal activity in preliminary trials, these antibiotics did not achieve agricultural usefulness, either because of accompanying toxicity to animals, humans, or crop plants or because of instability under weathering conditions, or, more importantly, because their use was not cost-effective. Actually, they were not effective enough as pesticides when applied at the rates comparable to their production costs. As described in Chapter 11, some antibiotics such as cycloheximide, chloramphenicol, and tetracycline were employed as agricultural antimicrobial agents. The efforts to find useful microbial metabolites were continued, but progress was rather slow (Misato et al., 1983, Isono, 1990).

It was in the 1970s that the microbial metabolite tetranactin was discovered and later proved to be the first useful pesticides. Today, five microbial products are being used as pesticides in agriculture: tetranactin, milbemycin, destomycin, hygromycin B, and avermectin (ivermectin). Avermectin and its derivative ivermectin, hygromycin B and its analog destomycin, and their

Table 13.2 Pesticidal activities of antibiotics[1]

Antibiotic	Producing microorganism	Pesticidal activity[2]	Anti-microbial activity[2]	Other activity
Amphotericin B	Streptomyces nodosus	I,M	F	
Antimycin	S. kitasawaensis	I,M,N	F,B	Respiratory inhibition
Aureothin	S. thioluteus	I,M,N	F	Antitumor
Beauvericin	Beauveria bassiana	I	F	
Bleomycin	S. verticillus	I	B	Antitumor
Colistin A	Bacillus colistinus	N	B	
Cycloheximide	S. griseus	M	F	
Destomycin	S. rimofaciens	I	B	
Gramicidin S	B. brevis	N	B	
Moroyamycin	Streptomyces sp.	M	B	
Neoaureothin	Streptoverticillium orinoci	N	F	Herbicidal
Patulin	Penicillium urticae	I	F	Mycotoxin
Piericidins A and B	Streptomyces mobaraensis	I,M,N	F	Inhibition of NADH oxidase
Polymyxin B$_1$	B. polymyxa	N	B	
Siccanin	Helminthosporium siccans	N	F	
Streptovitacin	S. griseus	I,M	B	Antitumor

[1]From Isono (1990); Otoguro et al. (1988); Fabre et al. (1988).
[2]Abbreviations: I, insecticidal; M, miticidal; N, nematicidal; F, antifungal; B, antibacterial.

antiparasitic activities were described in Chapter 4. The properties of three compounds are described briefly.

Tetranactin Tetranactin, produced by *Streptomyces aureus*, was discovered by Ando et al. (1971) in screening for insecticides. It is currently used in agriculture as a miticide. Tetranactin is a member of the family of macrotrolide, cyclic tetramers of furan carboxylic acid with hydroxylated substituents (Figure 13.1), which include tetranactin, trinactin, and nonactin (Haneda et al., 1974). Tetranactin shows antimicrobial activity against bacteria, yeasts, and filamentous fungi and is potently active against the azukibean weevil and spider mites, but has low acute toxicity to animals (Sagawa et al., 1972). It forms a metal complex, and is assumed to interfere the membrane function of mites. Resistant mites have emerged at a low rate; about 5% of a population became resistant after more than 5 years of use, and the resistant strains showed only a twofold higher minimal inhibitory concentration (MIC) value as compared with sensitive strains.

Milbemycin The milbemycins are a family of 20 components. They possess 16-membered, fused macrolide structures very similar to the aglycone of the avermectins (see Figure 13.1). They were discovered in screening acaricides from the cultured broth of *S. hygroscopicus* subsp. *aureolacvimosus* (Mishima, 1983).

Milbemycins show no antimicrobial activity but do show potent acaricidal and anthelmintic activity with some toxicity. Among the many components, milbemycin D has been approved and commercialized as an agricultural acaricide. It inhibits neurotransmission at the GABA receptor, which is the same site as reported for avermectins (Fisher and Mrozik, 1984).

Avermectins Avermectins, a family of fused, 16-membered macrolide produced by *S. avermitilis*, was discovered by a joint research effort of the Kitasato Institute, Japan and Merck Sharp and Dohme Research Laboratory, United States. The potent anthelmintic activity and usefulness of ivermectin as an antiparasite were described in Chapter 4.

Avermectin is being developed for use in crop protection. Because of its potent activity against acarines and a wide range of insects (Table 13.3) it is very likely that avermectin will be introduced in agriculture and horticulture (Dybas, 1989). The avermectin preparation formulated for agricultural use has a low acute toxicity (LD_{50} value): 650 and more than 2000 mg kg^{-1} on rats in oral and skin doses, respectively. Only slight phytotoxicity has been observed in a very few plants (ferns, carnation).

A series of avermectin derivatives have been synthesized chemically. Among these, 4'-*epi*-methylamino-4'-deoxyavermectin B1 was reported to show 1500 times greater potency against a variety of armyworms than did the mother compound (Dybas et al., 1989).

Tetranactin

Milbemycin D

Avermectin B$_{1a}$

Hygromycin B

Destomycin B

13.4 Screening Methods

General method: Inhibition of growth A conventional method for screening of insecticides and acaricides is to examine cultured broths for their ability to inhibit the growth of an indicator pest under laboratory conditions using larvae or adults. Reported examples are summarized in Table 13.4 and Figure 13.2. This is convenient when the test organism is rapid-growing, easy to culture, and detectable by microscope at low magnification or by the naked eye. Toxic compounds are also picked up, but can be eliminated in later steps.

Because cultivation conditions are different with individual pests and whether they are adults or larvae, the methods conventionally employed in insecticidal or miticidal assays are described here. When other methods such as the enzyme assay are employed as a means of the first selection of candidates, the methods described here are useful as the second or later screens.

Mosquito The mosquito is a member of the dipterous group of insects, which also includes the fly, horsefly, and 75,000 other kinds of insects. Adult and larva mosquitoes (*Culex pipiens*) are used as the test organism in many screenings of insecticidal activity.

Eggs are obtained in a small cup of water from adults 4 days after ad lib sucking of animal blood, e.g. mice. Eggs can be stored for week in the cold.

Table 13.3 Activity of avermectin against insect larvae, adult mites, and aphids

Mite species (contact effect against adult mites)	
Phyllocoptruta oleivora Ashmead (citrus rust mite)	0.02[1]
Tetranychus urticae Koch (two-spotted spider mite)	0.03
Panonychus ulmi Koch (European red mite)	0.04
Polyphagotarsonemus latus Banks (broad mite)	0.05
Tetranychus turkestani Ugarov & Nikolski (strawberry spider mite)	0.08
Tetranychus pacificus McGregor (Pacific spider mite)	0.16
Panonychus citri McGregor (citrus red mite)	0.24
Insect Species (foliar residue bioassay)	
Manduca sexta L. (tobacco hornworm)	0.02
Leptinotarsa decemlineata Say (Colorado potato beetle)	0.03
Heliothis virescens F. (tobacco budworm)	0.10
Epilachna varivestis Mulsant (Mexican bean beetle)	0.40
Trichoplusia ni Huebner (cabbage looper)	1.0
Heliothis zea Boddie (cotton bollworm)	1.5
Spodoptera eridania Cramer (southern armyworm)	6.0
Spodoptera frugiperda J.E. Smith (fall armyworm)	25.0

[1] Mortality assessed 72–96 h after exposure. All data are LC_{90} in ppm.
From Dybas, R.A. and Green, A. St. J. 1984. *Avermectins: Their Chemistry and Pesticidal Activity*, Proceedings, 1984 British Crop Protection Conference Pests and Diseases, Brighton, England 31:947–954. See also Dybas, 1989.

Figure 13.1 Agricultural microbial metabolites currently marketed.

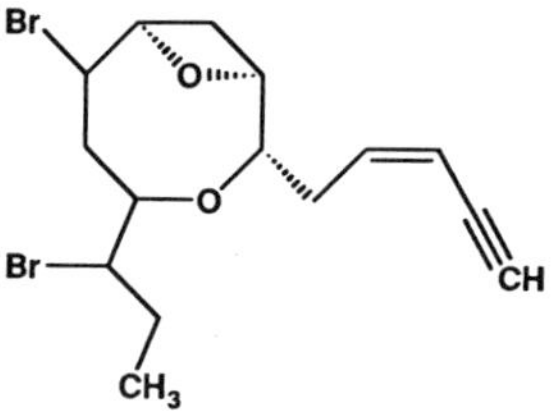

Z-Laureatin

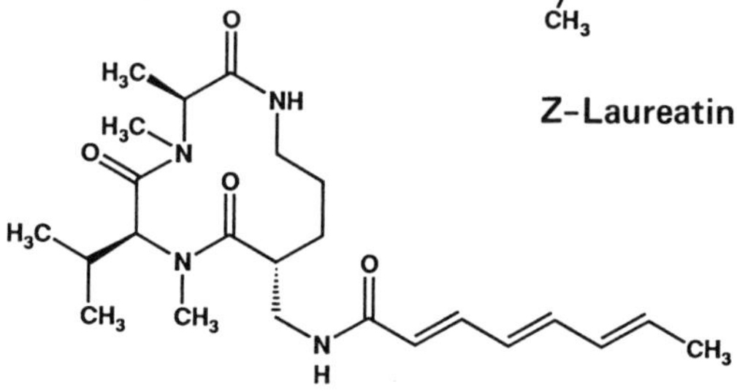

Aspochracin

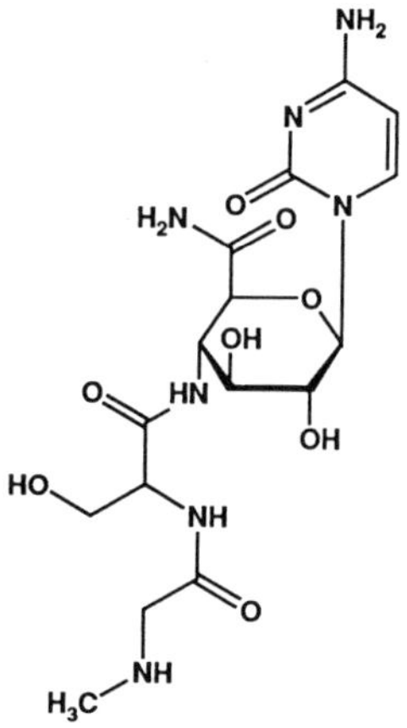

Aspiculamycin

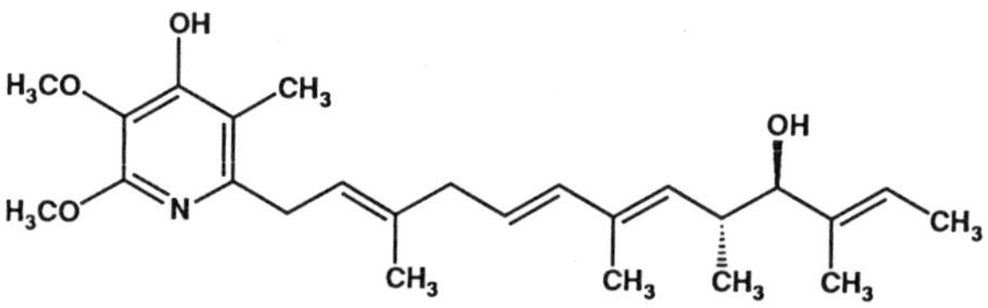

Piericidin A

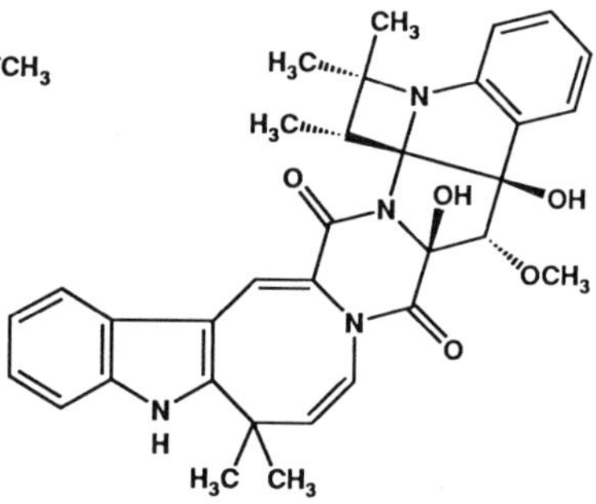

Okaramine B

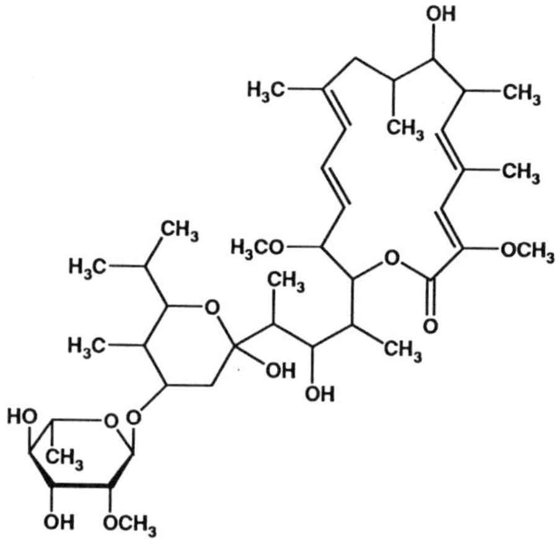

Leucanicidin

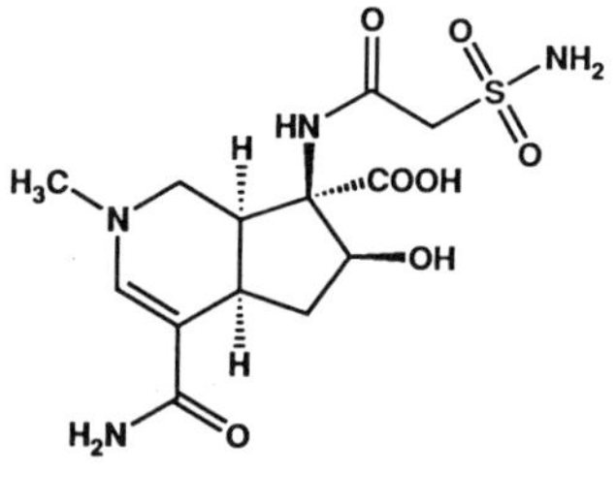

Altemicidin

For the culture of eggs, light-dark control is required. When 500 eggs are suspended in 3-day-old tap water in a vessel (35 × 25 × 6 cm) and incubated at 25°C or lower at a relative humidity of 90%, they hatch into larvae at a high rate. For better propagation of larvae, a small amount of dried yeast cells may be given as diet. Slow air bubbling and a supplement of a β-lactam antibiotic (50–100 μg ml^{-1}) are useful for suppression of bacterial contamination. Otherwise, a thin film may be formed on the surface of water. Under laboratory conditions, larvae grow up through first to fourth instar stages within 10–14 days. Water is renewed on every moulting, at 3–5 days. After the last instar stage, followed by pupae, they become adult mosquitoes. Adult mosquitoes are maintained on 3% sugar-wet cotton pieces in a cage.

Larvae of the second to third instar are convenient for use in screening. Usually 10 larvae are poured into test tubes or microplates using a pipet with a wide tip, mixed with a microbial culture broth, and incubated at 25°C. Larvacidal activity is measured after 1–2 days of incubation. During this incubation, no diet is necessary.

Azukibean weevil The azukibean weevil (*Callosobruchus chinensis* L.) is a member of the coleopterous group of insects, which includes the rice weevil, northern corn rootworm, gold beetle, ladybug, and 250,000 other kinds of insects. They are harmful to corn, citrus, and vegetables both before and after harvest. *C. chinensis*, together with mosquito larvae, was used as a test organism in the discovery of tetranactin.

Azukibean (500 g) is put in a 1-liter plastic pot. Adult weevils (500 worms) are then added, allowed to lay eggs at 25°–30°C at a relative humidity of 70%–80%. After 10 days, adult weevils are removed by sieving. Then azukibeans and eggs are incubated under the same conditions. Larvae, after hatching out, grow inside the beans for 3 weeks, and after a month, adult weevils appear outside the beans. They are separated from residual beans by sieving and are ready for use in insecticidal assays.

About 10–30 worms are placed on a filter paper in a plastic plate (9 cm in diameter). An active microbial culture supernatants is dripped onto the filter paper. Insecticidal activity is determined after 1–2 days.

The rice weevil (*Sitophilus zeamais* M.) is cultivated in a similar manner with rice particles as diet.

Armyworm Armyworms such as the beet armyworm and rice armyworm are members of the lepidopterous group of insects, which include the pink bollworm (cotton), common cutworm (cotton, corn, rice), rice stem borer (rice), oriental corn borer (corn) and 120,000 other kinds of insects. They are spread worldwide on cotton, rice, corn, fruits, and vegetables. The rice armyworm

◂
Figure 13.2 Pesticidal microbial metabolites discovered by various screening methods.

Table 13.4 Pesticidal compounds discovered using insects, mites, or nematodes as test organisms

Test organism	Pesticidal compound discovered	Pesticidal activity	Insecticide-producing microoganism	Known compound detected	Reference
Fly (*Musca domestica*)		I		Avermectin Actinomycin Antimycin Piericidin Valinomycin	Fabre et al., 1988
Mosquito (*Culex* sp.)	Aculeximycin	I	*Streptosporangium albidum*		Ikemoto et al., 1983 Murata et al., 1989
Mosquito (*Culex pipiens pallens*)		I	*Laurencia nipponica* (red alga)	Deoxyprepacifenol Z-laureatin Z-Isolaureatin	Watanabe et al., 1989
Silkworm larvae (*Bombix mori*)	Aspiculamycin	I,N,M	*Streptomyces toyocaensis*		Haneishi et al., 1974
	Aspochracin	I	*Aspergillus ochraceus*		Myokei et al., 1969
	Okaramine	I	*Penicillium simplicissimum*	Anthranylic acid	Hayashi et al., 1989
	Piericidin	I	*S. mobaraensis*		Tamura et al., 1963

Table 13.4 *(cont.)*

Test organism	Pesticidal compound discovered	Pesticidal activity	Insecticide-producing microorganism	Known compound detected	Reference
Armyworm (*Leucania separata*)	Leucanicidin	I	*S. halstedii*	Alanosin	Isogai et al., 1984
Azukibean weevil (*Callosobruchu chinensis*)	Tetranactin	M	*S. aureus*		Ando et al., 1971
Pine wood nematode (*Bursaphelenchus lignicolus*)	Jietacin	N	*Streptomyces* sp.	Piericidin Aureothin Neoaureothin Leucanicidin	Otoguro et al., 1988
Brine shrimp	Altemicidin	I	*S. sioyaensis*		Takahashi et al., 1989
Two-spotted spider mite (*Tetranychus urticae*)	Milbemycin	M	*S. hygroscopicus*		Mishima, 1983

[1] Abbreviations: I, insecticidal; M, miticidal; N, nematicidal. For structures, see Figures 13.1 and 13.2.

(*Pseudaletia separata* Wlk., probably synonymous with *Leucania separata* and *Mythimna separata*) was used as a test insect in the discovery of leucanicidin.

Adult armyworms (20 pairs) are grown in a steel net cage (27 × 23 × 40 cm) in which are put cotton pieces wet with 10% sugar solution. Eggs laid on the surface of multiply folded, paraffin-coated papers are collected into a petri dish. When eggs are incubated at 25°C with humidity, they hatch out and larvae grow with an artificial diet, which is better crushed into small pieces. At the late second-instar stage (2–3 cm in body length), each 5 larvae are separated and placed into a 200-ml-volume plastic cup, where they are allowed to grow under light-dark control until ready for use in insecticidal assays.

Microbial samples to be tested are dripped onto solid pieces of an artificial diet. Insecticidal activity is observed after 1–3 days.

Pest worms of related taxa, such as the rice cutworm (*Spodoptera litura*) and corn borer (*Ostrinia furnacalis, O. nubilalis*) are also cultivated on an appropriate artificial diet. However, the brown planthopper (*Nilaparvala lugens* Stal) and green rice leafhopper (*Nephotettix cincticeps* Uhler), both of which belong to other groups of important harmful insects, are maintained with rice, which is their natural habititat; no artificial diets are available.

Two-spotted spider mite Two groups of plantphagous mites, *Tetranychus* and *Panonychus*, are important as pests. No artificial diets are available for them. They are cultivated on their favorite plants under laboratory conditions: French beans (kidney beans) and lemon fruits. The two-spotted spider mite (*Tetranychus urticae*), grown on kidney bean leaves, was used in the discovery of milbemycin.

Conventionally, spider mites are allowed to lay eggs on kidney bean leaves for 1 day. Mites are removed with a wet soft brush, and the leaves are then cut into small pieces (2 × 2 cm) and placed on a wet filter paper. The leaf disks and filter papers are placed on a wet polyurethane sponge in a petri dish. A test sample is put on the leaf disk; alternatively, the leaf disk is dipped into a sample solution, air dried, and put back on the filter paper. The leaf discs are incubated at 25°–30°C under light-dark control with the relative humidity of 60%–80%. Miticidal activity is examined within a week through an inverted microscope.

Brine shrimp The brine shrimp is a free-living marine organism convenient for culture, although it has been employed in only a few cases as the test organism in screening of bioactive microbial metabolites. The merits of the use of this organism are as follows: (1) eggs are available commercially; (2) eggs are stored for months at room temperature; (3) they hatch out into larvae (1–2 mm in body length) in an artificial seawater at a high rate within 1–2 days; and (4) dried yeast cells, a commercial product, can be used as diet. It appears that brine shrimp will be employed more frequently if data are accumulated on the sensitivity of this organism to known insecticides and microbial metabolites. Takahashi et al. (1989) employed brine shrimp as the test organism and discovered altemicidin, a sesquiterpene that is very rare as an actinomycete metabolite. Their method is described next.

Brine shrimp eggs (100 mg, Warner Lambert, USA) were hatched at 27°C in 200 ml of synthetic seawater with forced aeration. After 1 day, swimming larvae were collected in a portion of artificial seawater to give 40–60 larvae ml^{-1}. An aliquot (20–30 larvae) was mixed with a cultured filtrate and incubated in a 24-well microplate for 48 h. The insecticidal activity was determined by measuring the mortality rate.

Nematode Free-living nematodes are convenient as test organisms used for screening of nematocidal compounds. They are also useful in the search for insecticidal and acaricidal compounds. They are used because they are easy to culture, although nematodes and insects or mites are far from each other in taxomic position. Other merits of the use of nematodes may be that insecticides and acaricides with a broad spectrum of activity are detected. Conventional methods for *Caenorhabditis elegans* and *Bursaphelenchus lignicolus* are described.

Usually, an *Escherichia coli* strain is grown at 37°C overnight on an agar slant containing a nutrient medium. *C. elegans* is inoculated onto the lawn of *E. coli* cells and allowed to grow at 20°–27°C. The worms are rinsed off into a physiological saline or a buffer (pH 7) and are then ready for use.

For the growth of *B. lignicolus* (pine wood nematode), a fungus, *Botrytis cinerea* is grown at 27°C on potato-glucose agar medium. The pine wood nematode is grown on the lawn of *B. cinerea* for about 10 days. Nematodes are rinsed into distilled water ($\approx$500 nematodes ml^{-1}) and are ready for use (Otoguro et al., 1988; see also Chapter 4).

Insecticide-resistant organisms Insecticide-resistant insects or mites are unique and convenient sources of test organisms. They are applicable to screening for nonclassical and novel types of pesticides. These compounds are useful for the control of resistant pests. The successful use of resistant strains in screening studies has not been reported, but the isolation of intriguing resistant strains has been described. Ivermectin-resistant *C. elegans* was obtained by laboratory selection (Schaeffer and Hanes, 1989). The IC$_{50}$ values of the wild (sensitive) and resistant strains of *C. elegans* were 3 and 75 nM, respectively. A variety of *Haemonchus contortus*, which is resistant to ivermectin (Egerton et al., 1988), and of *Trichostrongylus colubriformis*, resistant to ivermectin and benzimidazole (Shoop et al., 1990) were isolated. The latter two are common nematodes in sheep gastrointestinal tract.

Enzyme inhibition

Chitinase inhibitor Chitin, a linear β-1,4-polymer of *N*-acetylglucosamine, is a main component of cuticle in insects. Chitin synthesis is active in the course of ecdysis and is assumed to be regulated by chitinase as one mechanism. It is reasoned that the inhibition of either chitin synthase or chitinase results in the blockade of chitin turnover and finally in abnormal molting. This notion was supported by the finding that the chitin synthase inhibitor nikkomycin inhibited the growth of insects and mites, although this compound is not effective enough to use in agricultural fields. According to this reasoning, inhibitors of chitinase were searched for from microbial metabolites by Sakuda et al. (1987a). These authors found allosamidin, a water-soluble chitobiose derivative, from about 1700 cultures. Allosamidin inhibits endochitinase from the silkworm, but does not do so for those from yeasts and bacteria, for example, a commercial chitinase from *Streptomyces griseus*. When injected into the body of the last instar of silkworm, allosamidin caused the death of the insect.

Method of chitinase assay (Sakuda et al., 1987b)   The reaction mixture in a total volume of 2.0 ml contains 0.1 ml of crude chitinase from silkworm in 0.05 M citrate buffer (pH 5.0), 0.9 ml of 0.1 M citrate buffer (pH 5.0), 0.8 ml of sample solution or water, and 0.2 ml of methanol solution (25 mg ml^{-1}) of the chromogenic substrate chitin azure. After incubation at 35°C overnight, the blue color of the supernatant was measured against a no-inhibitor control.

Chitin is involved as a cell wall constituent of fungi. In this connection, it is of interest that demethylallosamidin induced the abnormality of the cell division of the yeast *Saccharomyces cerevisiae* (Sakuda et al., 1990).

Trehalase inhibitor   Trehalose, 2,2'-glucosylglucose, is widely distributed among most insects, microorganisms, plants, and animals. In some insects and microorganisms, trehalase is required for sugar transport, as a storage material, and as a source of readily available energy for insect flight (e.g., grasshopper). Trehalase probably participates in these steps directly or indirectly, and the inhibitors of trehalase activity are of interest as insecticides or insect growth (or behavior) regulators.

Validoxylamin A, a product of validamycin metabolism, is a potent inhibitor of trehalase from the tobacco cutworm (*Spodoptera litura*) (Asano et al., 1990). When injected into the instar larvae, validoxylamin A induced abnormal morphology, followed by the death of the insect.

Attempts were made to screen for trehalase inhibitors of microbial origin. Murao et al. (1991) isolated deoxynojirimycin, a known inhibitor of α-glucosidase. It inhibited the trehalases from a fungus and rabbit. The activity on insect enzyme was not reported. The same authors recently reported the discovery of trehalostatin, $C_{13}H_{22}O_{10}N_2$, molecular weight of 366, in the cultured broth of an actinomycete, *Amycolatopsis trehalostaticus*. This compound inhibited the trehalase from blowfly (*Aldrichina graham*) at a dose 3800 fold lower than from the fungus *Chaetomium* and 24 fold lower than did validamycin A.

Method of trehalase assay (Murao et al., 1991). The reaction mixture contained, in a total volume of 500 μl,: 50 μl of blowfly trehalase (640 μg ml^{-1} in 0.05 M phosphate buffer, pH 5.5), 50 μl of inhibitor in the same buffer, and 400 μl of the substrate trehalose (5 mM). After incubation at 37°C for 10 min, residual trehalase activity was estimated by measuring the resultant glucose by a glucose oxidase method.

[^{3}H]Ivermectin binding assay A [^{3}H]ivermectin binding assay system was established and used for studies of the mechanism of action of ivermectin (Shaeffer and Hanes 1989). The free-living nematode *C. elegans* was used as the receptor source because this nematode is extremely sensitive to this class of anthelmintic agents. This assay method is well applicable, although it has not yet been reported, to the screening of nematicides as well as insecticides and acaricides.

Schaeffer et al. (1990) isolated chochlioquinone A as a potent nematicide from a fermentation broth of *Helminthosporium sativum*. Using the [^{3}H]ivermectin binding assay, these authors found that chochlioquinone A was a competitive inhibitor of ivermectin binding (K_i, 30 μM). They suggested that chochlioquinone A and ivermectin might have a common mode of action, involving multiple binding sites. One of the binding sites might be independent of the α-aminobutyrate-chloride ion channel complex. A supporting observation for this hypothesis is that chochlioquinone A is less active against ivermectin-resistant *C. elegans*. The other nonavermectin type of anthelmintic agents tested (levamizol and paraherquamide, 1 mM at the maximum level) did not inhibit ivermectin binding.

Binding assay (Schaeffer et al., 1990) [^{3}H]Ivermectin (51.9 Ci mmol^{-1}) was prepared by catalytic hydrogenation of avermectin components at C-22, 23 positions with tritium gas. *C. elegans* was grown on *E. coli* cells on an agar slant and rinsed into 0.05 M HEPES buffer (pH 7.4). Membrane fraction was prepared by homogenization of the worms, followed by centrifugation. A pellet fraction obtained after a 20-min centrifugation at 28,000 $\times$ g was used as the receptor source (12.5 μg of protein ml^{-1}).

The membrane preparation (1.0 ml) was incubated with [^{3}H]ivermectin, and with or without an inhibitor, at 22°C for 45 min. A control run was carried out in the presence of a 500-fold molar excess of cold ivermectin. After incubation, the reaction mixture was filtered over Whatman GF/B filters containing 0.25% Triton X-100. The filters were then measured for radioactivity.

13.5 Anticoccidial Agents

Coccidiosis Poultry, in its veterinary or agricultural sense, includes not only chickens but also turkeys, ducks, and other avian species. Poultry production is often associated with agricultural activities. Thus, anticoccidial agents required in poultry production are also described here.

The economic importance of poultry in the total agricultural output varies among countries. In many countries, however, poultry production is increasing more and more, and is being carried out in a more regimented manner. In other words, it is becoming industrialized. It produces protein, either as eggs or meat, in a relatively high efficiency from raw cereals, thus accounting for the importance of this technology.

Coccidiosis is spread worldwide (Hofstad et al., 1984). It is one of the high-risk groups of poultry diseases that include bacterial (e.g., mycoplasmosis), fungal (aspergillosis), and viral (Newcastle disease) infections and tumors such as Marek's disease. Nowadays birds are grown in heavily populated houses, and an infectious disease once having emerged develops rapidly, so that early diagnosis is difficult (Onaga, et al., 1986). As a consequence, only a few reliable data are available on the extent of disease occurrence and distribution and the resulting economic effect. The mortality rate of coccidiosis is estimated to be 5%–10% in chicken. An approximation of the total economic losses from infectious diseases is 10%–20% in poultry production (Gordon and Jordan, 1982).

Coccidiosis in chickens, turkeys, geese, ducks, etc. is caused by a group of parasitic protozoa of intestinal epithelium. The major characteristics of the parasite follow (Long, 1982):

1. Those protozoa pathogenic in poultry belong almost entirely to the genus *Eimeria*. A rare exception is *Tyzzaria perniciosa*, which is most pathogenic in ducks. Members of the genus *Eimeria* require only one host for the completion of their life cycle.

2. The pathogens show strict host specificity. Seven *Eimeria* species are pathogenic in chickens, while nine other species of *Eimeria* are pathogenic only in turkeys. Coccidiosis is less serious in turkeys than in chickens.

3. They show species specificity in parasitizing sites of the intestine.

Since 1936, a number of synthetic compounds have been introduced into poultry, mainly for prophylactic uses. Among them are sulfonamides, quinolines, nitrofurans, carbanilides, etc. In the early 1950s, microbial metabolites such as tetracyclines joined them. A good success of coccidiosis control was achieved in the early 1970s by the introduction of polyether ionophores such as monensin, and lasalocid, followed by salinomycin, narasin, and maduramicin (Figure 13.3) (McDougald, 1982). They are useful as anticoccidial agents and as growth promoters.

The cidal effect of monensin on extracellular sporozoites is caused by the capability of the ionophore to act as a transmembrane cation carrier. This is exerted in an invasive stage of *Eimeria* (Long and Jeffers, 1982). The growth-promoting effect results from its antibacterial activity, which induces a change in the intestinal flora, shifting the microbial production of propionate and other volatile fatty acids in favor of chicken growth.

Polyether-resistant *Eimeria* spp. have emerged and increased. Therefore, many attempts have been carried out in the search for new anticoccidial agents that are more potent than the already commercialized anticoccidials or are active against resistant strains. New polyethers and nonionophore types of anticoccidial agents have been reported. Those showing promising activity are listed in Table 13.5; their structures are shown in Figure 13.3. Most of these were discovered using the conventional antimicrobial techiques.

CP-84657, recently isolated from *Actinomadura* sp. (Dirlam et al., 1990), is as potent as maduramicin, the most potent ionophore among those of current use. Kijimycin produced by a strain of *Actinomadura* sp. is superior to monensin and salinomycin (Takahashi et al., 1990). Among nonpolyethers, simaomicin α, a product of *Actinomadura* sp. (Taikwang et al., 1989), appears to be the most potent microbial anticoccidial ever reported. This is a polycyclic aromatic compound containing a xanthone unit, showing a broad anticoccidial spectrum. It is efficacious against chicken coccidia when fed at 1 g ton^{-1} of feed.

Screening method for anticoccidial agents Screening methods and evaluation of anticoccidial activity have been reviewed (Ryley, 1980). Ryley noted that (1) in vitro screens with a chick embryo system have a risk of false-negatives, and compounds active in vivo may be overlooked because of low activity in this system; (2) in vitro screens with a tissue culture system are superior to an embryo system because its sensitivity is one order of magnitude higher; and (3) the chick kidney cell screen is not satisfactory in evaluating the anti-*Eimeria* activity of unknown compounds but may be reliable for in vitro evaluation of known active compounds.

One of the merits of in vitro assay systems is the small quantity of compounds required for assay. Therefore, cultured cell systems are described here that were used for growth of *Eimeria* spp. Although these systems have limitations, they must be useful as in vitro screens in the search for anticoccidial agents of microbial origin to have been reported.

Table 13.5 Anticoccidial microbial products with practical usefulness or with feasible activity

Compound	Microbial source	Reference
Polyether		
Monensin	*Streptomyces cinnamonensis*	
Lasalocid	*S. lasaliensis*	
Salinomycin	*S. albus*	
Maduramycin	*Actinomadura rubra*	
CP-84,657	*Actinomadura* sp.	Dirlam et al., 1990
Kijimycin	*Actinomadura* sp.	Takahashi et al., 1990
Nonpolyether		
WS-5995	*S. auranticolor*	Ikushima et al., 1980
Frenolicin B	*S. roseofulvus*	Ōmura et al., 1985
Simaomicin	*A. madurae*	Taikwang et al., 1989

For structures, see Figure 13.3.

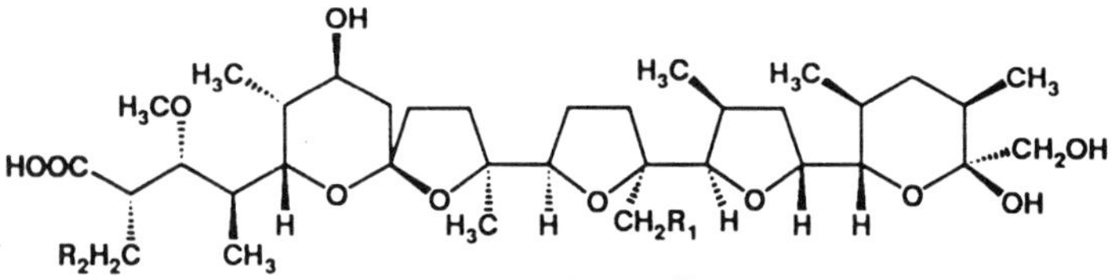

	R_1	R_2
Monensin A	CH_3	H
Monensin B	H	H
Monensin C	CH_3	CH_3

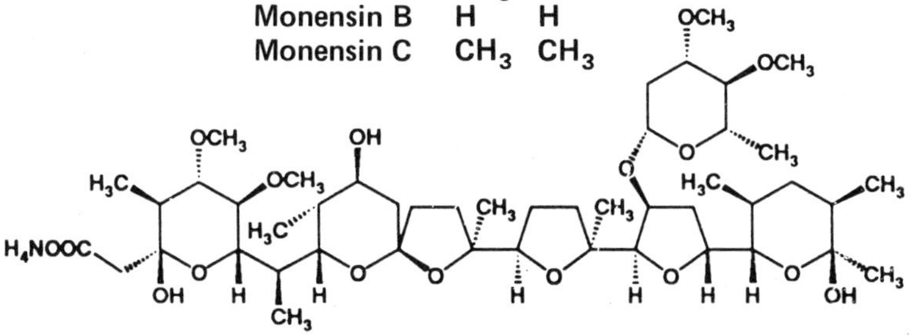

Maduramicin

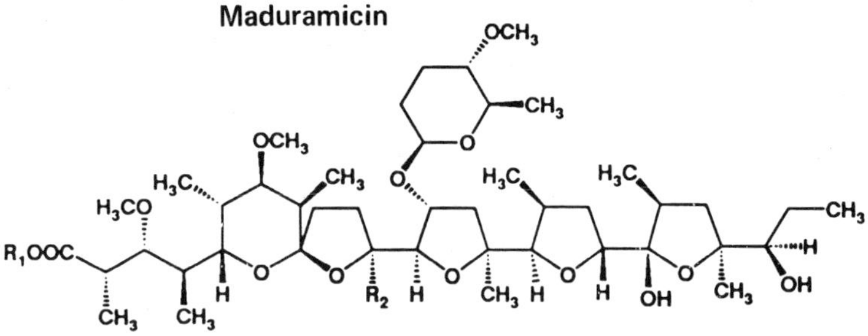

	R_1	R_2
CP-84657	H	CH_3
Portmicin	H	H

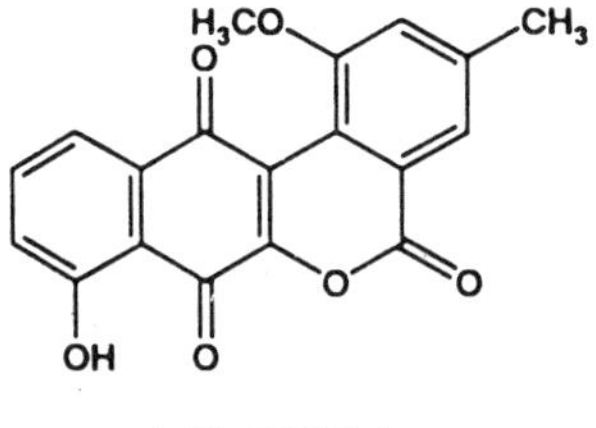

WS-5995A

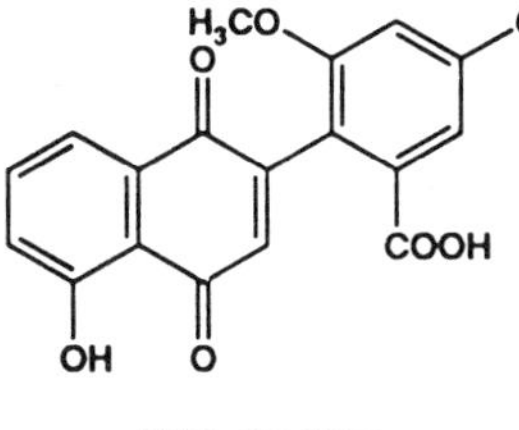

WS-5995B

Figure 13.3 Microbial metabolites with anticoccidial activity.

Chick kidney cell system (Ryley and Wilson, 1976; Ryley, 1980) Sterilized sporozoites were prepared from oocysts of *Eimeria tenella* that were obtained from infected chickens.

Chick kidney cells were grown in a medium consisting of Hanks saline

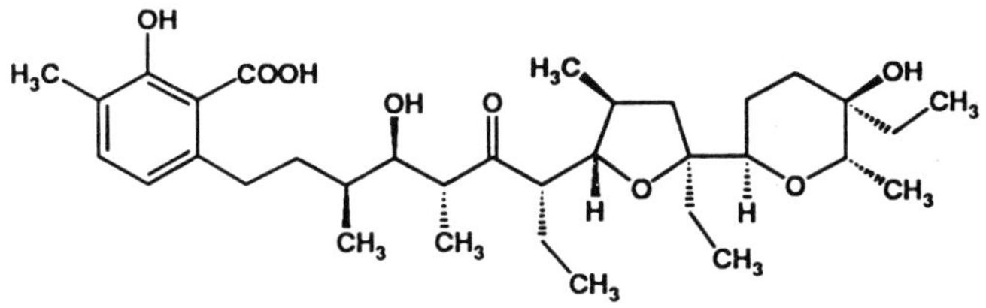

Chapter 13 Insecticides, Acaricides, and Anticoccidial Agents 257

Lasalocid

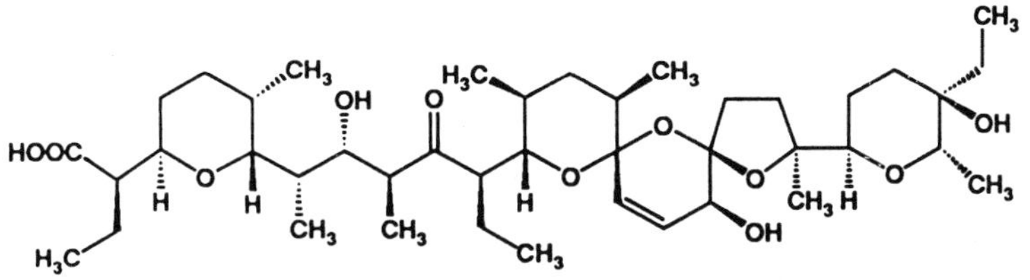

Salinomycin

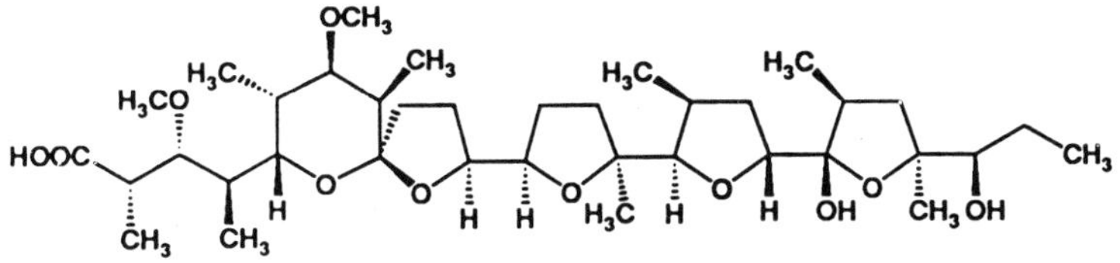

Kijimicin

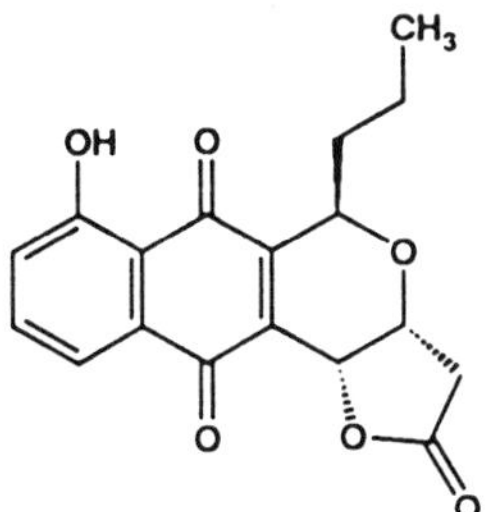

Frenolicin B

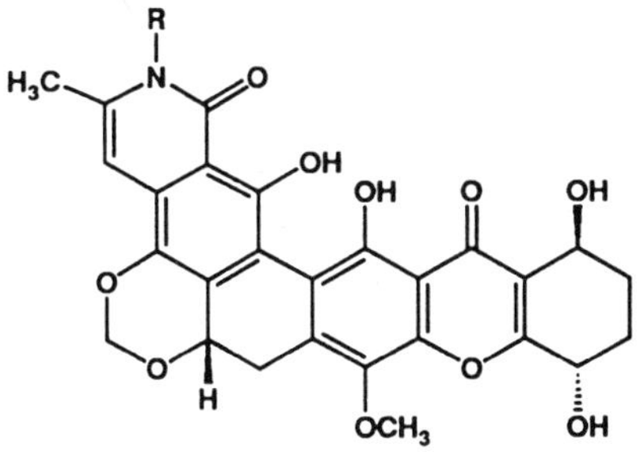

	R
Simaomicin α	CH$_3$
Simaomicin β	H

Figure 13.3 *(cont.)*

with 0.125% lactalbumin hydrolysate, 0.125% NaHCO$_3$, and 5% fetal calf serum (FCS) in an atmosphere of 5% CO$_2$:95% air at 37°C.

In testing in this cell system, the chick kidney cell cultures were grown in 25-welled plastic plates to produce a semiconfluent monolayer within 2 or 3 days; the medium then was drained off. A test sample (10 mg) was dissolved in 0.7 ml dimethylsulfoxide, and 3 ml culture medium was added to give a very fine suspension. This sample solution was used to prepare a series of threefold dilution in a final volume of 1.5 ml. Portions (0.5 ml) were added to wells containing the cell monolayer. Then, 0.5 ml of culture medium containing 1–1.5 × 10^6 sporozoites of *E. tenella* was added to each well.

The plates were incubated at 41°C for 4 days, the stage of the second schizogony, in an atmosphere of 5% CO_2:95% air. After discarding the medium, the monolayer was fixed with methanol before staining with 0.1% aqueous toluidine blue. The plates were examined with an inverted microscope.

Primary kidney cell system (Hofmann and Raether, 1990) Hofmann and Raether, (1990) have reported an improved new technique for the in vitro cultivation of *E. tenella* in primary chick kidney cells. In this system protozoa could be reproduced from sporozoites to oocysts; however the yield of oocysts was not very high.

Primary kidney cells (PKC) of 1- to 4 week-old chickens grown in culture flasks were inoculated with ultrapure sporozoites of *E. tenella*-Hoechst and *E. tenella*-Houghton. The sporozoites were purified by anion exchange chromatography, on an Econo-Column loaded with DEAE cellulose, from the mixture of sporozoites, sporocysts, and wall fragments. The 24-h parasite-free adaptation phase of the PKC in Williams E medium, plus 10% FCS and then the 168-h parasite-containing maintenance phase in medium 199 (Biochrom, Berlin), plus 2.5% FCS, established monolayers that enabled the routine development of all schizont generations as well as young oocysts.

BHK 21 cell system An established cell line is more convenient for a screening system. H. Onaga, of the Nippon Institute of Biological Science, Tokyo, established an in vitro system for *E. tenella* growth using BHK 21 cells, which was an established cell line from hamster kidney cells (Onaga and Ishii, 1978). This system supports *E. tenella* growth up to the first generation of schizogony.

For maintenance culture, a growth medium consisting of Eagle's MEM medium (Gibco), 10% FCS, 0.3% tryptose-phosphate broth (Difco), 0.07% $NaHCO_3$, 100 U of penicillin G ml^{-1}, and 100 µg of streptomycin ml^{-1}, was used.

BHK 21 cells (clone 13, Dainippon Pharmaceuticals Co., Tokyo, Japan) were grown in a plastic (or glass) bottle containing this growth medium at 37°C to confluent growth under an atmosphere of 5% CO_2:95% air. Medium was drained off, and cells were washed with phosphate-buffered saline [PBS(-)]. To the cell sheet, 0.3% trypsin solution in PBS(-) was added. After 1 min, the trypsin solution was removed and the bottle was allowed to stand upside down. After 1–2 min, growth medium was added to obtain a cell suspension (10^5 cells ml^{-1}).

For anticoccidial assays, the cells (1 ml) were poured into 24-welled plastic plates and incubated at 37°C for 4 days to give (semi)confluent cell growth. Culture medium was then drained off.

Sporozoites of *E. tenella* were suspended (1–5 × 10^5 ml^{-1}) in a modified growth medium. This medium contained no serum. A portion (1 ml) of this sporozoite suspension was added to the cell sheet and incubated at 41°C for 3 h, during which time sporozoites invaded the cells. The medium was discarded; serum-free growth medium with or without test samples was added and incubated further for 1–3 days.

After discarding the medium, the monolayer was stained with dyes and examined through an inverted microscope. Anticoccidial activity was based on the absence of schizonts without any cytotoxicity.

13.6 Concluding Remarks

The progress of pesticides have been made starting with excellent synthetic lead compounds or with active natural products, with the following criteria: potency in pesticidal activity, toxicity to animals and human, degradability in agricultural environments, and specificity in mode of action. Recently, efforts have focused on bioregulators rather than the traditional biocides. A chitinase inhibitor has appeared along this line.

From the foregoing point of view, microbial metabolites are good sources of potent and safe pest-controlling agents and their lead compounds. Additional properties required today for pest-controlling agents include: (1) stability under sunlight and weathering conditions; (2) less tendency to develop resistant strains; and (3) target-specific activity so as not to cause unexpected effects on non-target insects or mites. Microbial metabolites are likely to satisfy these requirements, because they are vast in number, extremely varied in activity and structure.

Finally, the importance of cooperation is mentioned. Because screening of every type of activity is a multidisciplinary science, the cooperation of researchers of different fields is beneficial. For those who have been engaged in the screening of antibacterial and antifungal agents, or for those who are specializing in developing the currently increasing pharmacologically active microbial metabolites, the knowledge and technique required for pesticide or anticoccidial screening are so different that they might feel that these agents are out of their scope and territory. Inversely, biologists in plant, insect, or parasite science are not always familiar with screening techniques. The very fact that this is true for both sides is one of the underlying factors making the progress in agrochemicals rather slow.

It is emphasized here that if insect biologists, plant physiologists, parasitologists, microbiologists, and chemists cooperate closely and over the long term, it will certainly be of great use for the discovery and development of potent, safe, and environmentally soft pesticides and anticoccidial agents.

The authors are grateful to Dr. Hiroshi Yoshida, Ageo Research Laboratory, Nippon Kayaku, Co., Saitama, Japan, for his valuable suggestions and discussions.

References

Ando, K., Oishi, H., Hirano, S., Okutomi, T., Suzuki, K., Okazaki, H., Sawada, M. and Sagawa, T. 1971. Tetranactin, a new miticidal antibiotic. I. Isolation, characterization and properties of tetranactin. *Journal of Antibiotics* (Tokyo) 24:347–352.

Asano, N., Takeuchi, M., Kameda, Y., Matsui, K. and Kono, Y. 1990. Trehalase inhibitors, validoxylamine A and related compounds as insecticides. *Journal of Antibiotics* (Tokyo) 43:722–726.

Dirlam, J. P., Belton, A. M., Bordner, B. J., Cullen, W. P., Huang, L. H., Kojima, Y., Maeda, H., Nishida, H., Nishiyama, S., Oscarson, J. R., Ricketts, A. P., Sakakibara, T., Tone, J. and Tsukuda, K. 1990. CP-84657, a potent polyether anticoccidial related to portimicin and produced by *Actinomadura* sp. *Journal of Antibiotics* (Tokyo) 43:668–679.

Dybas, R. A., 1989. Abamectin use in crop protection. In: Campbell, W. C. (editor), *Ivermectin and Abamectin*, pp. 288–308. Springer-Verlag, New York.

Dybas, R. A., Hilton, N. J., Babu, J. R., Presier, F. A. and Dolce, G. R., 1989. Novel second-generation avermectin insecticides and miticides for crop protection. In: Demain, A. L. et al. (editors), *Novel Microbial Products for Medicine and Agriculture*, pp. 203–212. Society for Industrial Microbiology, Washington, D.C.

Egerton, J. R., Suhayda, D. and Eary, C. H. 1988. Laboratory selection of *Haemonchus contortus* for resistance to ivermectin. *Journal of Parasitology* 74:614–617.

Fabre, B., Armau, E., Etienne, G., Legendere, F. and Tiraby, G. 1988. A simple screening method for insecticidal substances from actinomycetes. *Journal of Antibiotics* (Tokyo) 41:212–219.

Fisher, M. H. and Mrozik, H. 1984. The avermectin family of macrolide-like antibiotics. In: Ōmura, S. (editor), *Macrolide Antibiotics: Chemistry, Biology, and Practice*, pp. 553–606. Academic Press, Orland.

Georghiou, G. P., 1990. Overview of insecticide resistance. In: Green, M. B. et al. (editors), *Managing Resistance To Agrochemicals: From Fundamental Research To Practical Strategies*, pp. 18–41. ACS Symposium Series 421, American Chemical Society, Washington, D.C.

Gordon, R. F., Jordan, F. T. W., 1982. The poultry industry and the effect of diseases. In: Gordon, K. F. and Jordan, F. T. W. (editors), *Poultry Diseases*, pp. 1–8. Bailliere Tindall, London.

Haneda, M., Nawata, Y., Hayashi, T. and Ando, K., 1974. Tetranactin, a new miticidal antibiotic. VI. Determination of dinactin, trinactin, and tetranactin in their mixtures by NMR spectroscopy. *Journal of Antibiotics* (Tokyo) 27:555–557.

Haneishi, T., Arai, M., Kitano, N. and Yamamoto, S. 1974. Aspiculamycin, a new cytosine nucleoside antibiotic. III. Biological activities *in vitro* and *in vivo*. *Journal of Antibiotics* (Tokyo) 27:339–342.

Hayashi, H., Takiuchi, K., Murao, S. and Arai, M. 1989. Structure and insecticidal activity of new indole alkaloids, okaramines A and B, from *Penicillium simplicissimum* AK-40. *Agricultural and Biological Chemistry* 53:461–469.

Hofmann, J. and Raether, W. 1990. Improved techniques for the *in vitro* cultivation of *Eimeria tenella* in primary chick kidney cells. *Parasitology Research* 76:479–486.

Hofstad, M. S., Barnes, H. J., Calnek, B. W., Reid, W.M, and Yoder, H. W. Jr. (editors), 1984. *Diseases of Poultry*, 8th Ed. Iowa State University Press, Ames.

Ikemoto, T., Katayama, T., Shiraishi, A. and Haneishi, T. 1983. Aculeximycin, a new

antibiotic from *Streptosporangium albidum*. II. Isolation, physicochemical and biological properties. *Journal of Antibiotics* (Tokyo) 36:1097–1100.
Ikushima, H., Iguchi, E., Kohsaka, M., Aoki, H. and Imanaka, H. 1980. *Streptomyces auranticolor* sp. nov., a new anticoccidial antibiotic producer. *Journal of Antibiotics* (Tokyo) 33:1103–1106.
Isogai, A., Sakuda, S., Matsumoto, S., Ogura, M., Furihata, K., Seto, H. and Suzuki, A. 1984. The structure of leucanicidin, a novel insecticide produced by *Streptomyces halstedii*. *Agricultural and Biological Chemistry* 48:1379–1381.
Isono, K. 1990. Antibiotics as non-pollution agricultural pesticides. In: Gordon, B. et al. (editors), *Comments in Agricultural and Food Chemistry*, Vol. 2., pp. 123–142. Science Publishers, London.
Long, P. L. 1982. Parasitic diseases. In: Gordon, J. F. and Jordan, F. T. W. (editors), *Poultry Diseases*, pp. 166–197. Bailliere Tindall, London.
Long, P. L. and Jeffers, T. K., 1982. Studies on the stage of action of ionophorous antibiotics against *Eimeria*. *Journal of Parasitology* 68:363–371.
McDougald, L. R. 1982. Chemotherapy of coccidiosis. In: Long, P. L. (editor), *The Biology of the Coccidia*, pp. 373–427. University Park Press, Baltimore.
Misato, T., Ko, K. and Yamaguchi, I. 1983. Use of antibiotics in agriculture. In: Takahashi, N. et al. (editors), *Pesticide Chemistry: Human Welfare and the Environment*, Vol. 2, *Natural Products*, pp. 53–88. Pergamon Press, Oxford.
Mishima, H. 1983. Milbemycins, a family of macrolide antibiotics with insecticidal activity. In: Takahashi, N. et al. (editors), *Pesticide Chemistry: Human Welfare and the Environment*, Vol. 2, *Natural Products*, pp. 129–134. Pergamon Press, Oxford.
Murao, S., Sakai, T., Gibo, H., Nakayama, T. and Shin, T. 1991. A novel trehalase inhibitor, trehalostatin, produced by *Amycolatopsis trehalostatica*. *Agricultural and Biological Chemistry* 55:895–897.
Murata, H., Kojima, N., Harada, K. and Suzuki, M. 1989. Structure elucidation of aculeximycin. I. Further purification of glycosidic bond cleavage of aculeximycin. *Journal of Antibiotics* (Tokyo) 42:691–700.
Myokei, R., Sakurai, A., Chang, C.-F., Kodaira, Y., Takahashi, N. and Tamura, G. 1969. A new insecticidal metabolite of *Aspergillus ochraceus*. Part 1. Isolation, structure and biological activities. *Agricultural and Biological Chemistry* 33: 1491–1500.
Ōmura, S., Tsuzuki, K. and Iwai, Y., 1985. Anticoccidial activity of frenolicin B and its derivatives. *Journal of Antibiotics* (Tokyo) 38:1447–1448.
Onaga, H. and Ishii, T. 1978. Anticoccidial activity of monensin in chicks and cell culture. *Journal of the Japan Veterinary Medical Association* 31:592–596 (in Japanese; English abstract).
Onaga, H., Saeki, H., Hoshi, S. and Ueda, S., 1986. An enzyme- linked immunosorbent assay for serodiagonosis of coccidiosis in chicken: Use of a single serum dilution. *Avian Diseases* 30:658–661.
Otoguro, K., Liu, Z.-X., Fukuda, K., Li, Y., Iwai, Y., Tanaka, H. and Ōmura, S. 1988. Screening for new nematicidal substances of microbial origin. *Journal of Antibiotics* (Tokyo) 41:573–575.
Ryley, J. F. 1980. Screening for and evaluation of anticoccidial activity. *Advances in Pharmacology and Chemotherapy* 17:1–23.
Ryley, J. F. and Wilson, R. G. 1976. Drug screening in cell culture for the detection of anticoccidial activity. *Parasitology* 73:137–148.
Sagawa, T., Hirano, S., Takahashi, H., Tanaka, N., Oishi, H., Ando, K. and Togashi, K. 1972. Tetranactin, a new miticidal antibiotic. III. Miticidal and other biological properties. *Journal of Economic Entomology* 65:372–375.
Sakuda, S., Isogai, A., Makita, T., Matsumoto, S., Koseki, K., Kodama, H. and Suzuki, A. 1987a. Structure of allosamidins, novel insect chitinase inhibitors, produced by actinomycetes. *Agricultural and Biological Chemistry* 51:3251–3259.

Sakuda, S., Isogai, A., Matsumoto, S. and Suzuki, A. 1987b. Search for microbial insect growth regulators. III. Allosamidin, a novel insect chitinase inhibitor. *Journal of Antibiotics* (Tokyo) 40:296–300.

Sakuda, S., Nishimoto, Y., Ohi, M., Watanabe, M., Takayama, S., Isogai, A. and Yamada, Y. 1990. Effects of demethylallosamidin, a potent yeast chitinase inhibitor, on the cell division of yeast. *Agricultural and Biological Chemistry* 54:1333–1335.

Schaeffer, J. M., Hanes, H. W. 1989. Avermectin binding in *Caenorhabditis elegans*. Two-state model for the avermectin binding site. *Biochemical Pharmacology* 38:2329–2338.

Schaeffer, J. M., Fraizier, E. G., Bergstrom, A. R., Williamson, J. M., Liesch, J. M. and Goetz, M. A. 1990. Chochlioquinone A, a nematicidal agent which competes for specific [^{3}H]ivermectin binding sites. *Journal of Antibiotics* (Tokyo) 43:1179–1182.

Shoop, W. L., Egerton, J. R., Eary, C. H. and Suhayda, D. 1990. Anthelmintic activity of paraherquamide in sheep. *Journal of Parasitology* 76:349–351.

Taikwang, M. L., Carter, G. T. and Borders, D. B. 1989. Structure determination of simaomicins and : extremely potent, novel anticoccidial agents produced by *Actinomadura*. *Journal of the Chemical Society Chemical Communications* 1771–1772.

Takahashi, A., Kurasawa, S., Ikeda, D., Okami, Y. and Takeuchi, T., 1989. Altemicidin, a new acaricidal and antitumor substance. I. Taxonomy, fermentation, isolation and physico- chemical and biological properties. *Journal of Antibiotics* (Tokyo) 42:1556–1566.

Takahashi, Y., Nakamura, H., Ogata, R., Matsuda, N., Hamada, M., Naganawa, H., Takita, T., Iitaka, Y., Sato, K. and Takeuchi, T. 1990. Kijimycin, a polyether antibiotic. *Journal of Antibiotics* (Tokyo) 43:441–443.

Tamura, S., Takahashi, N., Miyamoto, S., Mori, R., Suzuki, S. and Nagatsu, J. 1963. Isolation and physilogical activities of piericidin A, a natural insecticide produced by *Streptomyces*. *Agricultural and Biological Chemistry* 27:576–582.

Watanabe, K., Umeda, K. and Miyakado, M. 1989. Isolation and identification of three insecticidal principles from the red alga *Laurencia mipponica* Yamada. *Agricultural and Biological Chemistry* 53:2513–2515.

Part 6
Chemical Screening

14

Chemical Screening

Akira Nakagawa

14.1 Introduction

A very large number of biologically active compounds from microorganisms have been discovered so far, mainly by screening programs based on such biological activities as antibacterial, antitumor, antiviral, etc. Nevertheless, the search for new bioactive compounds of microbial origin continues by the application of new screening methods, advancement of experimental techniques, and discovery of new sources.

Screening systems based on bioactivity measurements encounter the following difficult problems.

1. A great deal of research work and time is required for the establishment of a reliable screening program.

2. Variations originating from various factors such as growth media, inorganic ions, and high molecular weight metabolites are a hindrance to reliable measurements by biological assay techniques.

3. There is usually a great gap between the biological activities measured in vitro and in vivo of new compounds discovered by in vitro screening.

Taking into consideration that new and nearly inexhaustible numbers of microbial metabolites, not yet isolated, are present in microorganisms, Zähner et al. (1982), Zeeck et al., Umezawa et al., Ōmura et al., and other research groups have searched systematically for new bioactive compounds by chemical screening. The principle of chemical screening is, first, the isolation and detection of new metabolites by specific tests such as color reactions, using several reagents to detect functional groups present in molecules of known secondary metabolites. This step is followed by evaluation of the biological activities of the isolated metabolites. The main characteristics of this screening method are its convenience for simple detection and quick isolation of the new metabolites from microorganisms. In this chapter are described mainly methods of chemical screening using color reaction tests, along with a few other methods, and the characteristics of typical new compounds thus obtained. Table 14.1 summarizes some of the bioactive microbial metabolites that have been found by chemical screening.

14.2 Chemical Screening Using Color Reagents

The search for nitrogen-containing compounds produced by microorganisms has naturally been actively pursued because alkaloids from plants have played a significant role in medicine. Many alkaloids that have been obtained from fungi, such as nigragillin, echinulin, fumitremorgen, austamide, and rugulovasine, have been reported independently. Rosenberg et al. (1976) described screening methods for alkaloids from lower fungi using several color reagents, for example, Wagner's, Mayer's, and silicotungstic reagents. They also pointed out that microbial alkaloids were detected with the high frequency of 12%–40% from various type cultures of fungal species such as *Aspergillus, Fusarium, Penicillium,* and *Helminthosporium*. Color reagents and color reactions such as the Dragendorff, Sakaguchi, ninhydrin, Wood, diacetyl, Wagner, Mayer, and silicotungstic tests are well known as detection methods for nitrogen-containing compounds.

Dragendorff's reagent-positive metabolites Among the reagents detecting alkaloids, Dragendorff's reagent shows high sensitivity for nitrogen containing compounds. During the systematic chemical screening for new microbial alkaloids using Dragendorff's reagent, Ōmura et al., found several new alkaloids from the culture broths of streptomycetes. The screening method carried out by Ōmura et al. (1974) is shown in Figure 14.1.

Primary screening test A 5-day culture filtrate was divided into two parts. After adjusting the pH to 4.0 with 0.1 N HCl, one part of the culture filtrate was tested with Dragendorff's reagent according to Munier and Macheboeuf (Breed et al., 1957). The formation of precipitates indicated a positive result.

Secondary screening test If the primary test was positive, a secondary screening test was conducted on the reserved culture filtrate. Five milliliters of the filtrate was made alkaline with NH_4OH and extracted with 1 volume of butyl acetate. The butyl acetate layer was extracted with 1 ml of 0.1 N HCl and tested with Dragendorff's reagent. Titer (units) of the alkaloid accumulated in the culture medium was measured as the maximum dilution still yielding a positive reaction with the reagent.

By this method a piperidine alkaloid, pyrindicin, was found that posesses pharmacological activities such as narcoantagonistic and analgesic action, intestine relaxation, and inhibition of platelet aggregation induced by collagen. Alkaloids NA-337A and NA-337B were identified as the enzyme inhibitor THPP, which inhibits *N*-methyltransferase from rabbit lung, and an antiviral substance, abikoviromycin, respectively. An alkaloid TM-64 obtained from

Table 14.1 Bioactive compounds isolated from microorganisms by chemical screening

Compound	Color reagent	Producing microorganism	Biological activity	Reference
Nigrifactin	Dragendorff	*Streptomyces nigrifaciens*	A transient fall of blood pressure	Kaneko et al., 1968; Terashima et al., 1969, 1970a, 1970b
Pyrindicin	Dragendorff	*S. griseoflavus* var. *pyrindicus*	Narcoantagonistic, analgesic etc.	Ōmura et al., 1974; Onda et al., 1973
NA-337A (Abikoviromycin)	Dragendorff	*Streptomyces* sp. NA-337	A clearing of fat	Onda et al., 1975
NA-337B (THPP)	Dragendorff	*Streptomyces* sp. NA-337	Antiviral	Onda et al., 1975
1,3-Diphenethyl urea	Dragendorff	*Streptomyces* sp. AM-2498	Antidepressant	Iwai et al., 1978
Quinoline-2-methanol	Dragendorff	*Kitasatoa griseophaeus* PO-1228	Hypoglycemic	Ōmura et al., 1976
AM-6201 (Reductiomycin)	Dragendorff	*S. xanthochromogenus* AM-6201	Antitumor	Onda et al., 1982
AM-2504 (Dityromycin)	Dragendorff	*Streptomyces* sp. AM-2504	Antimicrobial, protein kinase C inhibitor	Ōmura et al., 1977a, Teshima et al., 1988
Staurosporine	Dragendorff	*S. staurosporeus*	Hypotensive	Ōmura et al., 1977b, Furusaki et al., 1978
Herquline	Dragendorff	*Penicillium herquei* Fg-372	Inhibition of platelet aggregation	Ōmura et al., 1979, Furusaki et al., 1980
Neoxaline	Dragendorff	*Aspergillus japonicus* Fg-551	Stimulation of central nervous system	Hirano et al., 1979, Konda et al., 1980
TM-64	Dragendorff	*Thermoactinomyces* sp. TM-64	Narcotic	Ōmura et al., 1975

Continued next page

Table 14.1 (cont.)

Compound	Color reagent	Producing microorganism	Biological activity	Reference
Antipain	Sakaguchi	Streptomyces sp. KC84-AG13	Antipapain	Umezawa et al., 1972b; Suda et al., 1972
KF77-AG6	Sakaguchi	S. filipinensis		Fujimoto et al., 1974
L-β-(5-hydroxy-2-pyridyl)-alanine	Ninhydrin Ferric chloride	S. chibaensis SF-1346	Antibacterial, antiinflammatory	Inouye et al., 1975
L-β-(3-hydroxy-ureido)-alanine	Ferric chloride	S. chibaensis SF-1346	Antibacterial, antiinflammatory	Inouye et al., 1975
Myxochelin A	Ferric chloride	Angiococcus disceformis	Antibacterial	Kunze et al., 1989
Dienomycin	Wood	Streptomyces sp. MC67-C$_1$	Antibacterial	Umezawa et al., 1970a, 1970b
Arglecin	Wood Diacetyl	Streptomyces sp. KB59-M$_1$	Antiarrhythmic	Tatsuta et al., 1971, 1972
Urdamycin	Ehrlich	S. fradiae sp. Tu 2717	Antimicrobial, antitumor, proliferation	Drautz et al., 1986
Elloramycin	Ehrlich	S. olivaceus sp. Tu 2353	Antimicrobial	Drautz et al., 1985
Streptazolin	Ehrlich	S. viridochromogenes sp. Tu 1678	Antimicrobial	Drautz et al., 1981
1,3-Dihydroxy-8-decene-5-one	Ehrlich	S. fimbriatus sp. Tu 2335		Keller-Schierlein et al., 1983
(E)-3-(1H-pyrrol-3-yl)-2-propenoic acid	Ehrlich	S. parvulus sp. Tu 2480		Keller-Schierlein et al., 1985
Shydrofuran	Ehrlich p-dimethylamino-benzaldehyde	S. termitum		Umezawa et al., 1971; Usui et al., 1971
Phosphoramidon	Ehrlich	S. tanashiensis		Umezawa et al., 1972a
2-Ethyl-5-(3-indolyl)oxazole	Ehrlich Barrollier	S. cinnamomeus		Noltermeyer et al. 1982
N-Acetyl-β-oxotryptamine	Barrollier Tetrazolium	S. ramulosus Tu 34		Chen et al., 1983

Table 14.1 (cont.)

Compound	Color reagent	Producing microorganism	Biological activity	Reference
N-Acetyl-α-hydroxy-β-oxotryptamine	Barrollier	S. ramulosus Tu 34		Chen et al., 1983
Ketalin	Tetrazolium Ehrlich	S. lavendulae	Ionophore	König et al., 1980a
Ferrithiocin	Tetrazolium	S. antibioticus Tu 1998	Antimicrobial	Naegeli and Zähner, 1980
Trihydroxy-6-methyl cyclohexanone	Tetrazolium	S. phaeochromogenes sp. venezuelae		Muller et al., 1986
KD16-U1	Tetrazolium	S. filipinensis		Tatsuta et al., 1974
Differolide	Tetrazolium	S. aurantiogriseus sp. Tu 3149	Enhancement of formation of aerial mycelium	Keller-Schierlein et al., 1986
3-Hydroxyquinoline-2-carboxylic acid		S. griseoflavus subsp. Go 3592	Growth stimulation against herpes simplex virus	Breiding-Mack and Zeeck, 1986
Amycin A, B		Streptomyces sp. DSM 3816	Antimicrobial	Grabley et al., 1990
Naphthomevalin		Streptomyces sp. Go 28	Antimicrobial	Henkel and Zeeck, 1991
SM 196A, B		Streptomyces sp. DSM 4769	Antimicrobial Growth inhibition of herpes simplex virus	Grarbley et al., 1991
Saphenic acid methyl ether		S. antibioticus		Geiger et al., 1988
6-Acetylphenazine-1-carboxylic acid		S. antibioticus		Geiger et al., 1988
Ethericin A,B		Aspergillus funiculosus Tu 680	Antimicrobial ionophore	König et al., 1978, 1980b
Aurachin A,B		Stigmatella aurantiaca Sg a15	Inhibition of NADH oxidation	Kunze et al., 1987

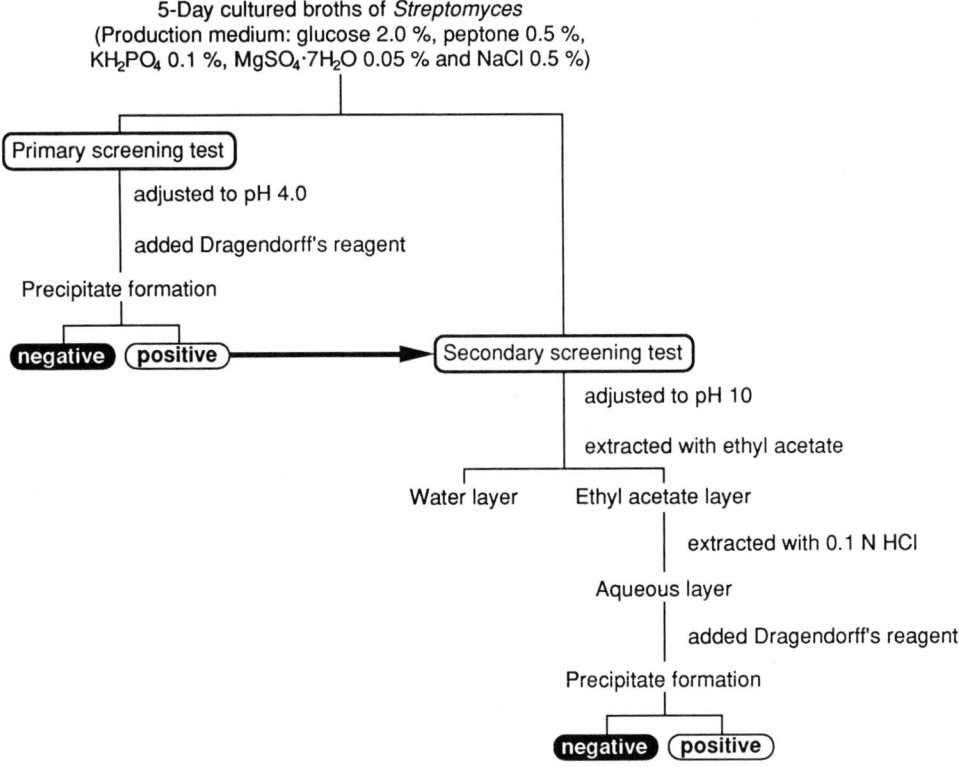

Figure 14.1 Screening procedure for microbial alkaloids using Dragendorff's reagent.

Thermoactinomyces possesses an indole skeleton containing a thiazole ring. Quinoline-2-methanol was isolated from *Streptomyces* sp. No. PO-1227, later classified into a new genus, *Kitasatoa*, and designated as *Kitasatoa griseophaeus* (Ōmura et al., 1976). The alkaloid shows hypoglycemic activity in the rat. On the other hand, Arai et al. (1976) have reported quinoline-4-methanol, an analog of quinoline 2-methanol, from *S. lavendulae* No. 314.

An alkaloid AM-6201, identified as reductiomycin (Shimizu et al., 1981a, 1981b), exhibits antitumor activity against Ehrlich ascites carcinoma in mice. The alkaloid AM-2504, later named dityromycin, was isolated from *Streptomyces*. It possesses a unique bicyclic depsipeptide structure and shows antimicrobial activity against gram-positive bacteria, especially *Bacillus* sp., *Corynebacterium paurometabolum*, and *Clostridium* sp. Because the production yield of dityromycin is low and is accompanied with two known antibacterial antibiotics, aureothricin and thiolutin, it would be difficult to find dityromycin selectively in a fermentation broth by ordinary screening methods using antimicrobial activity as an assay.

A new alkaloid AM-2282 named staurosporine is a stable compound first isolated from S. *staurosporeus*. The structure and stereochemistry of staurosporine were clarified by an X-ray crystallographic analysis. This alkaloid possesses strong hypotensive activity. Subsequently, it was found in another screening (Figure 14.2) as a potent platelet aggregation inhibitor from *Streptomyces* by Oka et al. (1986). The inhibitory activity against collagen-induced aggregation was more than 3000 times greater than that of aspirin. Further, the alkaloid was found to have a potent inhibitory activity against phospholipid/Ca^{2+}-dependent protein kinase (protein kinase C) by Tamaoki et al. (1986) (see Chapter 7). Staurosporine is now commercially available and is widely used as an inhibitor of protein kinase C. The discovery of staurosporine by chemical screening is a good example of the successful use of chemical techniques for finding a new compound with pharmacological activity from microbial sources.

Further, Ōmura et al. (1979) and Hirano et al. (1979) have discovered two new alkaloids, herquline and neoxaline, from *Streptomyces* in a screening using Dragendorff's reagent. Both alkaloids possess unique structures that have not been encountered previously in metabolites from streptomycetes. Herquline and neoxaline show weak inhibitory activity of blood platelet aggregation induced by adenosine diphosphate (ADP) and stimulation of the central nervous system, respectively.

Typical alkaloid reagent-positive metabolites Umezawa et al. (1972b) and Fujimoto et al. (1974) have found antipain and its related compound KF 77-AG 6, respectively, from the cultured broth of streptomycetes by chemical screening using Sakaguchi's reagent. KF-77-AG 6, which corresponds to the ureylene group-containing part of antipain, also shows positive reactions to other reagents such as Rydon-Smith, Wood, and diacetyl. Inouye et al. (1975) have screened metabolites that show a high sensitivity to the ninhydrin and ferric chloride reagents with the ultimate aim of finding analogs of physiologically active phenolic amines such as are present in epinephyrine, norepinephrine, and L-DOPA. As a result, two unusual amino acids, L-β-(5-hydroxy-2-pyridyl)-alanine and L-β-(3-hydroxyureido)-alanine were obtained from a fermentation broth of *Streptomyces chibaensis*. The antibacterial activities of L-β-(5-hydroxy-2-pyridyl)-alanine and its methylester were antagonized by L-tyrosine and those of L-β-(3-hydroxyureido)-alanine by L-glutamine. Both compounds also show moderate antiinflammatory activity, but so far no significant acute toxicity has been seen.

Kunze et al. (1989) isolated a new iron-chelating compound, myxochelin A, from the culture broth of the myxobacterium *Angiococcus disceformis* in the selecting compounds showing positive reaction to $FeCl_3$ on thin-layer chromatography (TLC). The production of myxochelin A was markedly increased by cultivation at low iron levels. This compound showed weak activity against some bacteria.

Taking into consideration that Wood's reagent is positive for purines, pyrimidines, and imidazole derivatives (Figure 14.3), Umezawa et al. (1970a,

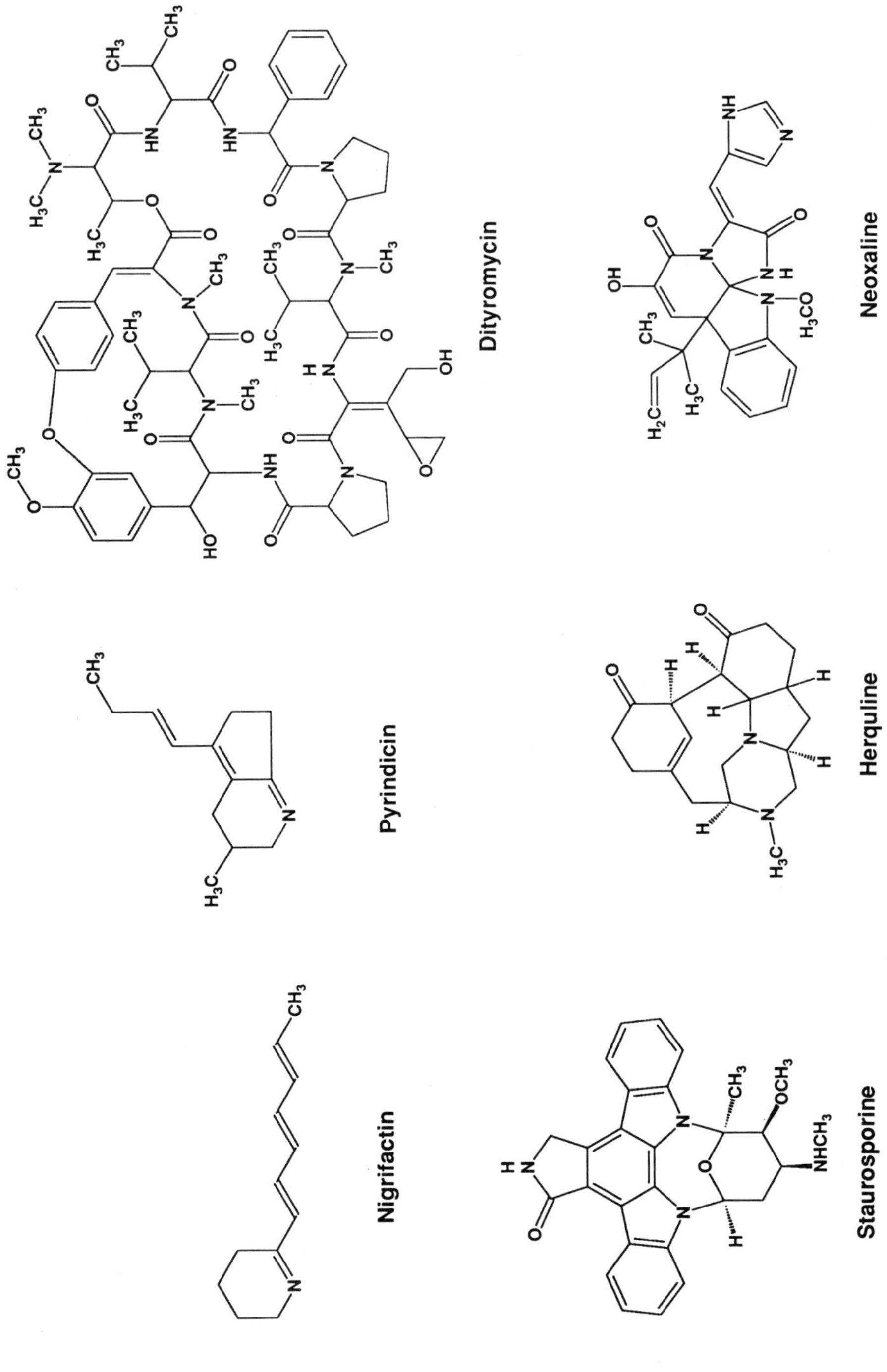

Figure 14.2 Structures of typical Dragendorff's reagent-positive metabolites.

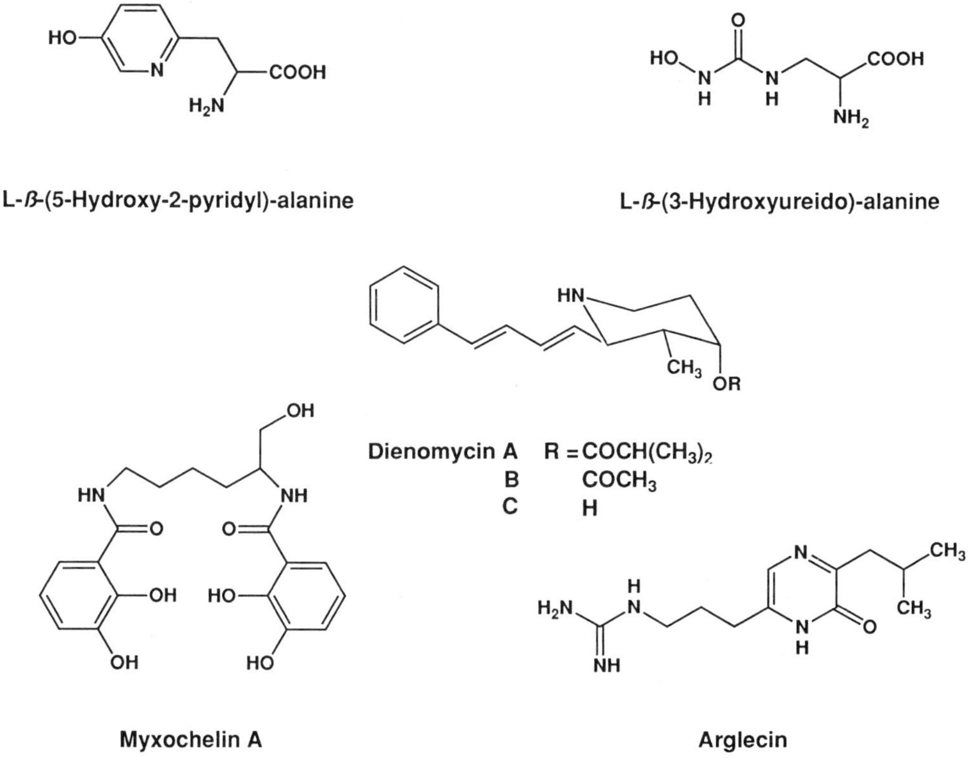

L-β-(5-Hydroxy-2-pyridyl)-alanine

L-β-(3-Hydroxyureido)-alanine

Dienomycin A R = COCH(CH$_3$)$_2$
B   COCH$_3$
C   H

Myxochelin A

Arglecin

Figure 14.3 Structures of typical alkaloid reagent-positive metabolites.

1970b) screened for compounds positive to this reagent and isolated the dienomycins. The dienomycins are derivatives of (4-phenylbutadienyl) pipecolin, the skeleton of which had not been previously found in plant alkaloids. Dienomycin A is active only against mycobacteria. On the other hand, Tatsuta et al. (1971, 1972) found arglecin as a compound positive against the Wood's and diacetyl reagents, both showing high sensitivity for guanidine and amidine moieties.

Chemical screening using the color reactions just described is aimed at finding new compounds possessing functional groups contained in the molecules of known pharmacologically active compounds and antibiotics. Although compounds with practical biological activity other than staurosporine have not been found by this procedure to date, in principle the method is a good one to use in searching for physiologically important compounds with desired structural skeletons, such as β-lactams and prostaglandins.

Ehrlich and Tetrazolium reagent-positive metabolites The groups of Zähner et al. (1982) have reported the discovery of novel compounds by systematic chemical screening using color reactions, based on the following considerations:

1. Even if new bioactive compounds are obtained from screening programs based on biological activity, the probability of their application and development for agricultural application or human or animal medicines is extremely low.

2. Numerous unknown secondary microbial metabolites, which have not been found by ordinary screening programs, have escaped detection.

These groups have screened for new compounds from streptomycetes by color reactions using the Ehrlich and Tetrazolium reagents, which are highly specific for the functional groups indole, urea, furan, and pyrrole that often occur in secondary metabolites (Figure 14.4). Drautz et al. (1986) found the urdamycins, modified benzanthraquinones, as substances positive to the Ehrlich reagent. The urdamycin antibiotics are structurally similar to kerriamycins and OM-4842 (aggreticin), which were isolated by Hayakawa et al. (1985) and by Ōmura et al. (1986), respectively. The urdamycins show antimicrobial activity, growth inhibition of L-1210 leukemia cells, and proliferation activity. The antibiotic complex shows especially strong growth inhibition of streptomycetes in its antimicrobial spectrum.

A tetracenomycin-related compound, elloramycin, isolated by Drautz et al. (1985) also shows a selective antimicrobial activity against gram-positive bacteria, especially streptomycetes. Elloramycin is characterized as an antibiotic possessing a tetracenone skeleton glycosidated with permethylated L-rhamnose. Streptazolin, isolated by Drautz et al. (1981), and (3S,8E)-1,3-dihydroxy-8-decene-5-one and (E)-3-(1H-pyrrol-3-yl)-2-propenoic acid and its amide, isolated by Keller-Schierlein et al. (1983, 1985), have been reported as Ehrlich-positive compounds from streptomycetes.

Based on the same idea, Umezawa et al. (1971) and Usui et al. (1971) have found sphydrofurans A_1, A_2, and A_3, which are an equilibrium mixture, and phosphoramidon, with a unique structural moiety, phosphoramide, respectively. Noltermeymer et al. (1982) isolated 2-ethyl-5-(3-indolyl) oxazole, which exhibits a red-violet color with the Ehrlich reagent and a blue-green color in the Barrollier test. The compound is identical to pimprimethine, is obtained from the cultured broth of *Streptoverticillium olivoreticuli*, and is an analog of pimprinin C; it is known to have antiepileptic and monoamine oxidase inhibitory activities. It is difficult to discover compounds of these types from culture broths by screening based on antibacterial activity alone.

The Zähner and Zeeck groups have reported several compounds with interesting structures and biological activities from streptomycetes by employing tetrazolium blue as a spray reagent. β-Oxotryptamine and α-hydroxy-β-oxotryptamine isolated by Chen et al. (1983) possess an indolyloxazol skeleton and seem to be direct precursors of the pimprinine described previously. This means that the β-oxotryptamine-producing strain, *S. ramulosus*, involves an enzyme able to oxidize the side chain of tryptophane or tryptamine analogs. Ketalin (7-methyl-2-methaleno-4-oxonona-1,3,8-triol), isolated by König et al. (1980a), shows no antibacterial activity but exhibits an ionophoric

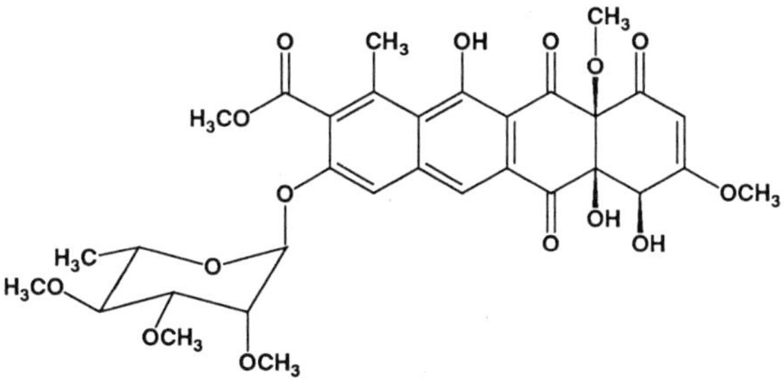

Elloramycin

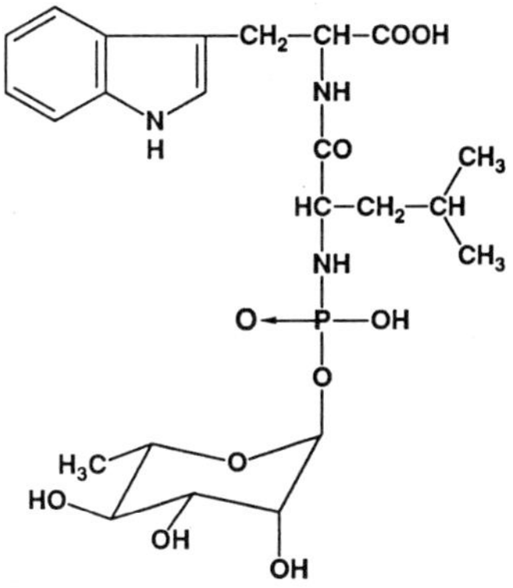

Phosphoramidon

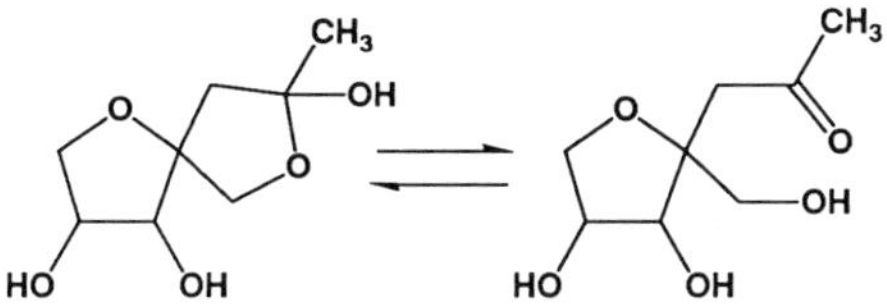

Sphydrofurans

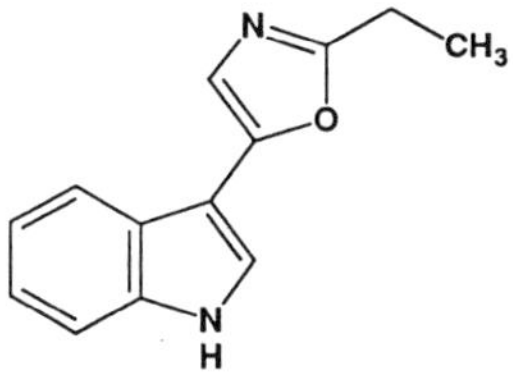

2-Ethyl-5-(3-indolyl)-oxazole (Pimprinethine)

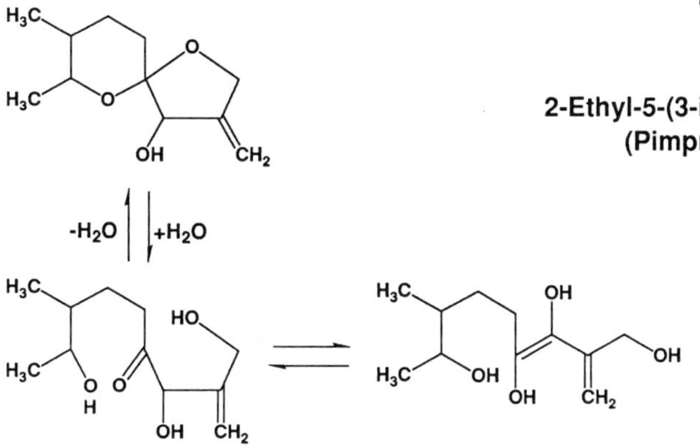

Ketalin

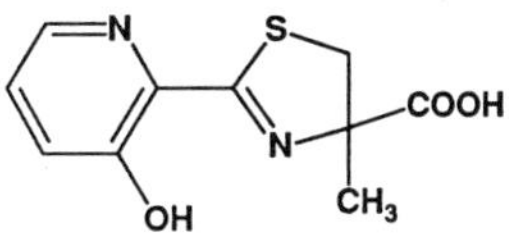

Ferrithiocin

Figure 14.4 Structures of Ehrlich and Tetrazolium reagent-positive metabolites.

activity comparable with that of valinomycin using artificial membranes. A striking finding is the similarity of ketalin to pheromones that occur in insects. This suggests the possibility of using ketalin as a lead compound convertible synthetically to pheromones.

Ferrithiocin, a new sulfur-containg metabolite, was isolated in the form of an iron complex as an indicator on thin-layer plates from S. *antibioticus* by Naegeli and Zähner (1980). The metabolite does not exhibit intrinsic antimicrobial activity against gram-positive and gram-negative bacteria, but increases the effectiveness of cephalosporin C against some bacteria. Differolide, with a racemic dilactone, enhances the formation of both aerial mycelium and spores of S. *glaucescens*, suggesting the possibility of using it as a growth factor for some *Streptomyces* (Keller-Schierlein et al., 1986). KD 16-UI, which was isolated after detection by a color reaction using triphenyl tetrazolium chloride reagent by Tatsuta et al. (1974) is related to the antibacterial antibiotics, epoxydon, and terremutin produced by fungi.

14.3 Other Chemical Screening Methods

As described, Zähner et al. have obtained new substances from a small series of streptomycetes with the help of chemical methods using thin-layer chromatography (TLC) with specific staining reagents for certain chemical groups. During the screening programs, they pointed out that the probability of finding new bioactive compounds by their approach is greater than by classical screening techniques. Breiding-Mack and Zeeck (1986) found 3-hydroxyquinoline-2-carboxylic acid (Figure 14.5), which on silica gel TLC plates exhibited a strong green-yellow fluorescence in UV light (366 nm) in metabolites from S. *griseoflavus*. Its calcium salt had not previously been isolated from microorganisms, but the free acid was reported as a saponification product from the cytostatic peptide antibiotic cinropeptin. The calcium salt exhibits no cytostatic activity and no antimicrobial activity, but shows growth stimulation against herpes simplex viruses.

Zeeck's group also reported new macrocyclic polyols, amycins A and B (Grabley et al., 1990), which are closely related to azalomycins. Grabley et al. (1991) isolated new compounds SM 196A and 196B, biologically active angucyclinones from *Streptomyces* sp. Henkel and Zeeck (1991) found naphthomevalin, a new dihydro-naphthoquinone, possessing antibacterial activity. On the other hand, Geiger et al. (1988) have isolated new phenazines, saphenic acid methyl ether and 6-acetylphenazine-1-carboxylic acid, from a strain of *Streptomyces antibioticus*.

The use of permeation-damaged cells, whether in the form of mutants or through the addition of an active substance, for example, EDTA, which damages permeation, increases the cell permeability of many substances. Using screening to circumvent the permeation barrier, König et al. (1978, 1980a) isolated new diphenylether antibiotics named ethericins A and B from a strain

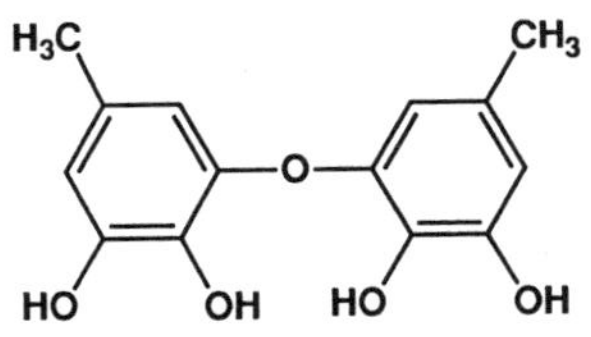

Ethericin A

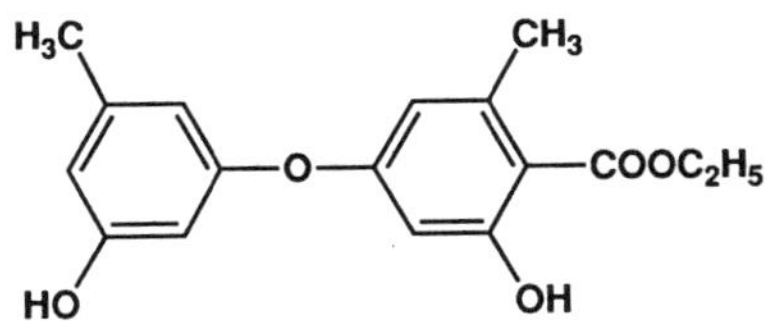

Ethericin B

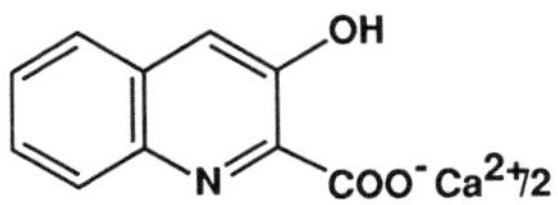

Calcium 3-hydroxyquinoline-
2-carboxylate

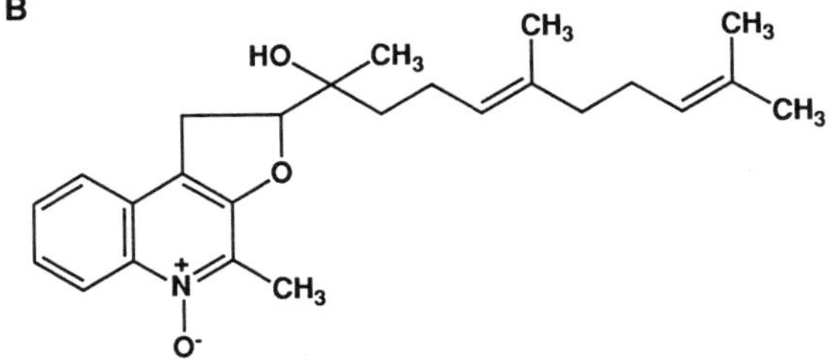

Aurachin A

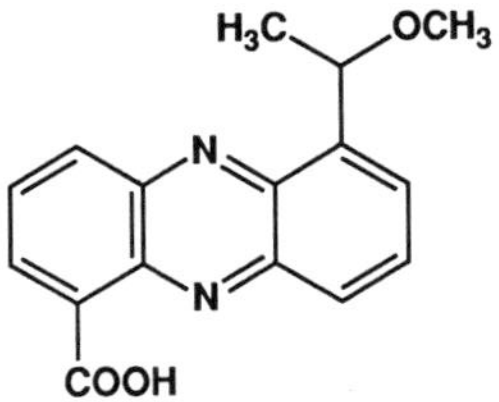

Saphenic acid methyl ether

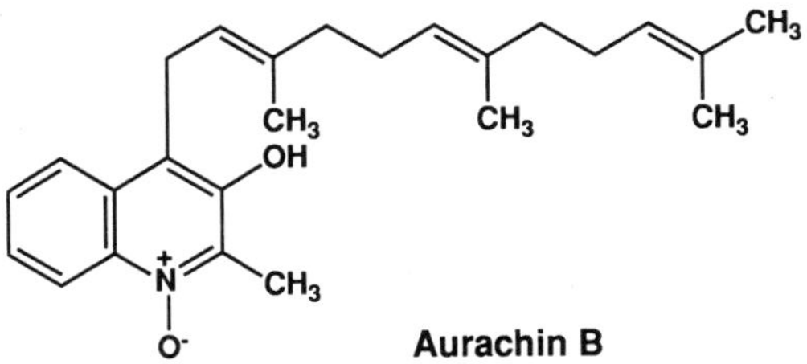

Aurachin B

Figure 14.5 Structures of metabolites isolated by other chemical screening methods.

of *Aspergillus*. Ethericin A is very likely an ionophore for bivalent cations, based on its structural features. It has a considerably stronger effect in combination with EDTA in nontoxic concentrations, which could result from a strengthening of complexes of the bivalent cations outside of cells. An increase in membrane permeability caused by ethericin would then lead to cell damage. Aspermutarubrol, isolated as a self-growth inhibitor from *Aspergillus* by Taniguchi et al. (1978), is identical to ethericin A.

Kunze et al. (1987) found the aurachins, new quinoline alkaloids, by utilizing ^{1}H-NMR (nuclear magnetic resonance) spectral data on cell extracts of *Stigmatella aurantiaca* strain Sg a15, which contained two structurally unrelated antibiotics, stigmatellin and a mixture of myxalamids. The aurachins consist of four components, A, B, C, and D, and possess activity against grampositive bacteria. The mode of action of the aurachins is inhibition of NADH oxidation in beef heart submitochondrial particles, as in the case of stigmatellin. The components C and D are compounds related to 2-heptyl-4-hydroxyqunoline-N-oxide, isolated from *Pseudomonas aeruginosa*, while components A and B possess a novel skeleton.

14.4 Concluding Remarks

The characteristics of the chemical screening methods are both simple and rapid ways for detection and isolation of new metabolites. During their experience with such screening, Zähner et al. (1982) suggested that the likelihood of finding new bioactive compounds by chemical screening seems to be greater than with ordinary screening employing biological activity. Among the new compounds obtained by chemical screening, our discovery of staurosporine is a successful example displaying those characteristic features of chemical screening for new biologically active substances.

It is also important to screen for structural analogs of known physiologically important substances, such as prostaglandins and polyamines. For maximum efficiency of this method of screening, it is necessary to evaluate the isolated compounds (on a broad basis) as far as their pharmacological activity is concerned. For this purpose cooperation among microbiologists, chemists, and pharmacologists is an essential key to the discovery and development of new bioactive compounds.

References

Arai, T., Yazawa, K., Mikami, Y., Kubo, A. and Takahashi, K. 1976. Isolation and characterization of satelite antibiotics, mimosamycin and chlorocarcins from *Streptomyces lavendulae*, streptothricin source. *Journal of Antibiotics* (Tokyo) 29:398–407.

Breed, R. S., Murry, E. Q. D. and Smith, N. R. 1957. *Bergey's Manual of Determinative Bacteriology*, 7th Ed., Williams & Wilkins, Baltimore.

Breiding-Mack, S. and Zeeck, A. 1986. Secondary metabolites by chemical screening.

I. Calcium 3-hydroxyquinoline-2-carboxylate from a *Streptomyces*. *Journal of Antibiotics* (Tokyo) 39:953–960.

Chen, Y., Zeeck, A., Chen, Z. and Zähner, M. 1983. Metabolic products of microorganisms. 222. β-Oxotryptamine derivatives isolated from *Streptomyces ramulosus*. *Journal of Antibiotics* (Tokyo) 36:913–915.

Drautz, H., Zähner, H., Kupfer, E. and Keller-Schierlein, W. 1981. 164. Stoffwechselprodukukte von mikroorganisme. Isolierung und Struktur von Streptazolin. *Helvetica Chimica Acta* 64:1752–1765.

Drautz, H., Reuschenbach, P., Zähner, H., Rohr, J. and Zeeck, A. 1985. Metabolic products of microorganisms. 225. Elloramycin, a new anthracycline-like antibiotic from *Streptomyces olivaceus*. Isolation, characterization, structure and biological properties. *Journal of Antibiotics* (Tokyo) 38:1291–1301.

Drautz, H., Zähner, H., Rohr, J. and Zeeck, A. 1986. Metabolic products of microorganisms. 234. Urdamycins, new angucycline antibiotics from *Streptomyces fradiae*. I. Isolation, characterization and biological properties. *Journal of Antibiotics* (Tokyo) 39:1657–1669.

Fujimoto, K., Tatsuta, K., Tsuchiya, T., Umezawa, S. and Umezawa, H. 1974. KF-77-AG6, an antipain-related metabolite. *Journal of Antibiotics* (Tokyo) 27:685.

Furusaki, A., Hashiba, N., Matsumoto, T., Hirano, A., Iwai, Y. and Ōmura, S. 1978. X-Ray crystal structure of staurosporine: a new alkaloid from a *Streptomyces* strain. *Journal of the Chemical Society Chemical Communications* 800 801.

Furusaki, A., Matsumoto, T., Ogura, H., Takayanagi, H., Hirano, A. and Ōmura, S. 1980. X-Ray crystal structure of herquline, a new biologically active piperazine from *Penicillium herquei* Fg-372. *Journal of the Chemical Society Chemical Communications* 698.

Geiger, A., Keller-Schierlein, W., Brandl, M. and Zähner, H. 1988. Metabolites of microorganism. 247. Phenazines from *Streptomyces antibioticus* strain TU 2706. *Journal of Antibiotics* (Tokyo) 41:1542–1551.

Grabley, S., Hammann, P., Raether, W., Wink, J. and Zeeck, A. 1990. Secondary metabolites by chemical screening. II. Amycins A and B, two novel niphimycin analogs isolated from a high producer strain of elaiophylin and nigericin. *Journal of Antibiotics* (Tokyo) 43:639–647.

Grableg, S., Hammann, P., Hütter, K., Kluge, H., Thiericke, R., Wink, J. and Zeeck, A. 1991. Secondary metabolites by chemical screening. 19. SM196A and B, novel biologically active angucyclinones from *Streptomyces* sp. *Journal of Antibiotics* (Tokyo) 44:670–673.

Hayakawa, Y., Furihata, K., Seto, H. and Otake, N. 1985. The structures of new isotetracenone antibiotics, kerriamycins A, B and C. *Tetrahedron Letters* 3475–3478.

Henkel, T. and Zeeck, A. 1991. Secondary metabolites by chemical screening. 15. Structure and absolute configuration of naphthomevalin, a new dihydronaphthoquinone antibiotic from *Streptomyces* sp. *Journal of Antibiotics* (Tokyo) 44:665–669.

Hirano, A., Iwai, Y., Masuma, R., Tei, K. and Ōmura, S. 1979. Neoxaline, a new alkaloid produced by *Aspergillus japonicus*. Production, isolation and properties. *Journal of Antibiotics* (Tokyo) 32:781–785.

Inouye, S., Shomura, T., Tsuruoka, T., Ogawa, Y., Watanabe, H., Yoshida, T. and Niida, T. 1975. L-β-(5-Hydroxy-2-pyridyl)-alanine and L-β-(3-hydroxyureido)-alanine from *Streptomyces*. *Chemical and Pharmaceutical Bulletin* (Tokyo) 23:2669–2677.

Iwai, Y., Hirano, A., Awaya, J., Matsuo, S. and Ōmura, S. 1978. 1,3-Diphenethylurea from *Streptomyces* sp. No. AM-2498. *Journal of Antibiotics* (Tokyo) 31:375–376.

Kaneko, Y., Terashima, T. and Kuroda, Y. 1968. A new alkaloid produced by *Streptomyces*. *Agricultural and Biological Chemistry* 32:783–785.

Keller-Schierlein, W., Wuthie, D. and Drautz, H. 1983. 122. Stoffwechselprodukte von Mikroorganismen. (3S,8E)-1,3-Dihydroxy-8-decen-5-on, ein Stoffwechselprodukt von *Streptomyces fimbriatus* (Millard und Burr 1926). *Helvetica Chimica Acta* 66:1253–1261.

Keller-Schierlein, W., Muller, A., Hagmann, L., Schneider, U. and Zähner, H. 1985. 61. Stoffwechselprodukte von Mikroorganismen, (E)-3-(1H-Pyrrol-3-yl)-2-propensaure und (E)-3-(1H-Pyrrol-3-yl)- 2-propensaureamid aus *Streptomyces parvulus*, Stamm Tu 2480. *Helvetica Chimica Acta* 68:559–562.

Keller-Schierlein, W., Bahnmuller, U., Dobler, M., Bielecki, J., Stumpfel, J. and Zähner, H. 1986. 196. Stoffwechselprodukte von Mikroorganismen. Isolierug und Strukturaufklarung von Differolid. *Helvetica Chimica Acta* 69:1833–1836.

Konda, Y., Onda, M., Hirano, A. and Ōmura, S. 1980. Oxaline and neoxaline. *Chemical and Pharmaceutical Bulletin* (Tokyo) 28:2987–2993.

König, W. A., Pfaff, K-P., Loeffler, W., Schanz, D. and Zahner, H. 1978. Stoffwechselprodukte von Mikroorganismen. 171. Ethericin A; Isolierung, Charakterisierung und Strukturauf-klarung eines neuen, antibiotisch wirksamen Diphenylethers. *Liebigs Annalen der Chemie* 1289–1296.

König, W. A., Krause, R., Loeffler, W. and Schanz. D. 1980b. Metabolic products of microorganisms. 196. The structure of ethericin B, a new diphenylether antibiotic. *Journal of Antibiotics* (Tokyo) 33:1270–1273.

König, W. A., Drautz, H. and Zahner, H. 1980a. Ketalin, ein Metabolit aus Streptomyces Tu 1 1668. *Liebigs Annalen der Chemie*: 622–628.

Kunze, B., Hofle, G. and Reichenbach, H. 1987. The aurachins, new quinoline antibiotics from myxobacteria: production, physicochemical and biological properties. *Journal of Antibiotics* (Tokyo) 40:258– 265.

Kunze, B., Bedorf, N., Kohl, W., Hofle, G., and Reichenbach, H. 1989. Myxochelin A, a new iron-chelating compound from *Angiococcus disciformis* (Myxobacterales). Production, isolation, physico-chemical and biological properties. *Journal of Antibiotics* (Tokyo) 42:14–17.

Muller, A., Keller-Schierlein, W., Bielecki, J., Rak, G., Stumpfel, J. and Zähner, H. 1986. 195. Stoffwechselprodukte von Mikroorganismen (2S,3R,4R,6R)-2,3,4-Trihydroxy-6-methylcyclo- hexanon aus zwei Actinomyceten-Stammen. *Helvetica Chimica Acta* 69:1829–1832.

Naegeli, H-V. and Zähner, H. 1980. 147. Stoffwechselprodukte von Mikroorganismen. Ferrithiocin. *Helvetica Chimica Acta* 63:1400–1408.

Noltermeyer, M., Sheldrick, G. M., Hoppe, H-V. and Zeeck, A. 1982. 2-Ethyl-5-(3-indolyl)oxazole from *Streptomyces cinnamomeus* discovered by chemical screening. Characterization and structure elucidation by X-ray analysis. *Journal of Antibiotics* (Tokyo) 35:549–555.

Oka, S., Kodama, M., Takeda, H., Tomizuka, N. and Suzuki, H. 1986. Staurosporine, a potent platelet aggregation inhibitor from a *Streptomyces* species. *Agricultural and Biological Chemistry* 50:2723–2727.

Ōmura, S., Hirano, A., Iwai, Y. and Masuma, R. 1979. Herquline, a new alkaloid produced by *Penicillium herquei*. Fermentation, isolation and properties. *Journal of Antibiotics* (Tokyo) 32:786–790.

Ōmura, S., Suzuki, Y., Kitao, C., Takahashi, Y. and Konda, Y. 1975. Isolation of a new sulfur-containing basic substance from a *Thermoactinomyces* species. *Journal of Antibiotics* (Tokyo) 28:609–610.

Ōmura, S., Iwai, Y., Hirano, A., Awaya, J., Suzuki, Y. and Matsumoto, K. 1977a. A new antibiotic, AM-2504. *Agricultural and Biological Chemistry* 41:1827–1828.

Ōmura, S., Iwai, Y., Hirano, A., Nakagawa, A., Awaya, J., Tsuchiya, H., Takahashi, Y. and Masuma, R. 1977b. A new alkaloid AM-2282 of *Streptomyces* origin. Taxonomy, fermentation, isolation and preliminary characterization. *Journal of Antibiotics* (Tokyo) 30:275–282.

Ōmura, S., Iwai, Y., Suzuki, Y., Awaya, J., Konda, Y. and Onda, M. 1976. Production of quinoline-2-methanol and quinoline-2- methanol acetate by a new species of Kitasatoa, Kitasatoa griseophaeus. Journal of Antibiotics (Tokyo) 29:797–803.

Ōmura, S., Tanaka, H., Awaya, J., Narimatsu, Y., Konda, Y. and Hata, T. 1974. Pyrindicin, a new alkaloid from a Streptomyces strain. Taxonomy, fermentation, isolation and biological activity. Agricultural and Biological Chemistry 38:899–906.

Ōmura, S., Nakagawa, A., Fukamachi, N., Miura, S., Takahashi, Y., Komiyama, K. and Kobayashi, B. 1986. OM-4842, a new platelet aggregation inhibitor from Streptomyces. Journal of Antibiotics (Tokyo) 41:812–813.

Onda, M., Konda, Y., Hinotozawa, K. and Ōmura, S. 1982. The alkaloid AM-6201 from Streptomyces xanthochromogenus. Chemical and Pharmaceutical Bulletin (Tokyo) 30:1210–1214.

Onda, M., Konda, Y., Narimatsu, Y., Ōmura, S. and Hata, T. 1973. Structure of pyrindicin. Chemical and Pharmaceutical Bulletin (Tokyo) 21:2048–2050.

Onda, M., Konda, Y., Narimatsu, Y., Tanaka, H., Awaya, J. and Ōmura, S. 1975. Revised structure for an alkaloid from Streptomyces sp. NA-337. Chemical and Pharmaceutical Bulletin (Tokyo) 23:2462–2463.

Rosenberg, H., Maheshwari, V. and Stohs, S. J. 1976. Alkaloid screening of fungi. Planta Medica 30:146–150.

Shimizu, K. and Tamura, G. 1981a. Reductiomycin, a new antibiotic. I. Taxonomy, fermentation, isolation, charac terization and biological activities. Journal of Antibiotics (Tokyo) 34:649–653.

Shimizu, K. and Tamura, G. 1981b. Reductiomycin, a new antibiotic. II. Structural elucidation by spectroscopic studies. Journal of Antibiotics (Tokyo) 34:654–657.

Suda, H., Aoyagi, T., Hamada, M., Takeuchi, T. and Umezawa, H. 1972. Antipain, a new protease inhibitor isolated from Actinomycetes. Journal of Antibiotics (Tokyo) 25:263–266.

Tamaoki, T., Nomoto, H., Takahashi, I., Kato, Y., Morimoto, M. and Tomita, F. 1986. Staurosporine, a potent inhibitor of phospholipid/Ca^{++} dependent protein kinase. Biochemical Biophysical Research Communication 135:397–402.

Taniguchi, M., Kaneda, N., Shibata, K. and Kamikawa, T. 1978. Isolation and biological activity of aspermutarubrol, a self- growth-inhibitor from Aspergillus sydowi. Agricultural and Biological Chemistry 42:1629–1630.

Tatsuta, K., Tsuchiya, T., Someno, T. and Umezawa, S. 1971. Arglecin, a new microbial metabolite, isolation and chemical structure. Journal of Antibiotics (Tokyo) 24:735–746.

Tatsuta, K., Tsuchiya, T., Umezawa, S., Naganawa, H. and Umezawa, H. 1972. Revised structure for arglecin. Journal of Antibiotics (Tokyo) 25:674–676.

Tatsuta, K., Tsuchiya, T., Mikami, N., Umezawa, S., Umezawa, H. and Naganawa, H. 1974. KD 16-U1, a new metabolite of Streptomyces: isolation and structural studies. Journal of Antibiotics (Tokyo) 27:579–586.

Terashima, T., Kuroda, Y. and Kaneko, Y. 1969. Studies on a new alkaloid of Streptomyces. Structure of nigrifactin. Tetrahedron Letters 2535–2537.

Terashima, T., Kuroda, Y. and Kaneko, Y. 1970a. Studies on alkaloids of Streptomyces. Part I. Screening of alkaloid- producing microorganisms and pharmacological activity of allkaloids. Agricultural and Biological Chemistry 34:747–752.

Terashima, T., Kuroda, Y. and Kaneko, Y. 1970b. Studies on alkaloid of Streptomyces. Part II. Pharmacological activity and isolation of nigrifactin and taxonomical studies of Streptomyces strain No. FFD-101. Agricultural and Biological Chemistry 34:753–759.

Teshima, T., Nishikawa, M., Kubota, I., Shiba, T., Iwai, Y. and Ōmura, S. 1988. The structure of an antibiotic, dityromycin. Tetrahedron Letters 1963–1966.

Umezawa, S., Tsuchiya, T., Tatsuta, K., Hiriuchi, Y., Usui, T., Umezawa, H., Hamada, M. and Yagi, A. 1970a. A new antibiotic, dienomycin. I. Screening method, isolation and chemical studies. Journal of Antibiotics (Tokyo) 23:20–27.

Umezawa, S., Tatsuta, K., Horiuchi, Y., Tsuchiya, T. and Umezawa, H. 1970b. Studies on dienomycins. II. Chemical structures of dienomycins A, B and C. *Journal of Antibiotics* (Tokyo) 23:28-34.

Umezawa, S., Tatsuta, K., Izawa, O., Tuchiya, T. and Umezawa, H. 1972a. A new microbial metabolite phosphoramidon (isolation and structure). *Tetrahedron Letters* 97-100.

Umezawa, S., Tatsuta, K., Fujimoto, K., Tsuchiya, T., Umezawa, H. and Naganawa, H. 1972b. Structure of antipain, a new Sakaguchi- positive product of *Streptomyces*. *Journal of Antibiotics* (Tokyo) 25:267-270.

Umezawa, S., Usui, T., Umezawa, H., Tschiya, T., Takeuchi, T. and Hamada, M. 1971. A new microbial metabolite, sphydrofuran. I. Isolation and the structure of a hydrolysis product. *Journal of Antibiotics* 24:85-92.

Usui, T., Umezawa, S., Tsuchiya, T., Naganawa, H., Takeuchi, T. and Umezawa, H. 1971. A new microbial metabolites, sphydrofuran. II. The structure of sphydrofuran. *Journal of Antibiotics* (Tokyo) 24:93-106.

Zähner, H., Drautz, H. and Weber, W. 1982. In Bulock, J. D., Nisbet, L. J. and Winstanley, D. J. (editors), Bioactive Microbial Products: Search and Discovery, pp. 51-70, Academic Press, New York.

Part 7
Sources, Fermentation, and Improvement of Producing Microorganisms

15

Selection of Microbial Sources of Bioactive Compounds

Yuzuru Iwai and Yōko Takahashi

15.1 Introduction

Soon after penicillin was rediscovered as an antibiotic and named the "yellow magic medicine," a new field of applied microbiology, i.e., that is, antibiotic screening, was introduced, especially by S.A. Waksman, who discovered streptomycin. A large number of bioactive metabolites have been isolated from various microbial sources, such as actinomycetes, bacteria, fungi, mushrooms, etc., but actinomycetes have proved to be the most important as antibiotic producers. The new antibiotics found in 1980 were derived from actinomycetes (80.4%), bacteria (11.4%), and fungi-mushrooms (8.3%); about 75% of those discovered during 1971–1980 were also from this group of microorganisms (Higashide and Yamamoto, 1982). An efficient way of finding new bioactive metabolites is by the discovery of new microorganisms, and many approaches have been used.

From a microbiological point of view, one of the most important investigations in the history of antibiotic screening was Weinstein's discovery of gentamicin (Weinstein et al., 1963) from a strain of *Micromonospora*, a genus of the so-called rare actinomycetes that had been very little investigated for antibiotic production before that time. With this study as a turning point, various kinds of rare actinomycetes were subjected to screening, resulting in the discovery of a variety of new antibiotics (Table 15.1).

In this chapter, we focus on the actinomycetes, describing the following subjects: (i) habitat variation of the actinomycetes; (ii) isolation methods for soil actinomycetes; (iii) procedures for selective isolation of rare actinomycetes; and (iv) microbial sources other than actinomycetes.

15.2 Habitat Variation of Actinomycetes

Variation of actinomycetes in soil   It is frequently said that the musty smell of the soil originates from its actinomycetes. Thus, the soil is a most important habitat for actinomycetes (Table 15.2). The differences in the source

Table 15.1 Chronological list of antibiotics isolated from actinomycetes[1]

Organism	≈1973	1974	1975	1976	1977	1978	1979	1980	1981	1982	1983	1984	1985	1986	Total
Streptomyces	1922	80	108	101	86	63	89	96	103	109	86	94	133	54	3124
Micromonospora	41	7	12	10	10	10	20	33	41	20	11	6	7	4	232
Nocardia	45	4	2	10	11	9	6	17	9	19	2	1	10		145
Actinoplanes	6	4	8	8	4	2	8	1	7	4	6	5	8		71
Streptoverticillium	19	2	2	2	4	4	2	1		1	3	4	19	3	66
Actinomadura		2	1	1			1	1		4	2	5	7	1	36
Streptosporangium	7	3	3	1	1	1	2	5	7		1	1	2		23
Saccharopolyspora					2		2	2	1	1	1	6	2		17
Chainia	8								1		1			6	15
Dactylosporangium						1	1	2	2	1	3	1	3		14
Ampullariella							1	2	4	1	1	1		1	11
Nocardiopsis					1		2	1				1	1	5	11
Pseudonocardia					2		2	1				1			6
Microellobosporia							3								6
Streptoalloteichus									3						5
Actinosporangium					2					2			1		5
Microtetraspora					2							2	1		4
Kitasatosporia								2	2						3
Micropolyspora									1	1			1		3
Actinosynnema	2												2		2
Saccharomonospora				1									2		2
Thermomonospora											1				1
Catenuloplanes							1								1
Sebekia													1	1	1
Kibdelosporangium													1		1
Total	2050	102	136	134	125	90	140	163	181	163	118	127	200	74	3803

[1]Reproduced from Okazaki and Enokita, 1986.

Chapter 15 Selection of Microbial Sources of Bioactive Compounds

Table 15.2 Numbers of microorganisms in various kinds of soil in Japan[1]

Soil No.	Soils examined[2]	Bacteria	Fungi	Actinoymycetes	Percent[3]
		Thousands per gram of dry soil			
1	Paddy soil, Niigata	1,120	100	400	25
2	Paddy soil, Miyagi	35,200	800	4,000	10
3	Silty clay loam, Miyagi	18,000	420	2,280	11
4	Paddy soil, Yamanashi	5,870	—	1,030	15
5	Field soil, Yamanashi	7,120	160	1,260	15
6	Fertilized paddy soil, Aichi	32,400	70	2,600	8
7	Air-dried soil, Aichi	7,380	20	530	7
8	Unfertilized VAS[4] Tokyo	43,700	480	3,320	7
9	Yoteiko acidic VAS, Hokkaido	2,900	140	460	13
10	Acidic VAS, Hokkdaido	40	2	5	16
11	Acidic muddy soil, Hokkaido	10,210	380	2,010	16
12	Field VAS, Gunma	3,680	170	1,650	30
13	Unfertilized VAS, Gunma	5,410	120	420	7
14	Unfertilized VAS, Tottori	1,920	160	520	20
15	Surface VAS, Tottori	70	2	20	22
16	Otoji-type humus, Ehime	300	50	90	21
17	Forest VAS, Tochigi	8,760	200	890	9
18	Low peat, Hokkaido	90	10	20	13
19	High peat, Hokkaido	40	6	10	16

[1] Modified from Nonomura and Ohara, 1959.
[2] Soil samples Nos. 1–5, comparatively fresh samples; soil samples Nos. 6–19, stored in vinyl bag at 5 C for a year.
[3] Ratio of actinomycetes to the total number of microorganisms.
[4] VAS, Volcanic ash soil.

give rise to the variation of strains and number of actinomycetes in soil. Although as many characteristic strains as possible should be isolated from the soil sample selected, the capacity of the agar plate for isolation is limited, and the number of strains detectable thus is restricted. Therefore the selection of the soil source tested greatly affects the efficiency of discovery of new bioactive metabolites.

Vertical distribution of actinomycetes in soil   "For the isolation of actinomycetes, collect soil at a 5–10 cm depth." Is this usual teacher's advice to the pupil correct? The vertical distribution of actinomycetes in soil has been studied only by Waksman et al. and a few other researchers (Hagedorn, 1976). We have recently studied the variety and distribution of microorganisms living underground (0–40 m) by taking advantage of the fortuitous borings for soil analysis on the grounds of The Kitasato Institute Medical Center in Kitamoto City, which is located 40 km northwest of the center of Tokyo, nearly at the middle of the Kanto Plain (Takahashi et al., 1990).

Fungi were isolated only in the upper layer (0–0.3 m), while the major population of actinomycetes and bacteria was found in the upper layer (0–1 m). Bacteria at 10^6 g^{-1} of dried soil were detected even at a depth of 40 m (Figure 15.1). The largest population of actinomycetes was observed in the surface layer, and their number gradually decreased as the depth increased. Most actinomycetes ($\approx$80%) existed in the surface layer (0–10 cm) (Figure 15.2). The percentage of characteristic strains decreases slightly in the lower layers, as shown in Figure 15.3. However, sampling of the lower layers may also result in isolating unique microorganisms.

Habitats other than soil   Some rare actinomycetes can be efficiently isolated from habitats other than soil, as shown in Tables 15.3 and 15.4. Among these sources are wet areas (water and mud of lakes, rivers, or seas) and various materials derived from living organisms.

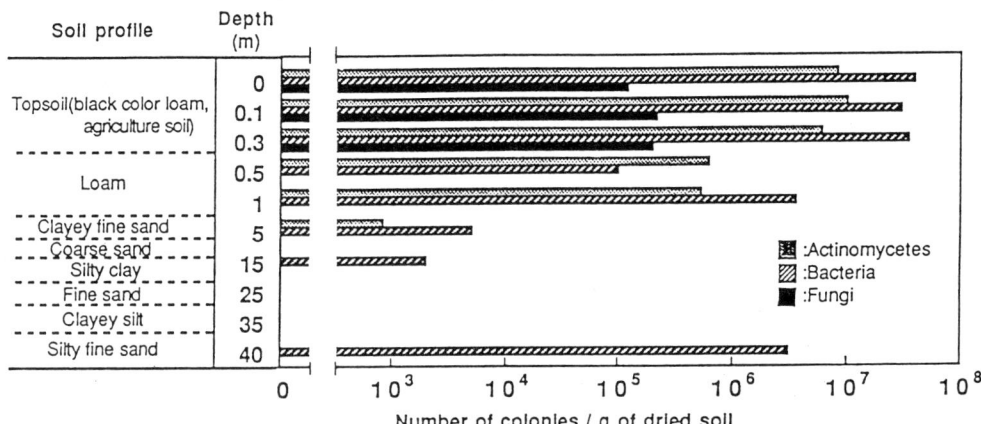

Figure 15.1   Vertical distribution of microorganisms in soil collected December 2, 1985 at former agricultural station, 121-1 Arai, Kitamoto-shi, Saitama prefecture.

Chapter 15 Selection of Microbial Sources of Bioactive Compounds   285

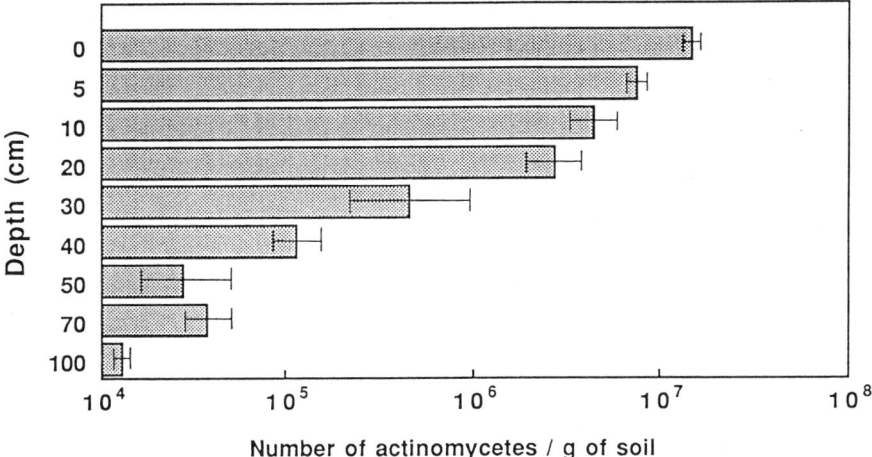

Figure 15.2 Vertical distribution of actinomycetes in soil (mean values of three soil samples).

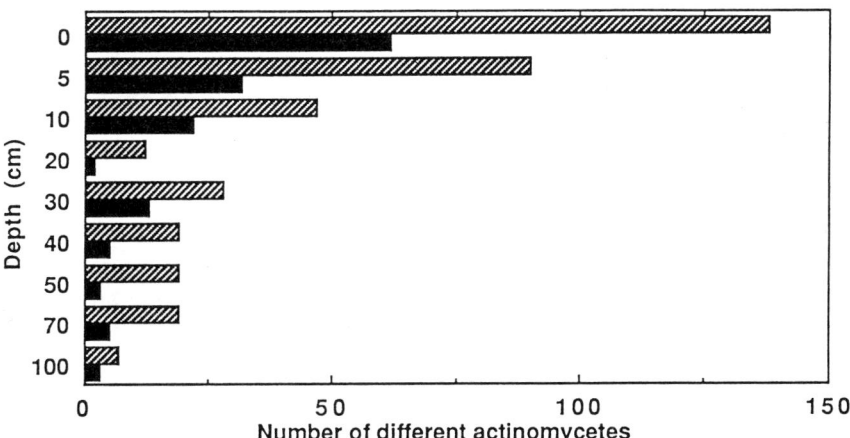

Figure 15.3 Vertical distribution of different actinomycetes in soil. Symbols: shaded bars, : number of apparently different actinomycetes isolated from soils of each layer; solid bars, : number of different actinomycetes found only specifically in that layer.

Okami's studies (Okami and Okazaki, 1972, Okami et al., 1976) on screening actinomycetes from marine sediments demonstrated that these actinomycetes, including some rare species, were derived from the terrestrial environment and selected under sea bottom conditions. Aplasmomycin, which was found in this screening is cultured in a special media similar to a marine environment that is prepared by dilution (1:6) of a yeast extract medium (yeast extract 0.4%, malt extract 1.0%, glucose 0.4%) with a 3% solution of NaCl and water.

Table 15.3 Mesophilic actinomycetes in soil samples taken from the Blelham Tarn drainage basin (colony-forming units $\times 10^4$ g^{-1})[1]

Sampling area	Micromonospora	Streptomyces	Nocardioforms
Wray Beck	5.8	59.3	3.8
Ford Wood Beck	1.5	35.9	0.3
Woodland Fish Pond Beck	0.02	1.2	0.4
Blelham Beck	4.7	28.4	1.2

[1] Modified from Cross, 1981.
Colloidal chitin agar was incubated at 25°C.

Table 15.4 Actinomycetes in Blelham Tarn and its streams (colony-forming units ml^{-1})[1]

Sample area	Micromonospora	Streptomyces	Nocardioforms
Wray Beck	42	4	30
Tock How Beck	76	76	362
Ford Wood Beck	78	6	70
Outflow Beck	54	7	50
Lake water (surface)	60	15	62

[1] Modified from Cross, 1981.
Colloidal chitin agar was incubated for 3 weeks at 25°C.

15.3 Isolation Methods for Soil Actinomycetes

The various isolation procedures reported to date are classified into two types: (1) pretreatment of soil samples and (2) control of culture condition on agar media. A combination of these is generally used for the isolation of soil actinomycetes (Nolan and Cross, 1988).

Pretreatment of soil samples Pretreatment with chemical agents (such as antibiotics, organic solvents, and surfactants), alkali, acid, and high temperature can reduce the number of undesirable microorganisms when screening for novel strains of antibiotic producers.

Pretreatment with antibiotics The procedure for preferential isolation of rare actinomycetes established in our laboratory is summarized in Figure 15.4. Rapid-growing microorganisms such as bacteria or *Streptomyces* strains are killed before spreading on an agar medium via incubation of the soil suspension with penicillin or other antibiotics.

Pretreatment with organic solvents We carried out the following experiment. Soil (1 g) was suspended in 5 ml of an organic solvent and allowed to stand for 60 min. After the solvent was discarded, the soil was airdried at

Chapter 15 Selection of Microbial Sources of Bioactive Compounds 287

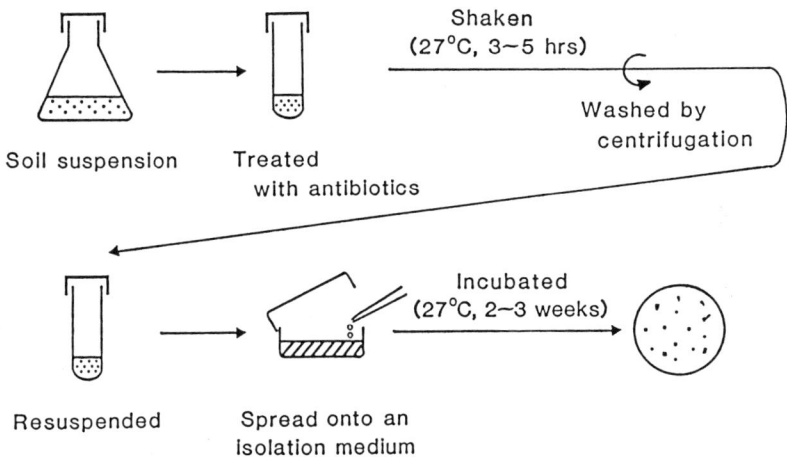

Figure 15.4 Selective isolation procedure for rare actinomycetes by antibiotic pretreatment.

room temperature. For reduction of fungi, chloroform or benzene were the most effective among the solvents tested (methanol, n-propanol, ethyl acetate, chloroform, cyclohexane, dichloromethane, methyl ethyl ketone, toluene, benzene and n-hexane).

Soil samples for screening are usually stored in the laboratory for some time before the isolation experiments and their preservation causes air-drying. Air-drying of a sample generally results in efficient recovery of the spores of actinomycetes because of their resistance to these conditions, and thus facilitates isolation of these microorganisms. However, some microorganisms cannot survive during the preservation period. For detection of these, a Japanese group has carried out an on-the-spot screening experiment using a special car designed for minimizing these risks.

Control of culture conditions on agar media To isolate the desired actinomycetes effectively, important factors are temperature, pH, addition of chemical agents such as antibiotics, etc. Specially designed culture conditions are employed.

Agar media for selective isolation of actinomycetes Various agar media for selective isolation of actinomycetes have been reported (Table 15.5). Among these, a chitin medium is one of the most useful media for isolation of a wide variety of actinomycetes (Lingappa and Lockwood, 1962), because as a rule actinomycetes can utilize chitin for their growth while fungi and bacteria cannot. The small white colonies of actinomycetes grown on a chitin medium generally appear to be similar to each other to the naked eye. Actinomycete strains are usually distinguished after transfer of the colonies from the chitin

Table 15.5 Media used for selective isolation of actinomycetes[1]

Actinomycete genus isolated	Major constituents of agar medium used
Actinomadura	Glucose, glycerol, L-Asp(NH$_2$), mineral salts, vitamins (AV medium)
Microbispora	AV medium
Micromonospora	Mineral salts, sodium propionate, thiamin (M3 medium)
Microtetraspora	Lower layer: glucose, L-Asp(NH$_2$), mineral salts Upper layer: casamino acids
Nocardia	M3 medium
Various genera	Colloidal chitin, mineral salts

[1] Modified from Williams and Wellington, 1982.

medium to a different medium that is suitable for the production of soluble pigments.

Thermophilic actinomycetes   There are some thermophilic actinomycetes, such as *Thermoactinomyces antibioticus* and *Saccharomonospora viridis* (the thermorubin- or thermoviridin-producing strain), although actinomycete strains are generally isolated around 27°C. *Thermoactinomyces antibioticus*, which produces TM-64 (an alkaloid), was isolated in our laboratory at 50°C from a soil sample (Ōmura et al., 1975).

Alkalophilic actinomycete   Horikoshi et al. have studied alkalophilic microorganisms isolated on agar media (pH 9–10) containing Na$_2$CO$_3$ or NaHCO$_3$. These studies gave rise to the discovery of several interesting alkaline enzymes, such as alkaline amylase from *Bacillus* No. A-40-2 and alkaline cellulase from *Bacillus* sp. The latter is an additive to a new type of laundry detergent (Horikoshi 1971). Based on the foregoing studies, antibiotic screening was also carried out, and Tanba et al. (1985) found the M119 complex, new 16-membered macrolide antibiotics, produced by an alkalophilic actinomycete, *Nocardiopsis* sp. M119.

While agar media have been very widely used in microbiological studies, they cannot be used for the study of acidophilic microorganisms because they do not remain solid under acidic conditions. Acidophilic microorganisms are also expected to be useful sources of bioactive metabolites, and thus they should be isolated efficiently if an agar-like material that remains solid under acidic conditions becomes available.

Addition of antibiotics   Antibiotics are generally used for suppressing the growth of rapid-growing actinomycetes, fungi and bacteria, as listed in Table 15.6. Antifungal agents such as cycloheximide or amphotericin B, sodium propionate, and rose bengal, which cause fungi to form smaller colonies, are used for inhibition of fungal growth. Because antibacterial agents interfere

Table 15.6 Antibiotics used in media for selective isolation of actinomycetes[1]

Actinomycete genus selectively isolated	Antibiotic used
Actinomadura	Bruneomycin, kanamycin, rifampicin, rubomycin, streptomycin
Actinoplanes	Novobiocin
Kitasatosporia	Novobiocin
Micromonospora	Gentamicin, novobiocin, tunicamycin
Nocardia	Chlortetracycline, cycloheximide, methacycline, mycostatin, tetracycline
Streptomyces	Cycloheximide, mycostatin, streptomycin
Streptoverticillium	Oxytetracycline
Thermoactinomyces	Cycloheximide, kanamycin, novobiocin rifampicin
Various genera	Polymyxin, polymyxin plus penicillin

[1] Modified from Williams and Wellington, 1982.

with the growth of actinomycetes as well as bacteria, the resistance relationship between *Micromonospora* and novobiocin (Table 15.6), can be efficiently used for their selective isolation. We recently found an efficient isolation method for *Kitasatosporia* strains using novobiocin, as described next.

Utilization of microbial resistance to antibiotics The antibiotic-producing strains are usually resistant to the antibiotics they produce. This phenomenon can be applied efficiently for concentrating the desired antibiotic-producing strains and also for selecting the high-producing ones.

Okami and Hotta have carried out very extensive studies on the relationship between the AG (aminoglycoside antibiotic)- multiresistant pattern of actinomycetes and the variety of AG they produce (Okami and Hotta, 1988). They found that highly multiresistant actinomycetes produced antimicrobial substances more frequently than nonmultiresistant strains. Based on this, he designed an interesting screening system: (1) Selection of AG-multiresistant actinomycetes; (2) classification of the isolated strains by AG-resistant patterns and their taxonomic characteristics; and (3) isolation of antibiotics from the selected new strain different from the known one. Two new antibiotics, dopsisamine and bagougeramine, were thus discovered from less than 1000 AG-resistant actinomycetes. This success may be rationalized by the recent concept that the genes for antibiotic biosynthesis and for antibiotic resistance form a cluster in the genome of the antibiotic-producing strains. (Details of Hotta's investigations are described in Chapter 1).

15.4 Procedures for Selective Isolation of Rare Actinomycetes

Compared with *Streptomyces* strains, rare actinomycetes show the following properties: (i) slower growth, (ii) more complex nutritional requirements, (iii) poorer sporulation, and (iv) instability toward preservation. In addition to these disadvantages, the amount of the desired antibiotic produced by a rare actinomycete frequently has been very limited, and its industrial production has been considered difficult. However, this may now be overcome by the rapid developments in chemical synthesis and gene engineering.

We next describe the outlines and examples of isolation procedures for the following six genera of rare actinomycetes which vary in morphological and chemotaxonomic criteria and are useful microbial sources for antibiotic screening: *Actinomadura, Actinoplanes, Kitasatosporia, Micromonospora, Nocardia,* and *Streptosporangium.* (see Table 15.1 through 15.10).

Selective isolation of *Actinomadura* by the combination of heat treatment and a medium containing rifampicin (Athalye et al., 1981)

Actinomadura strains form short spore chains or pseudosporangia on aerial mycelia.

Athalye et al. reported a selective isolation method for *Actinomadura* strains that combined heat treatment of the soil samples with an agar medium supplemented with rifampicin.

Method Soil (1 g), heated at 100°C for 15 min, is suspended in 10 ml of sterile 25%(vol/vol) Ringer's solution. The suspension is spread on an agar medium (1% glucose, 1% yeast extract, 0.05% K_2HPO_4, 100 μg ml^{-1} cycloheximide, 25 μg ml^{-1} rifampicin, 1.2% agar) and the plate is incubated at 30°C for 6 weeks.

A chemotactic method for isolation of *Actinoplanes* (Palleroni, 1980)

Actinoplanes strains usually form no true aerial mycelia, and their colonies are bright orange. Sporangia are formed on vegetative mycelia. Zoospores are produced in the matured sporangia.

Palleroni found that chloride and bromide ions were effective attractants for assembling zoospores. This chemotactic property was applied to the isolation of *Actinoplanes.*

Method A soil sample (0.5 g) is placed in each of the compartments of an isolation chamber (Figure 15.5). Sterile water is added to a depth of 2 mm in the connecting channel. A capillary filled with sterile 0.01 M phosphate buffer (pH 7) containing 0.01 M KCl is immersed in the chamber at least 1 mm below the water surface and incubated for 1 h at 28–30°C. Released from the sporangia and permitted motility, spores accumulate in the capillary by chemotaxis. The solution in the capillary is spread on a starch-casein agar medium (1% soluble starch, 0.1% casein, 0.05% K_2HPO_4, 0.5% $MgSO_4 \cdot 7H_2O$, and 1.5% agar).

Selective isolation of *Kitasatosporia* by the use of novobiocin (Takahashi et al., 1991)

Scanning electron micrographs of aerial mycelia and vegetative mycelia of a strain of *Kitasatosporia setae* are shown as Figure 15.6. The morphological properties and appearance of colonies of the genus *Ki-*

Chapter 15 Selection of Microbial Sources of Bioactive Compounds 291

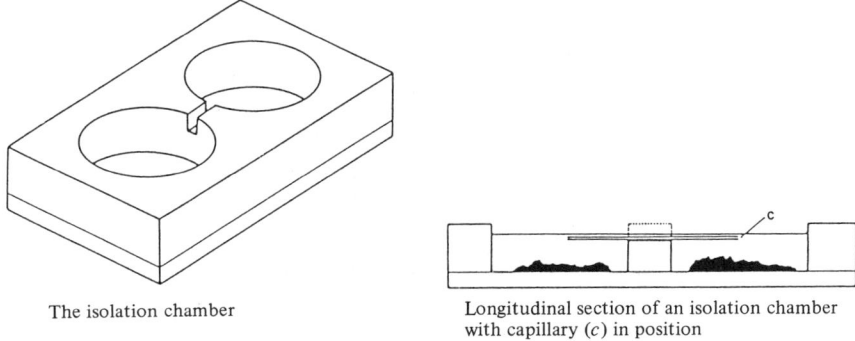

The isolation chamber

Longitudinal section of an isolation chamber with capillary (c) in position

Figure 15.5 Chamber (left) used for isolation of *Actinoplanes* strains by chemotactic method; longitudinal section of the chamber with capillary (c) in position is on right.

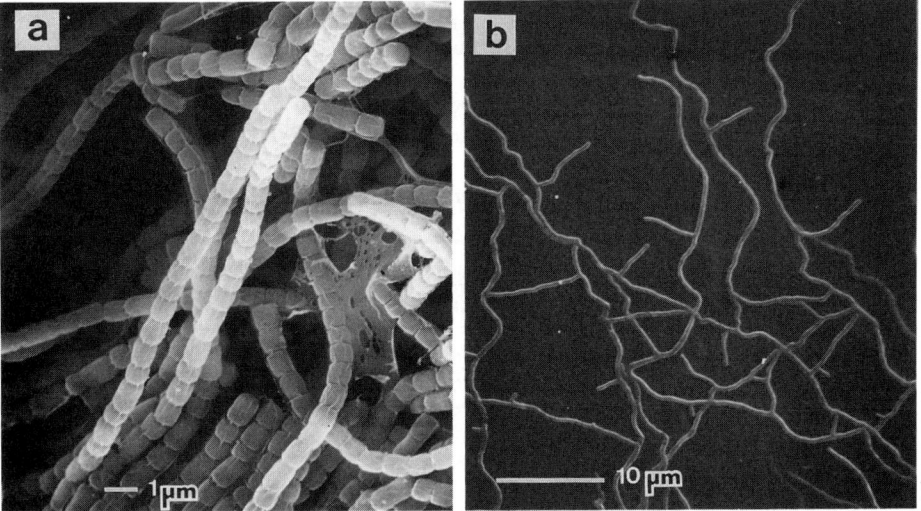

Figure 15.6 Scanning electron micrographs of aerial mycelia (a) and vegetative mycelia (b) of *Kitasatosporia setae* KM-6054 grown on inorganic salts starch agar for 14 days.

tasatosporia resemble those of *Streptomyces* strains, while the cell wall, containing LL-DAP (2,6-diaminopimelic acid), *meso*-DAP, glycine, and galactose, is entirely different from the *Streptomyces* type. It does not belong to any of the cell wall types I–IX in the chemotaxonomy of actinomycetes as proposed by Lechevalier and Lechevalier, 1970).

In a submerged culture of a *Kitasatosporia* strain, submerged spores and filamentous mycelia are observed (Figure 15.7). The former contain LL-DAP and the latter *meso*-DAP. The aerial mycelia contain LL-DAP, and the vegetative mycelia contain *meso*-DAP (Takahashi et al. 1983, 1984b).

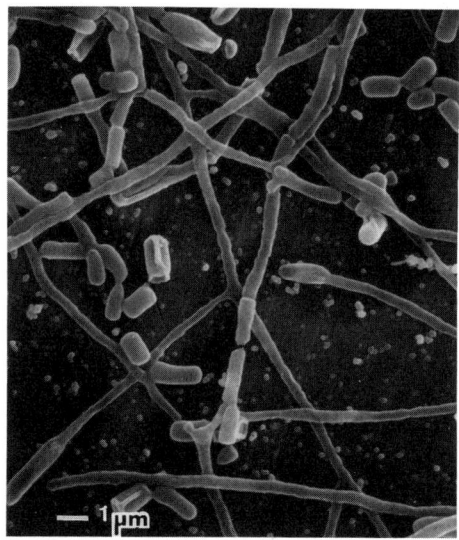

Figure 15.7 Scanning electron micrograph of *Kitasatosporia setae* KM-6054 grown in submerged culture for 72 h.

Table 15.7 Antibiotic sensitivities of some strains of *Kitasatosporia* and *Streptomyces*

Test organism	Minimum inhibitory concentration ($\mu g\ ml^{-1}$)				
	ABPC[1]	GM	NB	TC	Tyl
Kitasatosporia setae	>100	6.25	100	50	100
K. phosalacinea	6.25	6.25	>100	100	100
K. griseola	12.5	3.12	100	12.5	25
Streptomyces griseus	>100	6.25	0.8	25	>100
S. albus	25	1.6	20	100	>100
S. tanashiensis	>100	12.5	1.6	50	12.5
S. hygroscopicus	<0.8	1.6	<0.8	12.5	0.8

[1] Abbreviations: ABPC, ampicillin; GM, gentamicin; NB, novobiocin; TC, tetracycline; Tyl, tylosin.

We found that *Kitasatosporia* strains are resistant to novobiocin (Table 15.7) and isolated 270 actinomycetes from soil using a novobiocin-containing agar medium (1% starch, 1% glycerol, 0.2% $(NH_4)_2SO_4$, 0.1% K_2HPO_4, 0.1% $MgSO_4 \cdot 7H_2O$, 0.1% NaCl, 0.2% $CaCO_3$, 50 $\mu g\ ml^{-1}$ novobiocin, and 1.2% agar). Among them, 4 actinomycetes (1.5%) were found to belong to the genus *Kitasatosporia*. It is suggested that this procedure is effective for the enrichment of *Kitasatosporia* strains.

It is also worth noticing that *Kitasatosporia* strains produce bioactive compounds of diverse structures and biological activities (Table 15.8).

Chapter 15 Selection of Microbial Sources of Bioactive Compounds

Table 15.8 New bioactive compounds produced by *Kitasatosporia* strains

Antibiotic	Bioactivity	Producing organism	Reference
Setamycin	Antitrichomonal, antifungal	*K. setae* KM-6054 *K. griseola* AM-9660	Ōmura et al., 1982 Takahashi et al., 1984a
Phosalacine	Herbicidal	*K. phosalacinea* KA-338	Ōmura et al., 1984
Terpentecin	Antitumor	*K. griseola* MF 730-N6	Tamamura et al., 1985
Propioxatins	Enkephalinase B inhibitor	*K. setae* SANK 60684	lnaoka et al., 1986
FR-900494	Immunomodulator	*K. kifunense* No. 9482	Iwami et al., 1987
AB-110-D	Antibacterial	*K. papulosa* AB-110	Nakamura, et al., 1988
Cystargin	Antifungal	*K. cystarginea* RK-419	Uramoto et al., 1988
Tyrostatin	Proteinase inhibitor	*Kitasatosporia* sp. No. 55	Oda et al., 1989

Selective isolation of *Micromonospora* by the combination of alkaline treatment and a medium containing tunicamycin (Wakisaka et al., 1982) Most species of the genus *Micromonospora* lack aerial mycelia and form light yellow-orange or orange-red colonies. The surface of sporulating colonies turns black or dark brown, and they are sometimes waxy or viscid. Mono spores, formed only on vegetative mycelia, show no motility.

Wakisaka et al. (1982) designed a unique isolation procedure for *Micromonospora* by the combination of alkaline treatment of soil samples and an agar medium containing tunicamycin. The former treatment is effective for reduction of the growth of gram-negative bacteria, and the latter medium supports selective growth of *Micromonospora*, because tunicamycin is an inhibitor of almost all gram-positive bacteria except *Micromonospora*.

Method About 0.5 g soil is suspended in 5 ml of sterile saline containing 0.01% $MgSO_4 \cdot 7H_2O$ and 0.002% Tween 20, and stirred vigorously with glass beads by a mixer. The suspension is placed in a vacuum desiccator for 30 min to remove air. One milliliter of the suspension is mixed with 9 ml of 0.01 N NaOH and the mixture is allowed to stand for 10 min. The sample is neutralized with 0.1 N HCl and plated on an agar medium (1% glucose, 0.2% casamino acids, 0.2% yeast extract, 0.1% beef extract, 1.5% agar) containing 25 μg ml^{-1} of tunicamycin. The plates are incubated at 28°C for 14–21 days.

Paraffin bait technique for isolating *Nocardia* (Orchard et al., 1977)
Most species of the genus *Nocardia* form waxy or lustrous colonies but no aerial mycelia. However, some species produce immature aerial mycelia. The vegetative mycelia fragment into cocci or bacillary elements.

Method A soil sample (1 g) is added to 10 ml of sterile saline and shaken to give a suspension. The suspension is allowed to settle for 2 min, and the supernatant (1 ml) is pipetted into a tube containing 5 ml of McClung's C-free broth: 2 g $NaNO_3$, 0.8 g K_2HPO_4, 0.5% $MgSO_4 \cdot 7H_2O$, 10 mg $FeCl_3$, 8 mg $MnCl_2 \cdot 4H_2O$, 2 mg $ZnSO_4 \cdot 5H_2O$, 1 liter of distilled water, pH 7.2. Paraffin rods are suspended in the tube. After incubation at 37°C for 3 weeks, paraffin is scraped from the rods, broken up, and spread on the surface of DST agar (Oxoid) plates containing 50 μg ml^{-1} cycloheximide and 50 μg ml^{-1} nystatin. Chloroteracycline (45 μg ml^{-1}), demethylchlorotetracycline (5 μg ml^{-1}), and methacycline (10 μg ml^{-1}) are added singly or in combination. The plates are incubated at 25°C for 3 weeks.

Selective isolation of *Streptosporangium* by a combination of heat treatment and a growth medium containing vitamins (Nonomura and Ohara, 1969) The vegetative and aerial mycelia of *Streptosporangium* strains are well developed as with *Streptomyces* strains. Sporangia containing nonmotile spores are produced at the top of the aerial mycelia.

Method Soil samples are air dried at room temperature, then heated at 120°C for 1 h. The treated soil is directly scattered, or a soil suspension in sterile water is spread, on a vitamin-containing medium (AV-agar plates: 0.3 g L-arginine, 1 g glucose, 1 g glycerol, 0.3 g K_2HPO_4, 0.2 g $MgSO_4 \cdot 7H_2O$, 0.3 g NaCl, 0.5 mg thiamine, 0.5 mg riboflavin, 0.5 mg niacin, 0.5 mg pyridoxine, 0.5 mg inositol, 0.5 mg pantothenate, 0.5 mg PABA, 0.25 mg biotin, 10 mg $Fe_2(SO_4)_3$, 1 mg $CuSO_4 \cdot 5H_2O$, 1 mg $ZnSO_4 \cdot 7H_2O$, 1 mg $MnSO_4 \cdot 4H_2O$, 50 mg nystatin, 50 mg cycloheximide, 15 g agar, and 1 liter distilled H_2O).

Bacteria and most actinomycetes such as *Streptomyces*, *Nocardia*, and *Micromonospora* are killed by this heat treatment. However, because *Streptosporangium* and *Microbispora* strains tolerate heat, the isolation frequency of these genera is increased considerably.

15.5 Microbial Sources Other Than Actinomycetes

Early in the study of antibiotics, bacteria and fungi were believed to produce only a few limited types of antibiotics, for example, peptide antibiotics from bacteria or β-lactams from fungi. Now it is widely known that these organisms may also produce varied types of antibiotics. The following three approaches can lead to the discovery of new antibiotics.

Development of characteristic detection methods Some successful examples, such as monobactams or lactivicin, are described in Chapter 1 (Antibacterial Agents).

Development of special isolation methods   The lifecycles of myxobacteria are complicated, and their isolation is very difficult. However, this class of microorganisms is a rich source of antibiotics, and specially designed isolation methods have resulted in the discovery of several antibiotics, such as megovalicins, sorangicins, etc. (Yamanaka et al., 1987; Miyashiro et al., 1988).

Developments of new microbial sources   New information from other scientific fields sometimes suggests a new microbial source for antibiotics. This is the case with microorganisms in symbiosis with marine organisms. The bioactive compounds from marine sources have been now investigated worldwide, and it is noteworthy that the actual producer of bioactive material often is not the marine organism itself but the symbiotic microorganism.

Two new antibiotics, amphidinolides (Kobayashi et al., 1986) and andrimid (Fredenhagen et al., 1987), were reported to be produced by symbiotic microorganisms. The former is a new macrolide antitumor antibiotic produced by *Amphidinium* sp., a symbiont in a marine dinoflagellate, *Amphiscolops* sp. The latter is a new peptide antibiotic from *Enterobacter* sp., an intracellular bacterial symbiont isolated from the eggs of *Nilaparvata lugens*, a brown planthopper, and it is selectively active against *Xanthomonas campestris* pv. *oryzae*.

15.6   Concluding Remarks

Dr. Ōmura's group has been engaged in the screening for bioactive microbial metabolites for more than 15 years. Their discovery of more than 90 new bioactive compounds has been made with the close collaboration of microbiologists, biochemists, chemists, medical researchers, and others (Table 15.9).

Microorganisms may offer us not only new physiologically active materials and lead compounds but also information for chemical modification, as seen in the synthesis of amikacin. Kawaguchi et al. (1972) developed this antibiotic, active against kanamycin-resistant microorganism, on the basis of the fact that butirosin, whose side chain is L-($-$)-γ-amino-α-hydroxybutyric acid, is active against the kanamycin-resistant microorganisms.

The general consideration at present is that the discovery of new bioactive microbial products has become very difficult, when compared to the fruitful period of the 1960s; during this time, various new genera (Nonomura, 1981) of actinomycetes were found, resulting in the discovery of new and interesting bioactive compounds. On the other hand, as shown in Table 15.10, many blanks still remain to be filled with new genera that reflect the groupings of actinomycetes based on morphology and chemotaxonomy.

Future progress in microbiology, biology, and biochemistry will bring forth the design of specific techniques for isolating the microorganisms of these new genera and lead to filling in these blanks.

Table 15.9 Microbial products discovered by Ōmura's group at Kitasato

Microbial product	Producing organism	Bioactivity
Acetylpenicillide	Penicillium sp. FO-608	Acyl-CoA cholesterol acyltransferase inhibitor
Adechlorin	Actinomadura sp. OMR-37	Inhibition of adenosine deaminase
Adecypenol	Streptomyces sp. OM-3223	Inhibition of adenosine deaminase
Aggreceride	Streptomyces sp. OM-3209	Inhibition of platelet aggregation
Aggreticin	Streptomyces sp. OM-4842	Inhibition of platelet aggregation
Ahpatinin	Streptomyces sp. WK-142	Inhibition of acid protease
Alboleutin	Bacillus subtilis AF-8	Antifungal
A73A	Streptomyces viridifaciens MA-4864	Antibacterial
AM-6201	S. xanthochromogenus AM-6201	Anticancer
2'-Amino-2'-deoxyadenosine	Actinomadura corallina sp. nov[1]	Antimycoplasmal, antiviral
Asukamycin	S. nodosus subsp. asukaensis subsp. nov.	Antibacterial, anticoccidial
Atpenin	Penicillium sp. FO-125	Antifungal, inhibition of ATP generation
Aurantinins	Bacillus aurantinus sp. nov.	Antibacterial
Avermectins	S. avermitilis sp. nov[1]	Anthelmintic
Awamycin	Streptomyces sp. 80–217	Antitumor
Azureomycins	Pseudonocardia azurea sp. nov[1]	Antibacterial
Cerulenin	Cephalosporium caerulens sp. nov[1]	Antifungal, antibacterial, inhibition of fatty acid and "polyketide" syntheses
Cervinomycin	S. cervinus sp. nov[1]	Antibacterial
Chimeramycin	S. ambofaciens KA-448	Antibacterial
8-Chloroorobol	Streptomyces sp. OH-1049	Antioxidant
Clostomicin	Micromonospora echinospora subsp. armeniaca subsp. nov.	Antibacterial
3'-O-decarbamoyli-rumamycin	S. subflavus subsp. irumaensis subsp. nov[1]	Antifungal
Diazaquinomycin	Streptomyces sp. OM-704	Antibacterial, antifolate
1,3-Diphenethylurea	Streptomyces sp. AM-2498	Antidepressive
Dityromycin	Streptomyces sp. AM-2504	Antibacterial
Elasnin	S. noboritoensis KM-2753	Inhibition of human granulocyte elastase
Factumycin	S. lavendulae MA-4758	Antibacterial
Frenolicin B	S. roseofulvus AM-3867	Anticoccidial, antimycoplasmal
Furaquinocins	Streptomyces sp. KO-3988	Cytocidal
FO-740	Fusarium sp. FO-740	Acyl-CoA cholesterol acyltransferase inhibitor
Globopeptin	Streptomyces sp. MA-23	Antifungal
Glucopiericidinols	Streptomyces sp. OM-5689	Cytocidal

Chapter 15 Selection of Microbial Sources of Bioactive Compounds 297

Table 15.9 (cont.)

Microbial product	Producing organism	Bioactivity
Herbimycins	S. hygroscopicus AM-3672	Herbicidal, anticancer, anti-TMV, inhibition of tyrosine kinase
Herquline	Penicillium herquei Fg-372	Inhibition of blood platelet aggregation
Hitachimycin	Actinomycetales strain KM4927	Antiprotozoal, anticancer
	Streptomyces sp. KG-2245	
7-Hydro-8-methylpteroyl-glutamylglutamic acid	Promicromonospora sukumoe sp. nov[1]	Antibacterial, antifolate
3'-Hydroxydiadzein	Streptomyces sp. OH-1049	Antioxidant
8-Hydroxydiadzein	Streptomyces sp. OH-1049	Antioxidant
13-Hydroxy-glucopiericidin A	Streptomyces sp. OM-5689	Cytocidal
Irumamycin	S. subflavus subsp. irumaensis subsp. nov.	Antifungal
Irumanolide	S. subflavus subsp. irumaensis subsp. nov.	
Izupeptin	Streptomyces sp. AM-5289	Antibacterial
Jietacins	Streptomyces sp. KP-197	Antinematoda
Karabemycin	Streptomyces sp. AM-6424	Antibacterial
Kazusamycin	Streptomyces sp. 81–484	Antitumor
Kinamycins	S. murayamaensis sp. nov[1]	Antibacterial, Anticancer
KA-107	S. gedanensis KA-107	Inhibition of β-lactamase
KM-8	Streptoverticillium taitoensis sp. nov[1]	Antibacterial, Antifungal
KO-3599	Streptomyces sp. KO-3599	Cytocidal
Lactacystin	Streptomyces sp. OM-6519	Neurite outgrowth promoter
Leucomycins	Streptoverticillium kitasatoensis sp. nov.[1]	Antibacterial, antimycoplasmal
Luminamicin	Nocardioides sp. OMR-59	Antianaerobic
Lustromycin	Streptomyces sp. SK-1071	Antianaerobic
LA-1	Streptoverticillium kitasatoensis sp. nov.[1]	Antibacterial
Mederrhodin A	Streptomyces sp. AM-7161-pIJ2315	Antibacterial
Mishigamycin	S. mishigaensis OS-1804	Antifungal
Nanaomycins	S. rosa subsp. notoensis subsp. nov.[1]	Antifungal, antimycoplasmal, inhibition of anti-platelet aggregation
Neoxaline	Aspergillus japonicus Fg-551	Stimulation of central nervous system
NA-337	Streptomyces sp. NA-337	Fat clearing
O-2867	Streptomyces sp. O-2867	Antifungal
OH-2223	Streptomyces sp. OH-2223	Herbicidal
OM-173	Streptomyces sp. OM-173	Antimycoplasmal, antifungal
OS-3256B	S. candidus subsp. azaticus subsp. nov.[1]	Anticancer

Continued next page

Table 15.9 (cont.)

Microbial product	Producing organism	Bioactivity
OS-1000	*Streptomyces* sp. OS-1000	Milk clotting
Oxetin	*Streptomyces* sp. OM-2317	Herbicidal, glutamine synthetase inhibitor
Oxicenone	*Streptomyces* sp. KO-3599	Cytocidal
Pentalenolactone	*Streptomyces* sp. OM-2718	Antiviral
Phenazinomycin	*Streptomyces* sp. WK-2057	Antitumor
Phosalacine	*Kitasatosporia phosalacinea* sp. nov.[1]	Herbicidal, glutamine synthetase inhibitor
Phthoramycin	*Streptomyces* sp. WK-1875	Antifungal, herbicidal
Phthoxazolin	*Streptomyces* sp. OM-5714	Cellulose synthase inhibitor, herbicidal
Protylonolide	*S. fradiae* KA-427 No. 261	
Prumycin	*S. kagawaensis* sp. nov.[1]	Antifungal, anticancer
Purpactins	*Penicillium purpurogenum* FO-608	Acyl-CoA, cholesterol acyltransferase inhibitor
Pyrindicin	*S. griseoflavus* subsp. *pyrindicus* subsp. nov.	Intestine relaxation, platelet aggregation
Quinoline-2-methanol	*Kitasatoa griseophaeus* sp. nov.[1]	Hypoglycemic
Setamycin	*Kitasatosporia setae* gen. nov. sp. nov.[2]	Antibacterial, antitrichomonal
Setomimycin	*Streptomyces pseudovenezuerae* AM-2947	Antibacterial, anticancer
Sohbumycin	*Streptomyces* sp. 82–85	Anticancer
Sporamycin	*Streptosporangium pseudovulgare* PO-357	Anticancer
Staurosporine	*S. staurosporeus* sp. nov.[1]	Hypotensive, inhibition of protein kinase, inhibition of platelet aggregation
Takaokamycin	*Streptomyces* sp. AC-1978	Antibacterial
Thiotetromycin	*Streptomyces* sp. OM-674	Antibacterial
Tirandamycin C	*Streptomyces* sp. SK-2498	Antibacterial
TM-64	*Thermoactinomyces antibioticus* TM-64	Local anesthetic
Triacsins	*Streptomyces* sp. SK-1894	Inhibition of acyl-CoA synthetase
1*H*-1,2,4-Triazole-3-alanine	*Streptomyces* sp. KM-10329	L-Histidine antagonist
Trienomycins	*Streptomyces* sp. 83–16	Antitumor
Vineomycins	*S. matensis* subsp. *vineus*	Antibacterial, anticancer, inhibition of proline hydroxylase
Virantmycin	*S. nitrosporeus* AM-2722	Antiviral
Virustomycin	*Streptomyces* sp. AM-2604	Antiviral, antitrichomonal
1223A	*Scopulariopsis* sp. F-244	Inhibition of HMG-CoA synthase

[1] new species.
[2] new genus.

Table 15.10 Grouping of genera of actinomycetes by morphological and chemotaxonomic criteria

Sporangium	Aerial mycelium	Cell wall type[1]				
		I	II	III	IV	Others
+	+	□		Planobispora Planomonospora Streptoalloteichus Streptosporangium Spirillospora	Kibdelosporangium[2]	□
+	−		Actinoplanes Ampullariella Dactylosporangium Pilimelia	Frankia	□	Kineosporia
−	+	Intrasporangium Nocardioides Sporichthya Streptomyces Streptoverticillium	Glycomyces	Actinomadura Actinosynnema Microbispora Microtetraspora Nocardiopsis Saccharothrix Thermoactinomyces Thermomonospora	Actinokineospora Actinopolyspora Pseudoamycolata Pseudonocardia Saccharomonospora Saccharopolyspora	Kitasatosporia
−	−	Terrabacter	Catellatospora Micromonospora	Dermatophilus Geodermatophilus	Amycolata Amycolatopsis Mycobacterium Nocardia Rhodococcus Tsukamurella	□

[1] Classification by DAP isomers and sugars according to Lechevalier and Lechevalier (1970).
[2] The members of this genus contains sporangium-like bodies.
Boxes indicate blank spaces, or unknown genera.

References

Athalye, M., Lacey, J. and Goodfellow, M. 1981. Selective isolation and enumeration of actinomycetes using rifampicin. *Journal of Applied Bacteriology* 51:289–297.

Cross, T. 1981. Aquatic actinomycetes. A critical survey of the occurrence, growth and role of actinomycetes in aquatic habitats. *Journal of Applied Bacteriology* 50:397–423.

Fredenhagen, A., Tamura, S. Y., Kenny, P. T. M., Komura, H., Naya, Y., Nakanishi, K., Nishiyama, K., Sugiura, M. and Kita, H. 1987. Andrimid, a new peptide antibiotic produced by an intracellular bacterial symbiont isolated from a brown planthopper. *Journal of the American Chemical Society* 109:4409–4411.

Hagedorn, C. 1976. Influences of soil acidity on *Streptomyces* populations inhabiting forest soils. *Applied and Environmental Microbiology* 32:368–375.

Higashide, E. and Yamamoto, I. 1982. Recent advances in discovery of new antibiotics (I). In: Okami, Y. and Nara, T. (editors), *Microbial Products for Medical use, (II)*, pp. 128–157. Gakkai Shuppan Center, Tokyo.

Horikoshi, K. 1971. Production of alkaline enzymes by alkalophilic microorganisms. Part I. Alkaline protease produced by *Bacillus* No. 221. *Agricultural and Biological Chemistry* 35:1407–1414.

Inaoka, Y., Tamaoki, H., Takahashi, S., Enokita, R. and Okazaki, T. 1986. Propioxatins A and B, new enkephalinase B inhibitors. I. Taxonomy, fermentation, isolation and biological properties. *Journal of Antibiotics* (Tokyo) 39:1368–1377.

Iwami, M., Nakayama, O., Terano, H., Kohsaka, M., Aoki, H. and Imanaka, H. 1987. A new immunomodulator, FR-900494: Taxonomy, fermentation, isolation, and physico-chemical and biological characteristics. *Journal of Antibiotics* (Tokyo) 40:612–622.

Kawaguchi, H., Naito, T., Nakagawa, S. and Fujisawa, K. 1972. BB-K8, a new semisynthetic aminoglycoside antibiotic. *Journal of Antibiotics* (Tokyo) 25:695–708.

Kobayashi, J., Ishibashi, M., Nakamura, H., Ohizumi, Y., Yamasu, T., Sasaki, T. and Hirata, Y. 1986. Amphidinolide-A, a novel antineoplastic macrolide from the marine dinoflagellate *Amphidinium* sp. *Tetrahedron Letters* 27:5755–5758.

Lechevalier, M. P. and Lechevalier, H. A. 1970. Chemical composition as a criterion in the classification of aerobic actinomycetes. *International Journal of Systematic Bacteriology* 20:435–443.

Lingappa, Y. and Lockwood, J. L. 1962. Chitin media for selective isolation and culture of actinomycetes. *Phytopathology* 52:317–323.

Miyashiro, S., Yamanaka, S., Takayama, S. and Shibai, H. 1988. Novel macrocyclic antibiotics: Megovalicins A, B, C, D, G and H. I. Screening of antibiotics-producing myxobacteria and production of megovalicins. *Journal of Antibiotics* (Tokyo) 41:433–438.

Nakamura, Y., Ishii, K., Ono, E., Ishihara, M., Kohda, T., Yokogawa, Y. and Shibai, H. 1988. A novel naturally occurring carbapenem antibiotic, AB-110-D, produced by *Kitasatosporia papulosa* nov. sp. *Journal of Antibiotics* (Tokyo) 41:707–711.

Nolan, R. D. and Cross, T. 1988. Isolation and screening of actinomycetes. 1988. In Goodfellow, M., Williams, S. T. and Mordarski, M. (editors), *Actinomycetes in Biotechnology*, pp.1–32. Academic Press, Orlando.

Nonomura, H. and Ohara, Y. 1959. Distribution of actinomycetes in the soil. Part 3. Grouping of isolates and their frequency of isolation. *Yamanashi Hakkoken* 6:77–88.

Nonomura, H. and Ohara, Y. 1969. Distribution of actinomycetes in soil (IV). A culture method effective for both preferential isolation and enumeration of *Microbispora* and *Streptosporangium* strains in soil (Part I). *Journal of Fermentation Technology* 47:463–469.

Nonomura, H. 1981. Genera of *Actinomycetales* and the selective isolation methods. *Actinomycetologist* 39:3–10.

Oda, K., Fukuda, Y., Murao, S., Uchida, K. and Kainosho, M. 1989. A novel proteinase inhibitor, tyrostatin, inhibiting some pepstatin-insensitive carboxyl proteinases. *Agricultural and Biological Chemistry* 53:405–415.
Okami, Y. and Hotta, K. 1988. Search and discovery of new antibiotics. In Goodfellow, M., Wiliams, S. T. and Mordarski, M. (editors), *Actinomycetes in Biotechnology*, pp.33–67. Academic Press, Orlando.
Okami, Y. and Okazaki, T. 1972. Studies on marine microorgamisms. I. Isolation from the Japan Sea. *Journal of Antibiotics* (Tokyo) 25:456–460.
Okami, Y., Okazaki T., Kitahara, T. and Umezawa, H. 1976. Studies on marine microorganisms. ▽. A new antibiotic, aplasmomycin, produced by a streptomycete isolated from shallow sea mud. *Journal of Antibiotics* (Tokyo) 29:1019–1025.
Okazaki, T. and Enokita, R. 1986. Problems in classification and identification of rare actinomycetes. *Actinomycetologica* 49:15–20.
Ōmura, S., Takahashi, Y., Iwai, Y. and Tanaka, H. 1982. Kitasatosporia, a new genus of the order *Actinomycetales*. *Journal of Antibiotics* (Tokyo) 35:1013–1019.
Ōmura, S., Suzuki, Y., Kitao, C., Takahashi, Y. and Konda, Y. 1975. Isolation of a new sulfur-containing basic substance from a *Thermoactinomyces* species. *Journal of Antibiotics* (Tokyo) 28:609–610.
Ōmura, S., Murata, M., Hanaki, H., Hinotozawa, K., Ōiwa, R. and Tanaka, H. 1984. Phosalacine, a new herbicidal antibiotic containing phosphinothricin. Fermentation, isolation, biological activity and mechanism of action. *Journal of Antibiotics* (Tokyo) 37:829–835.
Orchard, V. A., Goodfellow, M. and Williams, S. T. 1977. Selective isolation and occurrence of nocardiae in soil. *Soil Biology and Biochemistry* 9:233–238.
Palleroni, N. J. 1980. A chemotactic method for the isolation of *Actinoplanaceae*. *Archives of Microbiology* 128:53–55.
Takahashi, Y,. Iwai, Y. and Ōmura, S. 1983. Relationship between cell morphology and the types of diaminopimelic acid in *Kitasatosporia setalba*. *Journal of General and Applied Microbiology* 29:459–465.
Takahashi, Y., Iwai, Y. and Ōmura, S. 1984a. Two new specics of the genus *Kitasatosporia*, *Kitasatosporia phosalacinea* sp. nov. and *Kitasatosporia griseola* sp. nov. *Journal of General and Applied Microbiology* 30:377–387.
Takahashi, Y., Kuwana, T., Iwai, Y. and Ōmura, S. 1984b. Some characteristics of aerial and submerged spores of *Kitasatosporia setalba*. *Journal of General and Applied Microbiology* 30:223–229.
Takahashi, Y., Seki, Y., Tanaka, Y., Ōiwa, R., Iwai, Y. and Ōmura, S. 1990. Vertical distribution of microorganisms in soil. *Actinomycetologica* 4:1–6.
Takahashi, Y., Seki, Y., Iwai, Y. and Ōmura, S. 1991 Taxonomic properties of five *Kitasatosporia* strains isolated by a new method. *Kitasato Archives of Experimental Medicine* 64:123–132
Tamamura, T., Sawa, T., Isshiki, K., Masuda, T., Homma, Y., Iinuma, H., Naganawa, H., Hamada, M., Takeuchi, T. and Umezawa, H. 1985. Isolation and characterization of terpentecin, a new antitumor antibiotic. *Journal of Antibiotics* (Tokyo) 38:1664–1669.
Tanba, H., Kawasaki, T., Adachi, K. and Mizobuchi, S. 1985. M 119-a, a new macrolide antibiotic produced by alkalophilic actinomycetes. In: *Abstracts Interscience Conference on Antimicrobial Agents and Chemotherapy*.
Uramoto, M., Ito, Y., Sekiguchi, R., Shin-ya, K., Kusakabe, H. and Isono, K. 1988. A new antifungal antibiotic, cystargin. *Journal of Antibiotics* (Tokyo) 41:1763–1768.
Wakisaka, Y., Kawamura, Y., Yasuda, Y., Koizumi, K. and Nishimoto, Y. 1982. A selective isolation procedure for *Micromonospora*. *Journal of Antibiotics* (Tokyo) 35:822–836.
Weinstein, M. J. Luedemann, G. M., Oden, E. M. and Wagman, G. H. 1963. Gentamicin, a new broad-spectrum antibiotic complex. *Antimicrobial Agents and Chemotherapy* 1–7.

Williams, S. T. and Wellington, E. M. H. 1982. Principles and problems of selective isolation of microbes. In: Bu'lock, J. D., Nisbet, L. J. and Winstanley, D. J. (editors), *Bioactive Microbial Products: Search and Discovery*, pp. 9–26. Academic Press, London.

Yamanaka, S., Kawaguchi, A. and Komagata, K. 1987. Isolation and identification of myxobacteria from soils and plant materials, with special reference to DNA base composition, quinone system, and cellular fatty acid composition, and with a description of a new species, *Myxococcus flavescens*. *Journal of General and Applied Microbiology* 33:247–265.

16

Fermentation Processes in Screening for New Bioactive Substances

Yoshitake Tanaka

16.1 Introduction

Among the many factors that affect screening for bioactive compounds from microorganisms, the following three are the most important: (1) devised fermentation, (2) unique microorganisms, and (3) selective and sensitive detection methods for the desired bioactivity (Ōmura, 1986).

In the general process of biological screening, fermentation studies are required in all three stages (early, middle, and late), each in different ways and with specific purposes. Fermentation in an early stage of primary screening is crucial among the three, because the overall success of a screening program depends much on initial results. Fermentation in the middle stages is necessary for duplication of initial results, as well as increased production, preliminary purification, and characterization of any compounds detected. Fermentation in a later stage of investigation is carried out on a large scale to provide a sufficient product (10–100 mg) in a pure state for full characterization. This fermentation in the latter stage appears to be closer to a conventional "antibiotic fermentation," in the sense that one selected microorganism is cultivated on a large scale, usually some tens to hundreds of liters.

One fact worthy of consideration is that fermentation in primary screening is different from a conventional "antibiotic fermentation" in both its characteristics and its purpose. The reasoning is as follows:

1. The microorganisms subjected to a screening program are very numerous, different from one another in taxonomic and physiological properties and in their production of secondary metabolites.

2. The biosynthetic mechanisms leading to the active compound and their regulation are unknown. Consequently, information on optimal cultural conditions is not available. Nevertheless, for convenience and simplicity they

are cultivated using a small number of culture media (usually one to two), at the same temperature, and under the same shaking conditions.

3. In a biological screening, although the bioactivity is known and detected at this stage, the chemical properties of microbial products are unknown. The converse is also true in a chemical screening for known structural types.

4. The essential requirements for fermentation in primary screening are that as many microorganisms as possible among those tested produce various secondary metabolites, with the hope that novel compounds be produced by one or of more of these organisms.

Taking into consideration these requirements and the characteristics just described, a variety of fermentation techniques can be employed in effective screening programs. This section will focus on the methods of fermentation concerned with new drug discovery.

16.2 Factors Affecting Antibiotic Biosynthesis

Antibiotics are produced under conditions with the characteristic for the production of other secondary metabolites (Ōmura and Tanaka, 1986): (1) antibiotic production is strain specific; (2) it is unstable, and the productivity tends to disappear on successive transfer of the producing organism and by mutational treatments; (3) antibiotic production follows growth associated kinetics in one medium, but is not growth associated in other media; (4) active antibiotic production often occurs in association with sporulation of the producing organism, which begins with nutritional limitation (such as carbon or nitrogen sources and inorganic phosphate) in cultivation media; and (5) they are produced as members of one or more families with many homologs, and distribution of the components is easily varied by a minor modification of the cultivation conditions.

Although the detailed mechanisms by which antibiotic biosynthesis is initiated and regulated are not well understood, these observations suggest that many factors influence antibiotic production. They are categorized into two types: biochemical (extracellular or nutritional and intracellular) and physicochemical (or operational).

Nutritional factors Nutritional ingredients in a fermentation medium are indispensable for the growth of microorganisms. Simultaneously, however, some of them affect antibiotic production negatively. Antibiotic production is either not initiated or proceeds at a very low rate when glucose, amino acids, and other carbon and nitrogen sources required for microbial growth are present at levels exceeding certain thresholds. Examples are summarized in Table 16.1.

Chapter 16 Fermentation in Screening for Bioactive Substances 305

Table 16.1 Regulation of antibiotic biosynthesis

Antibiotics	Producing microorganism	Regulatory factors	Mechanism of regulation
Antibiotics Derived from Amino Acids			
Penicillin	*Penicillium chrysogenum*	Glucose Methionine	Tripeptide synthesis (R), Induction
Cephalosporin	*Cephalosporium acremonium*	Glucose, NH_4^+, PO_4^{3-}	Ring-expandase (R)
	Streptomyces clavuligerus	NH_4^+, PO_4^{3-}	Cyclase (R,I), ring-expandase (R,I)
Cephamycin	*S. cattleya*	Glucose, NH_4^+, PO_4^{3-}	(NR)
Thienamycin	*S. cattleya*	PO_4^{3-}	(NR)
Actinomycin	*S. antibioticus*	Glucose	Aglycone synthetase (R)
Cycloserine	*S. garyphalus*	PO_4^{3-}	Cyclase (I)
Vancomycin	*S. orientalis*	PO_4^{3-}	(NR)
Bacitracin	*Bacillus licheniformis*	Glucose	Decline of pH
Neoviridomycin	*S. griseoviridus*	Glucose	(NR)
Echinomycin	*S. echinatus*	PO_4^{3-}	pH Control
Colistin	*B. polymyxa*	NH_4^+, PO_4^{3-}	Formation of diaminobutyric acid (R)
Gramicidin S	*B. brevis*	NH_4^+, PO_4^{3-}	(NR)
Antibiotics Derived from Sugars			
Streptomycin	*Streptomyces griseus*	PO_4^{3-}, A-factor	Phosphatase (I) Induction
Kanamycin	*S. kanamyceticus*	PO_4^{3-} Glucose	Phosphatase (I) N-acetyl hydrase (R)
Sagamicin	*Micromonospora sagamiensis*	Co^{2+}	Stimulation of N-methylation
Butirosin	*Bacillus vittelinus*	PO_4^{3-}	Phosphatase (R)
Lincomycin	*S. lincolnensis*	NH_4^+, PO_4^{3-}	(NR)
Antibiotics Derived from Shikimic Acid and Polyketide			
Chloramphenicol	*Streptomyces venezuelae*	NH_4^+ Chloramphenicol	(NR) Feedback inhibition to arylsulfatase
Leucomycin	*Streptoverticillium kitasatoensis*	NH_4^+ n-Butyric acid Glucose	Valine dehydrogenase 3-O-acylase (R) Induction
Spiramycin	*Streptomyces ambofaciens*	n-Butyric acid Glucose	3-O-Acylase (R) Induction

Continued on next page

Table 16.1 (cont.)

Antibiotics	Producing microorganism	Regulatory factors	Mechanism of regulation
Tylosin	S. fradiae	NH_4^+	Metabolism of succinate and valine (R)
		Glucose, PO_4^{3-}	Inhibition of methyl malonic acid synthesis
	cAMP	Stimulation of production	
Tetracycline	S. aureofaciens	PO_4^{3-}	Anhydro tetracycline oxygenase (R)
		Benzylthio-cyanic acid	Induction
Candicidin	S. griseus	PO_4^{3-}	PABA synthetase (R)
Nanaomycin	S. rosa subsp. notoensis	NH_4^+ PO_4^{3-}	Stimulation of production Aglycon synthesis (R)
Avermectin	S. avermitilis	Glucose, PO_4^{3-}, NH_4^+	(NR)
Daunomycin	S. griseus	L factor, I factor	Induction
Rifamycin	Nocardia mediterranei	B factor	Induction
Cerulenin	Caphalosporium caerulens	NH_4^+	(NR)
Patulin	Penicillium urticase	NO_3^-	m-Hydroxgbenzyl alcohol dehydrogenase (R)
Aflatoxin	Aspergillus parasiticus	Glucose, cAMP	Stimulation of production
Others			
Pyrromycin	Streptomyces alboniger	Glucose	3-O-α-Methyltransferase (R)
Novobiocin	S. niveus	Citric acid, NH_4^+	(NR)

[1] Abbreviations: (R), repression; (I), inhibition; (NR), Not reported.

Glucose, ammonia and inorganic phosphate   Table 16.1 illustrates that glucose, ammonium ions, and inorganic phosphate very often but not always suppress antibiotic production. The biochemical mechanisms of the negative effects have been elucidated in some cases.

The production of β-lactam antibiotics (cephalosporin C and penicillin N as the major components) by *Cephalosporium acremonium* (correctly *Acre-*

monium chrysogenum) is inhibited by glucose, ammonium ions, and inorganic phosphate when one or more of these nutrients are present at high concentrations in a medium. Analyses of bioconversion and enzymatic reactions have revealed that the three nutrients inhibit both the formation and activity of "ring expandase," an enzyme catalyzing the conversion of the 5-membered ring intermediate isopenicillin N into the 6-membered deacetoxycephalosporin C (Zanca and Martin, 1983).

Production of the 16-membered macrolide antibiotic tylosin by *Streptomyces fradiae* does not appear to be affected by 4% glucose added to a starch-based production medium. However, production was severely reduced when ammonium sulfate and potassium phosphate were added individually to the same medium. Ammonia is generated by microbial degradation of nitrogen-containing nutrients such as casein and peptone, and when released into a medium it inhibits tylosin production. Studies using blocked mutants, resting cells, and enzyme systems demonstrated that ammonia interferes with the formation of the three building units (acetate, propionate, and butyrate) used for aglycone biosynthesis (Tanaka et al., 1986).

Industrial production of chlortetracycline (CTc) is carried out using *Streptomyces aureofaciens* in a complex medium containing sucrose as the major carbon source. The pH is maintained below 6 by the addition of ammonia water. It is unlikely that sucrose and ammonia have an inhibitory effect on CTc formation under these conditions, although the addition of glucose or fructose to the sucrose-based medium is inhibitory. CTc production is highly susceptible to inorganic phosphate, so antibiotic formation begins only after exhaustion of inorganic phosphte in the medium. Under these conditions, inorganic phosphate is the rate-limiting factor for the growth of *S. aureofaciens*. Inorganic phosphate suppresses the formation of anhydrotetracycline oxygenase and perhaps some other enzymes involved in CTc biosynthesis (Behal et al., 1982).

Candicidin production by *Streptomyces griseus* is inhibited by inorganic phosphate because this nutrient suppresses the biogenesis of para-aminobenzoate, the starter unit for the synthesis of the 38-membered heptaene macrolide antibiotic (Gil *et al.*, 1985).

In general, inhibitory nutrients can be either glucose and other easily assimilated carbon sources, ammonia and other easily assimilable nitrogen sources, or inorganic phosphate and other phosphate-generating ingredients. Which one of the three classes exerts the major inhibitory action depends on the strain and the fermentation conditions employed. Namely, it depends on the mechanisms of biosynthesis and biodegradation of the antibiotic and their regulation. Because of these factors antibiotic production usually follows typical non-growth-associated kinetics.

Metal ions Several metal ions affect antibiotic production. The examples, summarized in Table 16.2, include both promotion and inhibition of antibiotic biosynthesis, which are caused at steps of transmethylation (Co^{2+}), biodegradation (Cu^{2+} and Mg^{2+}), and others.

Table 16.2 Influence of metal ions on antibiotic production

Antibiotic	Producing microorganism	Metal ion	Action
Putulin	*Penicillium urticae*	Mn^{2+}	Stimulation of production
Sagamicin	*Micromonospora sagamiensis*	Co^{2+}	Stimulation of N-methylation
A-Factor	*Streptomyces griseus*	Co^{2+}	Stimulation of production
Bialaphos	*S. hygroscopicus*	Co^{2+}	Stimulation of P-methylation
Lincomycin	*S. lincolnensis*	$Mg_3(PO_4)_2$	Reduction of NH_4^+ content
Candicidin	*S. griseus*	Zn^{2+}, Fe^{2+}, Mg^{2+}	Reduction of PO_4^{3-} content
Nourseothricin	*S. noursei*	Zn^{2+}	Stimulation of production
Gentamicin	*M. purpurea*	Ca^{2+}	Stimulation of N-methylation
β-Lactam	*S. clavuligerus*	Fe^{2+}	Reduction of inhibition by PO_4^{3-}

Precursors It is well known that the addition of amino acids, short chain fatty acids, and corresponding alcohols (e.g., propanol) often enhances antibiotic production, although the mechanism of enhancement has not always been reported. The enhancement can occur when the added compounds are converted to an intermediate or their precursors. If the pool size of the precursor is small, the enhancing effect can be pronounced.

Additions interacting with inhibitory substances or products A new fermentation technique was developed using NH4+- and phosphate-trapping agents to achieve 2- to 10-fold increases in the production of many types of antibiotics (Tanaka and Ōmura, 1988). The trapping agents include, as NH_4^+-trapping agents, magnesium phosphate and synthetic and natural zeolites such as mordenite; as phosphate-trapping agents, allophane, a noncrystalline clay of aluminosilicate, and magnesium carbonate. Fermentations were carried out in a conventional manner in shake flasks or in jar fermentors supplemented with one of the trapping agents before autoclaving. Ammonia and phosphate levels were lowered in the supplemented media, and antibiotic production increased markedly (Table 16.3) (Tanaka and Ōmura, 1988). The terms "ammonium ion-depressed fermentation" (Masuma et al., 1983) and "phosphate ion-depressed fermentation" (Masuma et al., 1986) were proposed for these techniques. The essential feature of the method is that ammonium ions or inorganic phosphate are decreased to a level at which they exert no inhibition but are sufficient to support microbial growth.

Table 16.3 Enhancement of production of antibiotics by ammonium ion- and phosphate-trapping agents

Trapping agents	Amount added (%)	Antibiotics	Maximum antibiotic titers (μg ml^{-1})	
			No addition	Addition
(NH$_4^+$)				
Magnesium phosphate	1.0	Leucomycin	700	3800
Sodium phospho-tungstate	0.5	Spiramycin	150	450
Natural zeolite	1.0	Tylosin	59	149
Magnesium phosphate	1.0	Cephalosporin	400	1600
NH$_4^+$-saturated zeolite	0.2	Nanaomycin	85	750
Natural zeolite	1.0	Cerulenin	40	280
(PO$_4^{3-}$)				
Allophane	0.5	Tylosin	50	130
Allophane	0.5	Nanaomycin	110	505
Allophane	0.5	Candicidin	180	530

Zeolite, kaoline, celite, and other natural minerals interact with microbial cells to suppress pellet formation. This can lead to enhancement of antibiotic production if increased uptake of oxygen and nutrients, which occurs with suppressed pellet formation, favors secondary metabolite biosynthesis. Cyclodextrins form inclusion complexes with, for instance, higher fatty acids, thereby rendering lipophilic compounds water soluble, volatile substances non volatile, unstable substances more stable, and compounds that are susceptible to microbial attack less so. Therefore it might be expected that the addition of cyclodextrins and their derivatives to a medium would increase antibiotic production. It turned out that lankacidin production is so affected (Sawada et al., 1987). The addition of β-cyclodextrin (0.2%–1.0%) promoted lankacidin production by about 10 fold. It was proposed that cyclodextrin stabilized lankacidin by forming an inclusion complex, which made lankacidin no more active as an feedback inhibitor of its own biosynthesis (Sawada et al., 1990).

Intracellular inducers   Antibiotic biosynthesis appears to be regulated by intracellular inducers of low molecular weight, such as those listed in Table 16.4.

Physicochemical factors   Physicochemical factors involved in fermentation technology, such as pH, temperature, and oxygen tension, were proposed as affecting antibiotic production following observations on fermentations using conical flasks (Higashide, 1984). These suggestions are valuable,

Table 16.4 Intracellular factors inducing antibiotic biosynthesis

Intracellular inducer	Producing microorganism	Antibiotic production induced
A Factor	*Streptomyces griseus*	Streptomycin
B Factor	*Nocardia mediterranei*	Rifamycin
I Factor	*S. griseus*	Daunomycin
L Factor	*S. griseus*	Daunomycin
Inducing material	*S. virginiae*	Virginiamycin

although rather qualitative. Quantitative analysis of the effect of a physicochemical factor on antibiotic production requires all other factors to be controlled at constant levels. A fermentation system equipped with various sophisticated monitoring devices may be able to do this, but such a system often is not available.

pH The pH of a culture medium, as well as mycelial growth, is an important and convenient measure to monitor antibiotic fermentation. The pH values of the media are adjusted to acidic, neutral, and alkaline ranges for growth of acidophilic, mesophilic, and alkalophilic microorganisms, respectively. Actual pH values within these growth-permissible pH ranges affect antibiotic production greatly. The pH of a medium can be controlled automatically in jar fermentors. With fermentations in flasks, pH is controlled by adding mineral acids, $CaCO_3$, phosphate salts, ammonia, and Na_2CO_3 to fermentation media. Under these conditions, actual pH values vary unavoidably, and the added materials have their own specific effects on antibiotic production, in addition to suppression of pH changes.

The inhibitory effect of glucose on bacitracin production results from a pH effect (see Table 16.1). Addition of short-chain fatty acids often results in a decrease in both mycelial growth and antibiotic production. This is probably also caused by an effect on intracellular pH.

The biochemical mechanisms of these pH effects are not known. It is assumed that growth at the suitable pH is favorable for the synthesis of a relevant enzyme, or that the medium pH coincides with the optimal pH of a key biosynthetic enzyme. Caution is needed in this interpretation, however, because intracellular pH values are independent of medium pH, lying within a range of 6 to 8, no matter whether cells are grown in an acidic medium or under highly alkaline conditions.

Buffering agents occasionally have negative effects on antibiotic production. Phosphate salts are often inhibitory. Therefore, when pH control is required around the neutral pH range, buffers such as MOPS (*N*-morpholinopropanesulfonic acid) and other synthetic amino acids are added to the medium. $CaCO_3$, one of the most common buffering agents used to avoid overly acidic conditions, was found to inhibit cerulenin production, shifting fermentation in favor of production of a by-product, helvolic acid (Iwai et al,. 1973).

Temperature   The effect of temperature on antibiotic production is similar to that of pH. Cryophilic, mesophilic, and thermophilic microorganisms must be cultivated under conditions of low (0°–15°C), physiological (20°–40°C), and high (>50°C) temperatures, respectively. Antibiotic production varies when growth temperatures are varied within any one of the three temperature ranges. A cryophilic strain of *Streptomyces* produced an antibacterial antibiotic A-60 at 15°C, but did not do so at 28°C, a permissible temperature, because an enzyme that inactivates the antibiotic was produced at 28°C (Ogata et al., 1977).

Aeration, agitation and pellet formation   Oxygen, absolutely necessary for aerobic growth of microorganisms, is dissolved into culture fluids by shaking culture flasks on a rotary (or reciprocal) shaking machine or by aeration and agitation in a jar fermentor. Carbon dioxide is dissolved together with oxygen when air is used, and both of these gases affect antibiotic production. Dissolved oxygen tension decreases to its lowest level when the growth rate is at its maximum. If the minimum level reaches zero, the antibiotic biosynthetic machinery appears to be damaged in some organisms, and antibiotic production does not start when the normal production phase is entered. According to Arai et al., (1976), when a strain of *Streptomyces* sp. was grown in a jar fermentor with agitation at 500 rpm, the minimum oxygen tension was sufficiently high to allow later coproduction of mimosamycin and chlorocarcins A, B, and C. When the agitation was reduced to 250 rpm, the minimum oxygen tension reached zero, and no production of these two antibiotics occurred. However, production of streptothricin did begin.

Tetracycline production by *S. aureofaciens* was susceptible to low oxygen tension but was not to CO_2. Erythromycin production by *Saccharopolyspora erythaeus* was enhanced by high dissolved CO_2, whereas penicillin production by *Penicillium chrysogenum* was inhibited by it.

Under conventional fermentation conditions the filamentous mycelia of fungi or streptomycetes are intertwined to form various sizes of "knit balls" called pellets. Pellet formation affects the viscosity of culture fluid and the physiological conditions of the cells involved in pellets. The formation of large pellets ≈3–10 mm in diameter results in a reduction of viscosity of the culture fluid, while oxygen transfer into the pellets is hindered. An anaerobic atmosphere exists inside large pellets. The formation of small pellets <3 mm in diameter also reduces somewhat the viscosity of the culture fluid, although oxygen transfer into the pellets is not limited. Culture fluid without pellet formation is usually highly viscous, and oxygen transfer in such culture fluids is impeded. Because pellet formation depends on agitation conditions, this is another way in which aeration and agitation affect antibiotic production.

Intermittent addition of water to a culture may be useful in reducing viscosity. The addition of zeolite, celite, kaolin, etc., suppressed pellet formation and increased antibiotic production by various *Streptomyces* strains.

Osmotic pressure Marine microorganisms, such as *Streptomyces tenjimariensis*, a producer of an aminoglycoside, istamycin, tolerated 5%–7% NaCl in culture media (Hotta et al., 1980). Some cultures required 3%–5% NaCl for growth, although many other marine microorganisms grew under conditions that also support land microorganisms. Terrestrial microorganisms grow in media containing less than 2% NaCl. Marine microorganisms that required an extremely low concentrations of organic nutrients for growth, that is, oligotrophy, were described by Kuznetsov et al. (1979). These grew slowly in media diluted to one-tenth to one-hundredth of the strength of a conventional medium.

16.3 Control of Fermentation Conditions

Variations in the fermentation environment often result in an alteration in antibiotic production. The alteration involves changes both in yields and in the composition of active components, both phenomena being associated with the time-dependent appearance and disappearance of some of the components actually produced. The newly appearing components may be novel compounds. Thus, new drug discovery is supported by an appropriate control of fermentation conditions. Successful examples of this approach to finding new bioactive compounds are described next.

Selection of nutrients for fermentation media Selection of suitable nutrients for fermentation media is commonly relied on to improve antibiotic production. Fermentation media are composed of carbon sources, nitrogen sources, inorganic metallic salts, and buffering agents such as $CaCO_3$. Complex nutrients are preferred because they often support higher yields and also for economic reasons; chemically defined media are rarely used. Table 16.5 exemplifies antibiotic production media reported in the literature, and illustrates various combinations of carbon sources (0–5%) and nitrogen sources (0–3%).

Ammonium salts or inorganic phosphates usually are not added to complex media because they are included in complex nutrients and because high concentrations of ammonium salts and inorganic phosphate do not favor antibiotic biosynthesis (as described under Facters Affecting Antibiotic Biosynthesis).

Media containing high concentrations of phosphate Despite the widely accepted notion that low-phosphate media are favorable, a high-phosphate medium containing 1%–2% inorganic phosphate was successfully employed in the fermentation yielding pyrrolnitrin, an antidermatophytic anti-

Table 16.5 Composition of typical media used for production of bioactive secondary metabolites

Avermectins[1]	Difficidins[2]	K-252[3]	Monacolin M[4]
Cerelose, 4.5%	Dextrin, 4%	Glucose, 0.5%	Glycerol, 7%
Peptonized milk, 2.4%	Distillers soluble, 0.7%	Soluble starch, 3%	Glucose, 3%
Autolyzed yeast, 0.25%	Yeast extract, 0.5%	Soybean meal, 3%	Meat extract, 3%
Polyglycol P2000, 2.5 ml liter^{-1}	$CoCl_2 \cdot 6H_2O$, 0.01%	Corn steep liquor, 0.5%	Peptone, 0.8%
		Yeast extract, 0.5%	$NaNO_3$, 0.2%
		$CaCO_3$, 0.3%	$MgSO_4 \cdot 7H_2O$, 0.01%
pH 7.0	pH 7.3	pH 7.2	pH 5.0

[1] From Burg et al., 1979.
[2] From Zimmerman et al., 1987.
[3] From Nakanishi et al., 1986.
[4] From Endo et al., 1986.

biotic (Arima et al., 1964). During the course of a study on the optimal cultivation conditions for pyrrolnitrin production by *Pseudomonas fluorescens*, it was found by Imanaka in Arima's laboratory (University of Tokyo, Tokyo), that pyrrolnitrin production became more reproducible and higher in the presence of 2% inorganic phosphate. This effect was found after the accidental addition of 10 times more inorganic phosphate than had been planned to a fermentation medium.

Following on this finding, Imanaka, after rejoining the Fujisawa Co., started systematic screening projects using high-phosphate media. Imanaka and coworkers thus discovered many bioactive compounds of bacterial, actinomycete, and fungal origins (Table 16.6). These include the monocyclic β-lactam nocardicin (Aoki et al., 1976) and the immunostimulant FK-156 (Gotoh et al., 1982). Other Japanese groups, encouraged by Fujisawa's success, employed similar methods, leading to the discovery of carpetimycin, a thermostable β-lactamase inhibitor, cationomycin, a polyether (Nakamura et al., 1981), and oxetanocin, an inhibitor of reverse transcriptase (Shimada et al., 1986) (see Table 16.6). Selected compounds are described later in some detail.

Table 16.6 Bioactive substances discovered in fermentation media containing various salts added at high concentrations

Addition (%)	Antibiotics discovered
Phosphate salt (0.5–3.6)	Pyrrolnitrin, Herbicidin, FR-31564, FR-32863, FR-33289, FR-900130, FR-900148, Herbicolins, Malioxamycin, Carpetimycin, *epi*-Thienamycin, Candiplanecin, Cationomycin, 3-Demethoxy-3-ethoxy- tetracenomycin C, AT-265, 6643-X, FK-156, Demethoxyrapamycin, Muraceins, Penitricin, SN-7, Mitomycin analogs, Oxetanocin, PB-5266A, B & C
Ammonium salt (0.5–1.0)	Octapeptin D, Gilvocarcins, 4-*epi*-Cetocycline, AN-3, 8006-I, Safracins, Ancovenin, AN-7A, 7B & 7C, AN-1, I5B2, Phenacein, Monacolins A & J, Biphenomycins A & B, Dihydrofusarubins, Pereniporins A & B, Herbimycin B, Karabemycin
Magnesium salt (0.3–1.25)	A-60, Senacarcin A, Neocarzinostatin chromophore, Y-T0678 H, Chicamycin
Phosphate salt (0.5–3.6) plus magnesum salt (0.25–1.0)	Bicyclomycin, Nocardicin A, FR-900098, FR-900137, Echinosporin, Oxirapentyn
Phosphate salt (0.5–1.0) plus ammonium salt (0.5–1.0)	Valienamine, Fengycin, Amiclenomycin peptides

Sources: *Journal of Antibiotics* (Tokyo) (1975–1990); *Agricultural and Biological Chemistry* (1960–1988); *Antimicrobial Agents and Chemotherapy* (1970–1988).

According to Imanaka, high phosphate favors antibiotic production in three ways: as a constituent atom of an antibiotic molecule, as a pH controller, and as an agent to suppress production of other known bioactive compounds. Hall and Hassall (1970) demonstrated that different antibacterial compounds were produced by a single microorganism in low- and high-phoshate conditions. It is likely that similar events occurred in Imanaka's screening media.

Nocardicins (Aoki et al., 1976) Nocardicins are produced by a soil isolate, designated *Nocardia uniforms* subsp. *tsuyamaensis* ATCC 21806. The producing organism is more likely to be a strain of *Actinosynnema*, because synnemata characteristic to this genus was observed later. The antibiotic activity was detected using a β-lactam-hypersensitive mutant of *E. coli*, strain Es-11. Nocardicin was the first monocyclic β-lactam (monobactam) antibiotic (Figure 16.1).

The fermentation was carried out in a high-phosphate medium (glycerol, 3%; cottonseed meal, 2%; dried yeast cells, 2%; KH_2PO_4, 2.18%; $Na_2HPO_4 \cdot 12H_2O$, 1.43%; and $MgCl_2 \cdot 6H_2O$, 0.5%) at 30°C for 4 days.

Cationomycin (Nakamura et al., 1981) The polyether antibiotic cationomycin (see Figure 16.1) is produced by a strain of *Actinomadura azurea*. It is active against gram-positive bacteria and shows anticoccidial activity in vivo with low toxicity. It is an ionophore with its highest affinity being for Ca^{2+}.

The fermentation was conducted at 28°C for 9 days in a high-phosphate medium (glycerol, 1.5%; oat meal 3%; dried yeast cells, 0.5%, KH_2PO_4, 0.5%; $Na_2HPO_4 \cdot 12H_2O$, 0.5%; $MgCl_2 \cdot 6H_2O$, 0.1%).

An immunostimulant, FK-156 (Gotoh et al., 1982) FK-156 was discovered in the cultured broths of two soil actinomycetes, *Streptomyces olivaceogriseus* ATCC 31427 and *S. violaceus* ATCC 31481. It is an adjuvant-active and immunostimulating peptide with structural similarity to peptidoglycan (see Figure 16.1). It shows protection against bacterial infection, enhancement of humoral antibody formation, and anticancer activity in experimental animals.

The fermentation was carried out at 30°C for 3 days in a high-phosphate medium (soluble starch, 2%; cottonseed meal, 0.5%; wheat germ, 0.5%; dried yeast cells, 0.25%; corn steep liquor, 0.25%; KH_2PO_4, 0.5%; $Na_2HPO_4 \cdot 12H_2O$, 1.25%).

Oxetanocin (Shimada et al., 1986) Oxetanocin was found in the culture broth of *Bacillus megaterium* NK 84-0218. It is a new nucleoside antibiotic having an oxetane ring as the sugar moiety (see Figure 16.1). It is active against restricted genera of gram-positive bacteria but is inactive against fungi. It also shows activity against Herpes simplex virus (type 2) and the human immunodeficiency virus in vitro.

The producing organism was cultured at 37°C for 43 h in a high-phos-

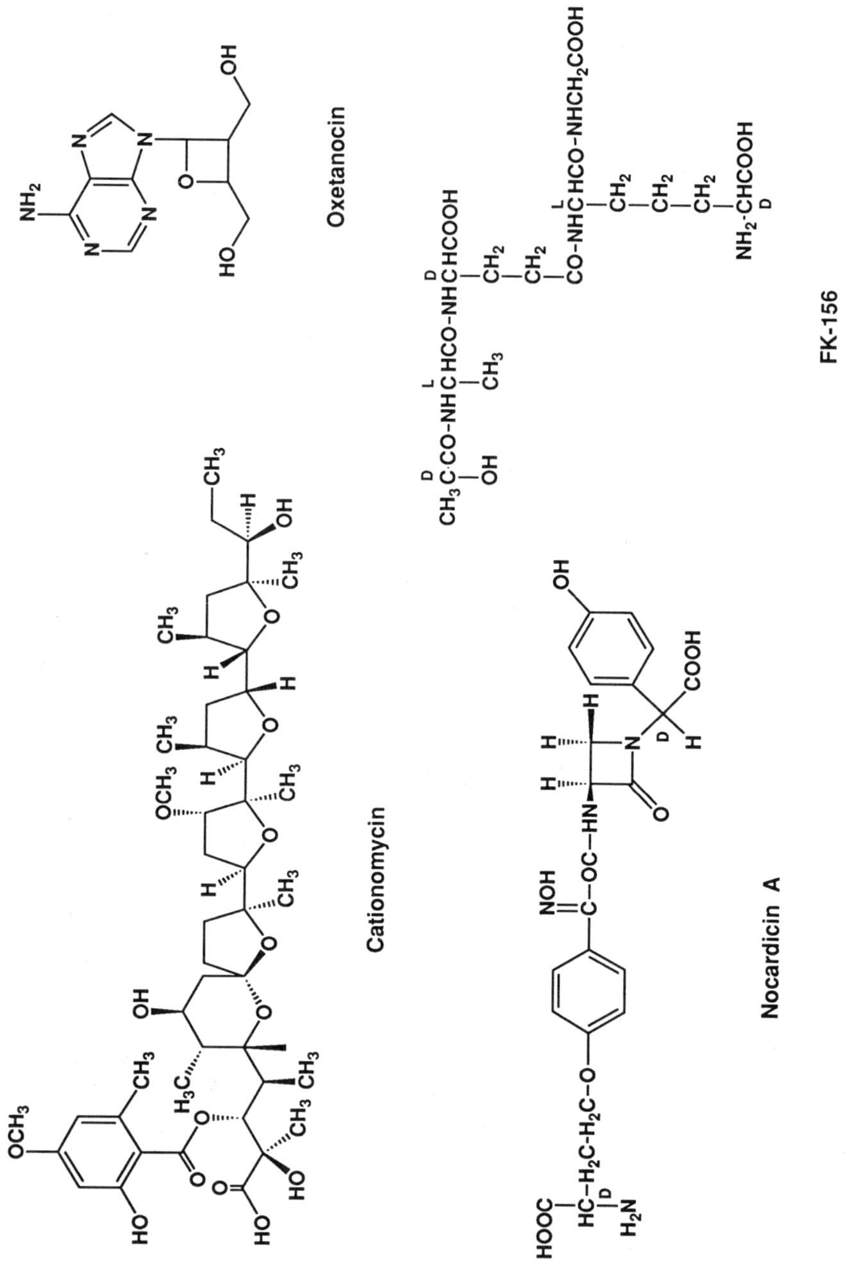

Figure 16.1 Structures of bioactive compounds produced in high-phosphate media.

phate medium (soluble starch, 2%; soybean meal, 1.5%; KH_2PO_4, 0.3%; Na_2HPO_4, 0.2%; $MgSO_4 \cdot 7H_2O$, 0.05%; $CoCl_2 6H_2O$, 0.0002%; $FeSO_4 \cdot 7H_2O$, 0.0002%; pH 6 before sterilization).

Media containing high contents of other salts A high-phosphate medium is regarded as yielding extreme fermentation conditions considering the well-documented inhibitory action of inorganic phosphate on antibiotic biosynthesis. Extreme fermentation conditions are also caused by other nutrients. The addition of ammonium sulfate, ammonium phosphate, NaCl, or magnesium sulfate in high concentrations to antibiotic fermentation media has led to the finding of new compounds (see Table 16.6). Antibiotic A-60 (Ogata et al., 1977) was produced only in the presence of greater than 2% magnesium sulfate. The mechanism by which antibiotic biosynthesis was stimulated greatly by this salt was not determined.

Ammonium ion-depressed and phosphate-depressed fermentations
In contrast to the high ammonium salt and high-phosphate media just described, Tanaka and Ōmura (1988) developed ammonium ion-depressed and phosphate ion-depressed fermentations for antibiotic screening. They utilized ammonium ion-trapping agents and phosphate-trapping agents. As described in 16.2, the production of many antibiotics is enhanced markedly when magnesium phosphate or allophane is added to a fermentation medium. If an antibiotic is produced only in the presence of a trapping agent, it may be a new compound. Based on this reasoning, screening for bioactive compounds was conducted in culture media containing magnesium phophate (ammonium ion-trapping agent) and magnesium carbonate or allophane (phosphate- trapping agent), and eight new bioactive compounds, listed in Table 16.7, were discovered. They are nanaomycin derivatives (Iwai et al., 1983), globopeptin

Table 16.7 New antibiotics found at Kitasato in screening using ammonium-depressed or phosphate-depressed fermentations

Addition (%)	Nutritional limitation	New antibiotics	Producing microorganism
MgP[1] (0.5)	NH_4^+	Nanaomycin αA, αB, αE	*Streptomyces* sp. OM-173
		Thiotetromycin	*Streptomyces* sp. OM-674
		Diazaquinomycin	*Streptomyces* sp. OM-704
		Globopeptin	*Streptomyces* sp. MA-23
MgC[2] (0.5)	PO_4^{3-}	Triacsins	*Streptomyces* sp. SK-1894
		Ahpatinin	*Streptomyces* sp. WK-142
Allophane[3] (0.5%)	PO_4^{3-}	Jietacins	*Streptomyces* sp. KP-197
		Phthoxazolin	*Streptomyces* sp. OM-5714

[1] MgP, Magnesium phosphate, $Mg_3(PO_4)_2$ $8H_2O$.
[2] MgC, Magnesium carbonate, $4MgCO3$ $Mg(OH)_2$ $5H_2O$.
[3] Allophane, Alumino silicate, Al_2O_3 $SiO2$ nH_2O (n = 2–6), a noncrystalline clay.

(Tanaka et al., 1987), diazaquinomycin (Ōmura et al., 1982), thiotetromycin (Ōmura et al., 1983a), triacsin (Ōmura et al., 1986a), ahpatinin (Ōmura et al., 1986b), phthoxazolin (Ōmura et al.,1990) and jietacin, (Ōmura et al., 1987). The structures of these compounds are shown in Figure 16.2.

Diazaquinomycin (Ōmura et al., 1982) Diazaquinomycin is produced by *Streptomyces* sp. OM-704. It is active against gram-positive bacteria and inhibits thymidilate synthase. It is the first naturally occurring antifolate so far reported. Diazaquinomycin was discovered by an antifolate assay using *Enterococcus faecium*, a folate-requiring bacterium.

The fermentation was carried out in a medium containing an ammonium ion-trapping agent (0.5% glucose, 1% corn steep liquor, 1% oatmeal, 1% Pharmamedia, 0.5% K_2HPO_4, 0.5% $MgSO_4 \cdot 7H_2O$, and 0.001% each $FeSO_4 \cdot 7H_2O$, $MnCl_2 \cdot 4H_2O$, $ZnSO_4 \; 7H_2O$, $CuSO_4 \cdot 5H_2O$, and $CoCl_2 \cdot 2H_2O$, pH 7.0).

Nanaomycins (Ōmura et al., 1974, Iwai et al., 1983) Nanaomycins A–E were discovered in 1974 in a culture of *Streptomyces rosa* subsp. *notoensis* KA-301. They are active against gram-positive bacteria, mycoplasma, and dermatophytic fungi. Nanaomycin A has found veterinary use as an excellent agent against ringworm of cattle. Later, new methyl ester derivatives of the nanaomycin family, nanaomycins αA, αB, and αE, were found in 1983 in the culture broth of *Streptomyces* sp. OM-173. They show antimicrobial spectra similar to those of the nonmethylated series A–E.

Streptomyces sp. OM-173 was cultivated in the presence of an ammonium ion-trapping agent, magnesium phosphate (MgP) (2% glycerol, 2% starch, 0.2% corn steep liquor, 1% soybean meal, 0.5% meat extract, 0.3% yeast extract, 0.3% $CaCO_3$, and 1% $Mg_3(PO_4)_2 \cdot 8H_2O$, pH 7.0).

The fermentation of the former nanaomycin producer, strain KA-301, in a MgP-containing medium did not result in the production of the methyl ester series of nanaomycins.

Triacsins (Ōmura et al., 1986a) The triacsins are produced by *Streptomyces* sp. SK-1894. They were discovered in a screening program for inhibitors of lipid metabolism using acyl-CoA synthetase I-deficient and fatty acyl-CoA synthetase-deficient mutants of *Candida lipolytica*. The triacsins specifically inhibit acyl-CoA synthetase from bacteria and animal cells. They also inhibit acylation of proteins and glycoconjugates. WS-1228 A and WS-1228 B, structurally related to the triacsins, were reported as hypotensive vasodilators. No correlation was found between the acyl-CoA I-inhibiting activity and the hypotensive vasodilating activity of the triacsins. The fermentation was conducted at 27°C for 60 h in a medium containing basic magnesium carbonate, a phosphate-trapping agent (1% glucose, 1% soluble starch, 0.3% corn steep liquor, 1% oatmeal, 1% Pharmamedia, and 0.5% basic magnesium carbonate).

Chapter 16 Fermentation in Screening for Bioactive Substances 319

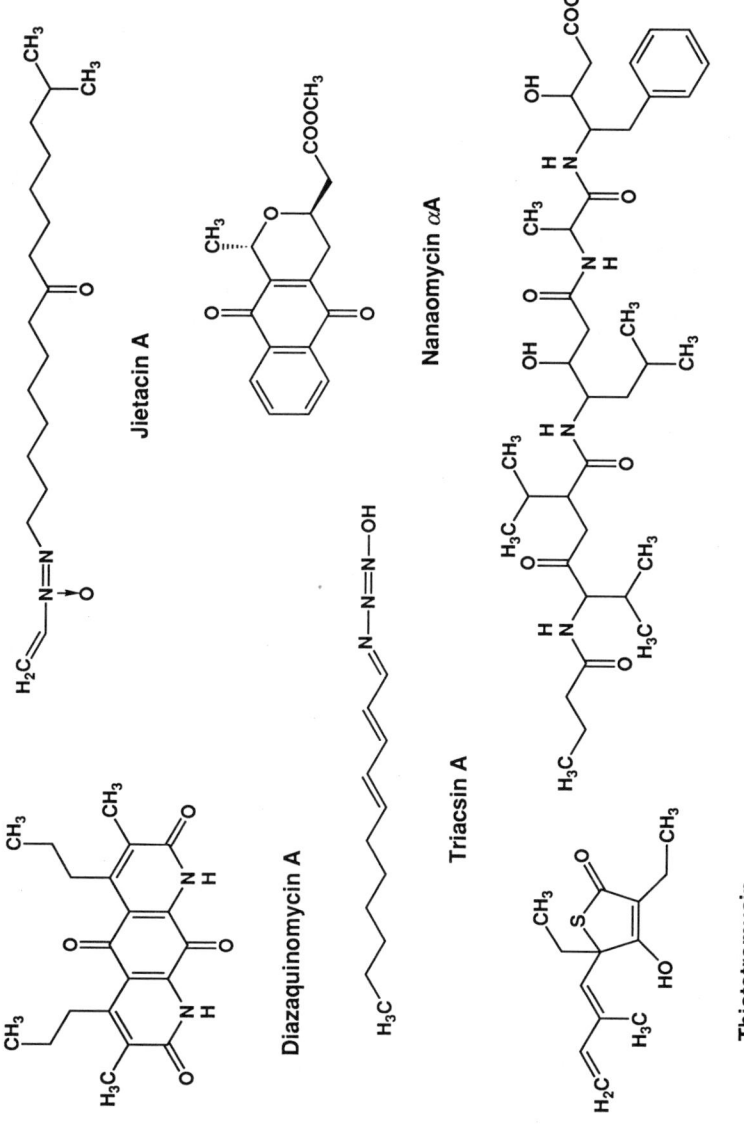

Figure 16.2 Structures of bioactive compounds produced by ammonium ion- or phosphate ion-depressed fermentation.

Jietacin (Ōmura et al., 1987) Jietacin is produced by *Streptomyces* sp. KP-197. It was named after Jie-tai temple, an old temple in Peking, where the soil sample was collected. The nematocidal activity in the cultured broth of KP-197 was detected using the pine nematode *Bursaphelenchus lignicolus*. It shows potent in vitro activity comparable to that of avermectin.

Jietacin production was higher and more reproducible in a medium containing a phosphate-trapping agent, allophane (2% glycerol, 2% soybean meal, 0.3% NaCl, 0.01% citrulline, 0.002% $CoCl_2 \cdot 6H_2O$, and 0.5% allophane, pH 7.0).

Media with an alkaline pH Microorganisms able to grow at an alkaline pH have been described. They produce extracellular enzymes with optimal pH values in the alkaline range. Actinomycetes and fungi are among these alkalophilic microorganisms. They were surveyed for their ability to produce new antibiotics under alkaline conditions. This trial led to the discovery of a new antibacterial and antifungal peptide, No. 1907, from *Paecilomyces* sp. No 1907 grown at pH 11 (Sato et al., 1980).

Pseudo-mixed cultures Of interest is an attempt to obtain new antibiotics by a mixed culture of two microorganisms. It is expected that one strain produces a biologically inactive metabolite that is converted by the second microorganism to an active antibiotic. However, a systematic and large-scale screening program using this method may be hampered by various difficulties, such as differences in the growth rate of the individual microorganisms. In fact, no such attempt has been described in the literature. An alternative method was reported by Watanabe et al., (1982), who employed a spent medium.

A *Fusarium* strain was grown in a medium for 4 days, and the supernatant fluid, filter-sterilized and supplemented with 2% glucose, was inoculated with various microorganisms. The bacterial cultures showing antibacterial activity after growth in the used medium, but inactive when grown in a conventional medium, were selected. Among them, one *Gluconobacter* strain was found to produce a new trienic antibiotic, AB-315.

16.4 Microbial Conversions

Together with chemical modification, biological modification of a natural compound provides an additional route to new bioactive compounds. Both the chemical and biological properties of an antibiotic can be modified. Biological modification utilizes enzymes from various sources and intact cells both of the producing microorganisms and of foreign microorganisms having desirable capabilities. Mutant cells and enzymes from them are also utilized. These may be immobilized for repeated use and longer periods of reaction. Three approaches have been successful and promising.

Directed biosynthesis A natural compound is incubated with intact cells or enzymes that cause hydroxylation, methylation, acylation, glycosylation, etc.; when a substrate is known to be a biosynthetic intermediate leading to a final antibiotic molecule, an analog of the substate is added to the culture to yield an analog of the antibiotic.

In a late step of penicillin biosynthesis, α-aminoadipate is attached to 6-aminopenicillanic acid (6-APA) to produce penicillin N. Phenylacetate or its corresponding alcohol, added to the penicillin fermentation culture, substitutes for α-aminoadipate yielding penicillin G. Addition of a 4-hydroxy derivative of phenylacetic acid results in the production of the corresponding 4-hydroxylated penicillin ∇. A similar approach was used to obtain an amino acid analog of cyclosporine A, a peptide antibiotic with potent immuno-suppressing activity (Hensens et al., 1992). Blasticidin S is a nucleoside antibiotic with antifungal activity. When a fluorinated base analog, 5-fluoro-cytosine, was added to the producing culture, a corresponding base analog of blasticidin S was produced (Kawashima et al., 1987).

Hybrid biosynthesis

Using blocked mutants A mutant defective in the biosynthesis of an antibiotic is more likely to incorporate an analog of a biosynthetic intermediate added to the culture because of the lack of competition with the natural counterpart. The analogs to be added can be synthesized or isolated from other antibiotic molecules of the same or related classes. This approach to new antibiotics was called mutational biosynthesis.

A DOS- (deoxystreptamine)-requiring mutant of a neomycin-producing *Streptomyces fradiae* was incubated with a DOS analog, streptamine or epistreptamine, to give rise to new hybrid aminoglycosides named hybrimycins (Shier et al., 1969). On incubation with chemically derived 8-fluoroerythronolide, an aglycone-requiring mutant of the erythromycin producer *Saccharopolyspora erythaeus* produced 8-fluoroerythromycins (Toscano et al., 1983). These fluoroerythromycins are more acid-stable than the naturally occurring erythromycins.

Using enzyme inhibitors The preparation of blocked mutants is a tedious task. Therefore, instead of using blocked mutants, an enzyme inhibitor with a specific site of inhibition was successfully employed by Ōmura to obtain hybrid macrolides.

The antibiotic cerulenin specifically inhibits fatty acid biosynthesis. It also inhibits the synthesis of polyketides, such as the aglycones of the macrolides, at a low, non-growth-inhibitory concentration (Ōmura, 1981). Thus, a macrolide-producing culture grown with a low level of cerulenin mimics an aglycone-requiring mutant, retaining an intact capability to synthesize other parts of the macrolide molecule.

A spiramycin producer, *S. ambotaciens*, was grown in a production medium. When cerulenin (40 µg ml^{-1}) was added once a day, no spiramycin-related compounds were produced because of the lack of aglycone biosynthesis. However, when protylonolide, a 16-membered lactonic precursor of the aglycone moiety of tylosin produced by a mutant of a tylosin producer, *S. fradiae*, was added to this culture, new macrolide antibiotics named chimeramycins were produced. These are hybrids of the two macrolide molecule, possessing the aglycone moiety of tylosin and the three sugars mycaminose, mycarose and forosamine, all corresponding to those of spiramycin (Figure 16.3; Ōmura et al., 1983b).

Recombinant DNA technology   Recent progress in recombinant DNA technology allows one to predict with high certainty the possibility of creating new antibiotics. The ability of a microorganism to synthesize one antibiotic is combined with that for another antibiotic at the gene level. The expression of the combined ability in one microorganism results in the production of hybrid-type antibiotics. The first demonstration of this approach is described in Chapter 17.

16.5   Concluding Remarks

Microorganisms as a whole are highly efficient in their ability to produce many kinds of antibiotics and other bioactive compounds. As individuals, however, they are different from one another, and appear to be very opportunistic with respect to antibiotic production. They are different in the way they respond to fermentation environments. They are so different that a superficial look at their behavior in antibiotic production might lead one to suspect that there are no rules or direction followed when they produce bioactive compounds under given fermentation conditions. Such complexity reflects strict and elegant regulatory mechanisms in the biosynthesis and biodegradation of antibiotics, which operate in different ways in individual microorganisms. The details of their metabolic regulation are not known at present. Unfortunately, "microorganisms are so shy, that they never show the mechanism to us by themselves," according to D. Perlman.

If the fermentation conditions employed in a screening program are favorable for a few microorganisms among a group of hundreds of soil isolates, then these will produce their antibiotics for us. Possible approaches to this goal have been described in this chapter. It may be said that antibiotic discovery is a function of the variety of fermentation conditions, when the set of microorganisms and detection methods are fixed.

The author expresses his thanks to Dr. Rokuro Masuma, The Kitasato Institute, for his help and valuable discussions in the preparation of this chapter.

Chapter 16 Fermentation in Screening for Bioactive Substances 323

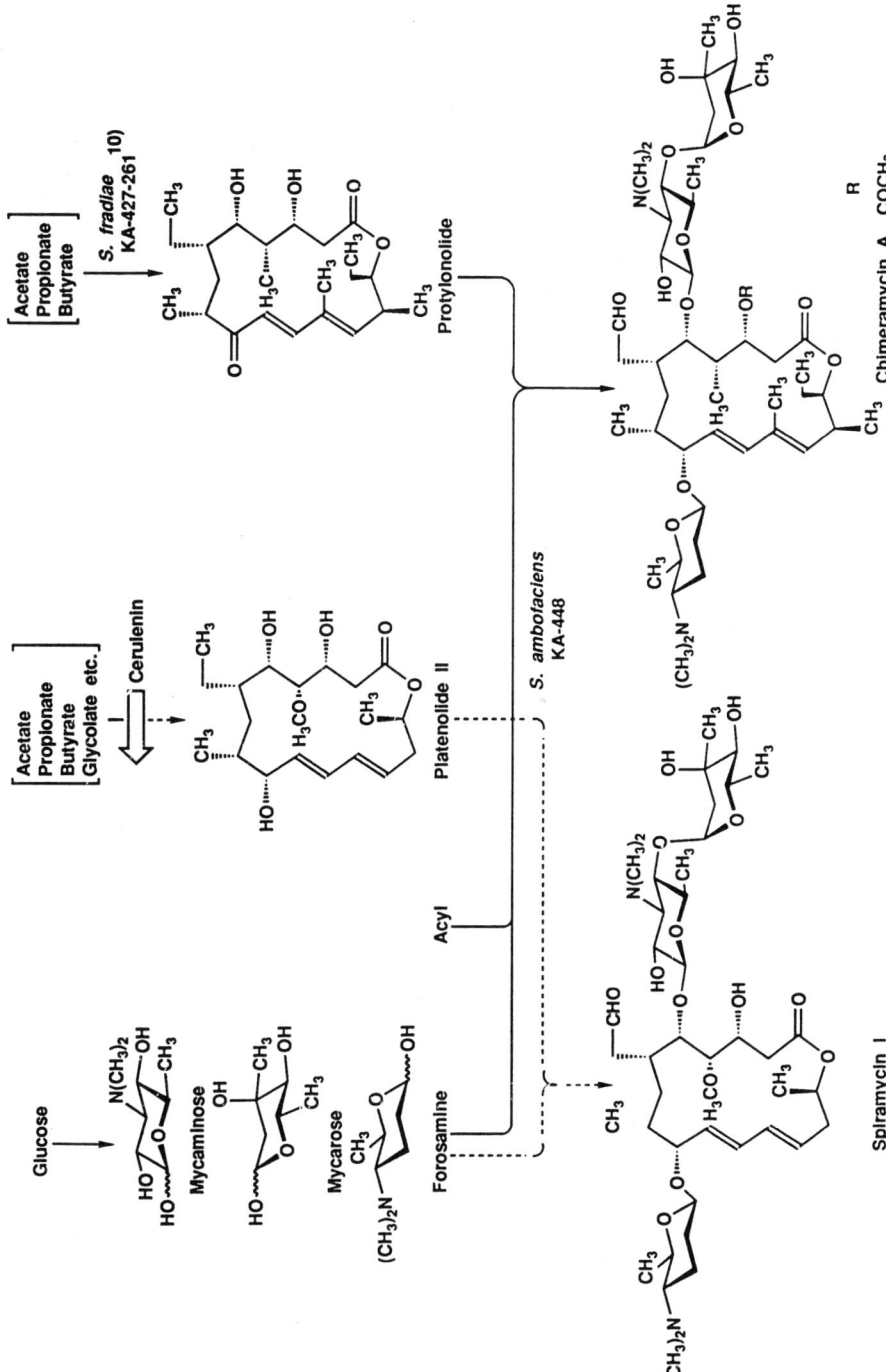

Figure 16.3 Hybrid biosynthesis of chimeramycins by a spiramycin-producing strain of *Streptomyces ambofaciens* KA-488.

References

Agathos, S. N. and Domain, A L. 1986. Dissolved oxygen levels and the *in vivo* stability of gramicidin S synthetase. *Applied Microbiology and Biotechnology* 24:319–322.
Aoki, H., Sakai, H., Kohsaka, M., Konomi, T., Hosoda, J., Kubochi, Y., Iguchi, E. and Imanaka, H. 1976. Nocardicin A, a new monocyclic β-lactam antibiotic. 1. Discovery, isolation, and characterization. *Journal of Antibiotics* (Tokyo) 29:492–500.
Arai, T., Yazawa, K. and Mikami, Y. 1976. Isolation and characterization of sattelite antibiotics, mimosamycin and chlorocarcins from *Streptomyces lavendulae*, streptothricin source. *Journal of Antibiotics* (Tokyo) 29:398–407.
Arima, K., Imanaka, H., Kohsaka, M., Furuta, A. and Tamura, G. 1964. Pyrrolnitrin, a new antibiotic substance, produced by *Pseudomonas*. *Agricultural and Biological Chemistry* 28:575–576.
Behal, X., Gregrova-Prusakova, J. and Hostalek, Z. 1982. Effect of inorganic phosphate and benzyl thiocyanate on the activity of anhydrotetracycline oxygenase in *Streptomyces aureofaciens*. *Folia Microbiologica* 27:102–106.
Burg, R. W., Miller, B. M., Baker, E. E., Birnbaum, J., Currie, S. A., Hartman, R., Kong, Y. L., Monaghan, R. L., Olson, G., Putter, I., Tunac, J. B., Wallick, H., Stapley, E. O., Ōiwa, R. and Ōmura, S. 1979. Avermectins, new family of potent anthelimintic agents: Producing organism and fermentation. *Antimicrobial Agents and Chemotherapy* 15:361–367.
Endo, A., Komagata, D. and Shimada, H. 1986. Monacolin M, a new inhibitor of cholesterol biosynthesis. *Journal of Antibiotics* (Tokyo) 39:1960–1673.
Gil, J. A., Naharro, G., Villanueva, J. R., Martin, J. F. 1985. Characterization and regulation of para-aminobenzoic acid synthase from *Streptomyces griseus*. *Journal of General Microbiology* 131:1279–1287.
Gotoh, T., Nakahara, K., Iwami, M., Aoki, H. and Imanaka, H. 1982. Studies on a new immunoactive peptide, FK-156. I. Taxonomy of the producing strain. *Journal of Antibiotics* (Tokyo) 35:1280–1285.
Hall, M. J. and Hassall, C. H. 1970. Production of monamycins, novel depsipeptide antibiotics. *Applied Microbiology* 19:109–112.
Hensens, O. D., White, R. F., Goegelman, R. T., Inamine, E. S. and Patchett, A. A. 1992. The preparation of [2-deutero-3-fluoro-D-Ala[8]]cyclosporine A by directed biosynthesis. *Journal of Antibiotics* (Tokyo) 45:135–135.
Higashide, E. 1984. The macrolides: properties, biosynthesis, and fermentation. In: Vandamme, (ed.), *Biotechnology of Industrial Antibiotics*, pp. 452–509. Madrcel Dekker, New York.
Hotta, K., Saito, N. and Okami, Y. 1980: Studies on new aminoglycoside antibiotics, istamycins, from an actinomycete isolated from a marine environment. *Journal of Antibiotics* (Tokyo) 33:1502–1509.
Iwai, Y., Awaya, J., Kesado, T., Yamada, H., Ōmura, S. and Hata, T. 1973. Selective production of cerulenin by *Cephalosporium caerulens* KF-140. *Journal of Fermentation Technology* 51:575–581.
Iwai, Y., Kimura, K., Takahashi, Y., Hinotozawa, K., Shimizu, H., Tanaka, H. and Ōmura, S. 1983. OM-173, new nanaomycin-type antibiotics produced by a strain of *Streptomyces*. *Journal of Antibiotics* (Tokyo) 36:1268–1274.
Kuwashima, A., Seto, H., Ishiyama, T., Kato, M., Uchida, K. and Otake, N. 1987. Fluorinated blasticidin S. *Agricultural and Biological Chemistry* 51:1183–1184.
Kuznetsov, S. I., Dubinia, G. A. and Lapteva, N. A. 1979. Biology of obigotrophic bacteria. *Annual Review of Microbiology* 33: 377–387.
Masuma, R., Tanaka, Y. and Ōmura, S. 1983. Ammonium-depressed fermentation of tylosin by the use of a natural zeolite and its significance in the study of biosynthetic regulation of the antibiotic. *Journal of Fermentation Technology* 61:607–614.

Masuma, R., Tanaka, Y., Tanaka, H. and Ōmura, S. 1986. Production of nanaomycin and other antibiotics by phosphate- depressed fermentation using phosphate-trapping agents. *Journal of Antibiotics* (Tokyo) 39:1557-1564.

Nakamura, G., Kobayashi, K., Sakurai, T. and Isono, K. 1981. Cationomycin, a new polyether ionophore antibiotic produced by *Actinomadura* nov. sp. *Journal of Antibiotics* (Tokyo) 34:1513-1514.

Nakanishi, S., Matsuda, Y., Iwahashi, K. and Kase, H. 1986. K- 252b, c, and d, potent inhibitors of protein kinase C from microbial origin. *Journal of Antibiotics* (Tokyo) 39:1066-1071.

Ochi, K. 1987. Metabolic initiation of differentiation and secondary metabolism by *Streptomyces griseus*: Significance of the stringent response (ppGpp) and GTP content in relation to A factor. *Journal of Bacteriology* 169:3608-3619.

Ogata, K., Osawa, H. and Tani, Y. 1977. Production of an antibiotic by *Streptomyces* a-60. *Journal of Fermentation Technology* 55:285-289.

Ōmura, S. 1981. Cerulenin. In: Lowenstein, J. M. (editor), *Methods in Enzymology*, Vol. 72, pp. 520-532. Academic Press, New York.

Ōmura, S. 1986. Philosophy of new drug discovery. *Microbiological Review* 50:259-279.

Ōmura, S. and Tanaka, Y. 1986. Macrolide antibiotics. In: Pape, H. and Rehm, H. J. (editors), *Biotechnology-A Comprehensive Treatise in 8 Volumes*, Vol. 4, pp. 359-391. Verlag-Chemie Weinheim.

Ōmura, S., Tanaka, H., Koyama, Y., Ōiwa, R., Katagiri, M., Awaya, J., Nagai, T. and Hata, T. 1974. Nanaomycins A and B, new antibiotics produced by a strain of *Streptomyces*. *Journal of Antibiotics* (Tokyo) 27:363-365.

Ōmura, S., Iwai, Y., Hinotozawa, K., Tanaka, H., Takahashi, Y. and Nakagawa, A. 1982. OM-704 A, a new antibiotic active against gram-positive bacteria, produced by *Streptomyces* sp. *Journal of Antibiotics* (Tokyo) 35:1425-1429.

Ōmura, S., Iwai, Y., Nakagawa, A., Iwata, R., Takahashi, Y., Shimizu, H. and Tanaka, H. 1983a. Thiotetromycin, a new antibiotic. Taxonomy, production, isolation, and physicochemical and biological properties. *Journal of Antibiotics* (Tokyo) 36:109-114.

Ōmura, S., Sadakane, N., Tanaka, Y. and Matsubara, H. 1983b. Chimeramycins, new macrolide antibiotics produced by hybrid biosynthesis. *Journal of Antibiotics* (Tokyo) 36:927-930.

Ōmura, S., Tomoda, H., Xu, Q. M., Takahashi, Y. and Iwai, Y. 1986a. Triacsins, new inhibitors of acyl-CoA synthetase produced by *Streptomyces*. *Journal of Antibiotics* (Tokyo) 39:1211-1218.

Ōmura, S., Imamura, N., Kawakita, K., Mori, Y., Yamazaki, Y., Masuma, R., Takahashi, Y., Tanaka, H., Huang, L. Y. and Woodruff, H. B. 1986b. Ahpatinins, new acid protease inhibitors containing 4-amino-3-hydroxy-5-phenylpentanoic acid. *Journal of Antibiotics* (Tokyo) 39:1079-1085.

Ōmura, S., Otoguro, K., Imamura, N., Kuga, H., Takahashi, Y., Masuma, R., Tanaka, Y., Tanaka, H., Su, X. H. and Yao, E. T. 1987. Jietacins A and B, new nematocidal antibiotics from a *Streptomyces* sp. *Journal of Antibiotics* (Tokyo) 40:623-629.

Ōmura, S., Tanaka, Y., Kanaya, I., Shinose, M. and Takahashi, Y. 1990. Phthoxazolin, a specific inhibitor of cellulose biosynthesis, produced by a strain of *Streptomyces* sp. *Journal of Antibiotics* (Tokyo) 43:1034-1036.

Rinehart, K. L., Jr. and Sebek, O. K. 1979. Mutasynthesis of antibiotics. In: Schlessinger, D. (editor), pp. 307-321. American Society for Microbiology, Washington, D. C.

Sato, M., Beppu, T. and Arima, K. 1980. Properties and structure of a novel peptide antibiotic No. 1907. *Agricultural and Biological Chemistry* 44:3037-3042.

Sawada, H., Suzuki, T., Akiyama, S. and Nakao, Y. 1987. Stimulatory effect of cyclodextrins on the production of lankacidin-group antibiotics by *Streptomyces* sp. *Applied Microbiology and Biotechnology* 26:522-526.

Sawada, H., Suzuki, To, Akiyama, S. and Nakao, Y. 1990. Mechanism of stimulatory effect of cyclodextrins on lakacidin-producing *Streptomyces*. *Applied Microbiology and Biotechnology* 32:556–559.

Shimada, N., Hasegawa, S., Harada, T., Tomisawa, T., Fujii, A. and Takita, T. 1986. Oxetanocin, a novel nucleoside from bacteria. *Journal of Antibiotics* (Tokyo) 39:1624–1625.

Shier, W. T., Rinehart, K. L., Jr. and Gottlieb, D. 1969. Preparation of four new antibiotics from a mutant of *Streptomyces fradiae*. *Proceedings of the National Academy of Sciences of the United States of America* 63:198–204.

Spagnoli, R., Cappelletti, L. and Toscano, L. 1983. Biological conversion of erythronolide B, an intermediate of erythromycin biogenesis, into new hybrid macrolide antibiotics. *Journal of Antibiotics* (Tokyo) 36:365–375.

Tanaka, Y. and Ōmura, S. 1988. Regulation of biosynthesis of polyketide antibiotics. In: Okami, Y., Beppu, T. and Ogawara, H. (editors), *Biology of Actinomycetes '88*, pp. 418–423. Japan Science Society Press, Tokyo.

Tanaka, Y., Taki, A., Masuma, R. and Ōmura, S. 1986. Mechanism of nitrogen regulation of protylonolide biosynthesis in *Streptomyces fradiae*. *Journal of Antibiotics* (Tokyo) 39:813–821.

Tanaka, Y., Hirata, K., Takahashi, Y., Iwai, Y. and Ōmura, S. 1987. Globopeptin, a new antifungal peptide. *Journal of Antibiotics* (Tokyo) 40:242–244.

Toscano, L., Tioriello, G., Spagnoli, R., Cappelletti, L. and Zanuso, G. 1983. New fluorinated erythromycins obtained by mutasynthesis. *Journal of Antibiotics* (Tokyo) 36:1439–1450.

Watanabe, T., Izaki, K. and Takahashi, H. 1982. New polyenic antibiotics active against Gram-positive and -negative bacteria, II. Screening of antibiotic producers and taxonomical properties of *Gluconobacter* sp. W-315. *Journal of Antibiotics* (Tokyo) 35:1148–1155.

Zanca, D. M. and Martin, J. F. 1983. Carbon catabolite regulation of conversion of penicillin N into cephalosporin C. *Journal of Antibiotics* (Tokyo) 36:700–708.

Zimmerman, S. B., Schwartz, C. D., Monaghan, R. L., Pelak, B. A., Weissberger, B., Gilfillan, E. C., Mochales, S., Hernandez, S., Currie, S. A., Tejera, E. and Stapley, E. O. 1987. Difficidin and oxydifficidin: Novel broad spectrum antibacterial antibiotics produced by *Bacillus subtilis*. *Jornal of Antibiotics* (Tokyo) 40:1677–1681.

17

Genetic Engineering of Antibiotic-Producing Microorganisms

Haruo Ikeda

17.1 Introduction

Many antibiotics are produced by strains of the genus *Streptomyces*. Although they are classified as gram-positive organisms, the streptomycetes exhibit a more complicated life cycle than do *Bacillus* and also show greater differentiation. The genomic size is about 6000 kbp, 1.6 times larger than that of *E. coli* (Genthner et al., 1985). Genetic manipulation in *Streptomyces* has been established by D.A. Hopwood and his coworkers (Hopwood et al., 1985a). So far, treatment with mutagens and the screening of resulting clones has been repeated and adopted as a procedure for the breeding of antibiotic-producing microorganisms and has produced good results.

The intentional breeding of strains is performed by using genetic manipulation. These techniques make possible: (1) the increase of antibiotic productivity by the gene dosage effect of a cloned gene; (2) the production of new analogs of antibiotics by the introduction of heterogeneous biosynthetic genes; (3) stable production of antibiotics by introducing whole genes for antibiotic biosynthesis into a more convenient strain; and the elucidation of control systems for antibiotic biosynthesis. It is important for cloning trials that the genes for antibiotic biosynthesis form a cluster in the chromosome.

The preparation of protoplasts and introduction of the recombinant DNA into them (transformation and transfection) are fundamental methods for cloning in *Streptomyces*. These two protocols are described as follow.

17.2 Preparation Protoplasts (Hopwood et al., 1985a)

Medium Tryptic soy broth (BBL or Difco) or YEME (34% sucrose, 1.0% glucose, 0.3% yeast extract, 0.3% malt extract, 0.5% peptone, 5 mM MgCl$_2$, no pH adjustment) are used for growing the cultures. The addition of 0.25

to 1.2% of glycine to the growth medium causes an increase of lysozyme sensitivity of the cell wall. However, the concentration needed varies from strain to strain. The optimum concentration should be determined individually.

P buffer (10.3% sucrose, 0.025% K_2SO_4, 0.202% $MgCl_2$, 0.2% trace element solution, 0.005% KH_2PO_4, 0.368% $CaCl_2$, 25 mM TES (N-tris(hydroxymethyl)methyl-2-aminoethane sulfonic acid) buffer, pH 7.2) is used for the making and washing of the protoplasts. L buffer (the composition is similar to P buffer except that the concentrations of $MgCl_2$ and $CaCl_2$ are 2.5 mM) is also used for the formation of protoplasts.

Procedure To a baffled flask(500 ml) add 50 ml of growth medium and an appropriate volume of 20% glycine. Add about 0.1 ml of a spore or mycelial suspension; incubate for 21–72 h at 25° to 35°C in an orbital shaker. Pour the culture into a 50-ml screw-cap tube and sediment the mycelium by centrifugation at 1000 x g for 10 min; and discard the supernatant and suspend the mycelium in 25 ml of 10.3% sucrose. The washed mycelium is sedimented at 1000 $\times$ g for 10 min. Repeat the washing and sedimentation steps; suspend the mycelium in 8 ml of L or P buffer containing 1 mg of lysozyme per milliliter and incubate at 30°C for 15–120 min; mix at 15-min intervals; add 10 ml of P buffer and triturate a few times with a pipette; filter through cotton wool and transfer to two 10-ml screw cap tubes; sediment the protoplasts by centrifugation at 1000 $\times$ g for 10 min; discard the supernatant solution and suspend each pellet in 10 ml of P buffer; and centrifuge at 1000 $\times$ g for 10 min. Repeat washing and sedimentation steps. The washed pellet is resuspended in a small volume of P buffer to make about 10^9 protoplasts ml^{-1}. Protoplasts may be stored at -70°C in P buffer, although the transformation or transfection efficiency using the stored protoplasts would be decreased about 10-fold compared to that of a fresh batch.

17.3 Transformation and Transfection (Hopwood et al., 1985a)

Medium and solutions Protoplast regeneration medium R2YE (10.3% sucrose, 0.025% K_2SO_4, 1.012% $MgCl_2$, 1.0% glucose, 0.01% casamino acids, 2% trace elements solution, 0.005% KH_2PO_4, 0.2944% $CaCl_2$, 0.3% L-proline, 25 mM TES buffer pH 7.2, 0.02 N NaOH, 0.5% yeast extract, 2.2% agar); P buffer; 25% polyethylene glycol #1000 in T buffer (2.5% sucrose, 0.025% K_2SO_4, 0.2% trace element solution, 0.1 M $CaCl_2$, 50 mM Tris-maleate pH 8.0); or P buffer.

Procedure Sediment about 1 $\times$ 10^9 protoplasts by centrifugation at 1000 $\times$ g for 7 min and discard the supernatant. Resuspend the protoplasts in the drop of buffer left after pouring off the supernatant. Add DNA sample and immediately add 0.5 ml of 25% PEG in T buffer or P buffer. Mix by pipetting up and down. In the case of transfection, add phage DNA sample and then add 20% PEG in P buffer. After 1 min, add 0.5 ml of P buffer and mix. Plate out 0.1 ml on each of 10 plates of protoplast regeneration medium and incubate at 30°C. In the case of transfection, plate out 3 ml of half-strength R2YE made by P buffer containing 1×10^8 of *Streptomyces* spores and incubate at 30°C. A phage plaque is observed on the next day. If transformants are to be selected by antibiotic resistance, overlay with 2.5 ml of soft agar containing antibiotic at 8–21 h (8–16 h for neomycin, 16–21 h for thiostrepton).

Some *Streptomyces* strains produce very few protoplasts by lysozyme treatment. However, achromopeptidase can be applied to these strains (Ogawa et al., 1983). Transformation or transfection efficiency is increased by adding the polycation protamine (Matsushima and Baltz, 1985), heparin (Birmingham et al., 1986), or positively charged liposomes (Rodicio and Chater, 1982). In particular, transfection efficiency has been dramatically increased (100 fold) by adding the positively charged liposome.

17.4 Methodology and Strategy of Cloning Genes for Antibiotic Synthesis

Five methods have been developed for cloning genes for antibiotic biosynthesis, as follow.

Detection of an individual gene product by cloning in a standard host (Gil and Hopwood, 1983; Jones and Hopwood, 1984) This procedure is suitable for the gene product(s) for which there is a convenient and sensitive detection system. The genomic library can be prepared in *Streptomyces lividans* 1326 or its derivatives, because these are efficient hosts for transformation.

Complementation of blocked mutants in the producing strain (Feitelson and Hopwood, 1983; Feitelson et al., 1985; Baily et al., 1984; Malpartida and Hopwood, 1984, 1986; Ohnuki et al., 1985; Distler et al., 1987) This is the most common procedure used, and the antibiotic nonproducing mutants, which are genetically blocked in a region of antibiotic biosynthesis, must be isolated. After making the genomic library by shotgun cloning in which an antibiotic nonproducing mutant (blocked mutant) is used as a host in the transformation, antibiotic-producing clones appear to result from complementation of the mutation.

Mutational cloning in the antibiotic producing strain (Chater and Bruton, 1985) This method was used with temperate actinophage ϕC31 of *S. coelicolor* A3(2) as a cloning vector (Figure 17.1). The vector phage was lacking the attachment site (*att*), which was recombined with the *att* site of the chromosome so that only the vector phage is unable to lysogenize the host strain. However after the recombinant phages, which contain a part of the chro-

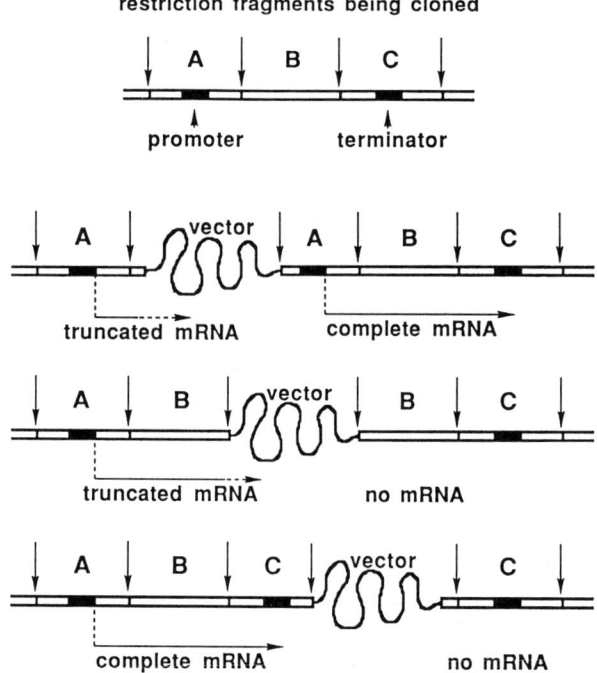

Figure 17.1 Use of φC31 for mutational cloning.

mosome of the antibiotic-producing strain, have infected the strain, a homologous region of the chromosome of the antibiotic-producing strain is able to recombine with a region of the recombinant phage genome, forming lysogens. Bacause the vector phage was inserted into an antibiotic-resistant gene, lysogens were detected by selecting for antibiotic resistance. Because the recombinant phage that was inserted into the target DNA fragment did not contain both the promoter and the terminator regions infected and lysogenized, a gene corresponding to the inserted DNA of the recombinant phage was inactivated (Chater and Bruton, 1983).

Cloning for antibiotic resistance followed by analysis of linked DNA for biosynthetic genes (Murakami et al., 1986; Stanzak et al., 1986)
Antibiotic-producing strains show self-resistance to their own antibiotics. It is considered that expression of resistance is synchronized with that of antibiotic biosynthesis. Thus, the gene(s) for antibiotic resistance should be located close to the antibiotic biosynthetic genes. Cloning of these genes is possible, although the DNA fragment cloned would be relatively large in size.

Probing gene libraries with synthetic oligonucleotides matched to partial sequences of antibiotic biosynthesis (Fishman et al., 1987; Samson et al., 1985) If at least one enzyme involving antibiotic biosynthesis is purified, oligonucleotides corresponding to the amino acid sequence of the N terminus of the enzyme can be synthesized. Such a clone could be screened by the plaque or colony hybridization technique using the radioactive oligonucleotide as a probe with appropriate genomic libraries, which are constructed by using a lambda phage or cosmid vector in E. coli. In this method, the enzyme involved in the antibiotic biosynthesis must be purified to homogeneity, and the antibiotic-blocked mutants should be defective in their restriction systems because it is more difficult to introduce the recombinant DNA propagated in E. coli (unmodified DNA) into restriction-positive microorganisms.

17.5 Structure of Gene Cluster Antibiotic Biosynthesis

Several antibiotic biosynthetic genes were obtained. However, almost all of them were a part of the region of a gene cluster. Because all the genes for biosynthesis of actinorhodin have been isolated, the structure of the cluster of actinorhodin biosynthetic genes is described here.

Genes for actinorhodin biosynthesis (Figure 17.2) are classified as actI-VII and its biosynthetic steps from I,III → VII → IV → VI → V to actinorhodin (Rudd and Hopwood, 1979). Some subclones were prepared from primary recombinant plasmid pIJ2300 and pIJ2301, and they were then introduced into seven types of actinorhodin-blocked mutants. The results of the complementation experiments indicate that the arrangement of the genes for actinorhodin biosynthesis is on the 26-kb DNA fragment (Malpartida and Hopwood, 1986). Further, each gene step was analyzed by the mutational cloning technique and at least four transcriptional units are located on the cluster of actinorhodin biosynthetic genes. ActI, actVII and actIV were polycistronic and the direction of transcription was from left to right on the restriction map. ActII and actIII have an independent unit for transcription. ActV and VI were expected to be formed from one or more units for transcription; the transcription was from right to left on the restriction map.

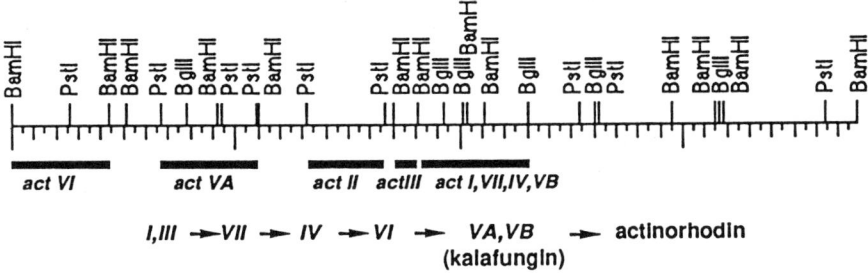

Figure 17.2 Restriction map of gene cluster of actinorhodin biosynthesis.

17.6 Genes for Regulation of Antibiotic Production

Some compounds that are related to regulation of differentiation or antibiotic production have been reported (Khokhlov et al., 1967; Scribner et al., 1973; McCann and Pogell, 1979). One of them, A factor, was produced by streptomycin-producing *Streptomyces* strains, and the production of both streptomycin and sporulation were stimulated by an extremely small amount of A factor.

A 9-kb DNA fragment containing a gene for A factor biosynthesis (*afsA*) was obtained from streptomycin-producing *Streptomyces bikiniensis* (Horinouchi et al., 1984b). After analysis of the DNA fragment, it was found that a 1.2-kb DNA region on a primary 9-kb DNA fragment was necessary for the expression of the A factor biosynthetic gene, and the biosynthesis of A factor was performed by one enzyme protein that contained 280 amino acids.

In addition, a regulatory gene for A factor production (*afsB*) was obtained from *Streptomyces coelicolor* A3(2) as a 10 kb DNA fragment (Horinouchi et al., 1983). The *afsB* gene was not complementary to *afsA* mutants; therefore, the *afsB* gene was determined to be a regulatory gene for A factor production. Interestingly, the transformant carrying the *afsB* gene was stimulated in the production of both actinorhodin and undecylprodigiosin in *Streptomyces lividans* (Horinouchi and Beppu, 1984a). Although the messenger RNAs corresponding to *act* genes were barely detectable in *Streptomyces lividans*, the transformant carrying the *afsB* gene was able to synthesize the complementary mRNA of the *act* genes. Thus, the *afsB* gene involved regulation of not only A-factor production but also actinorhodin and undecylprodigiosin production.

Further, the *afsC* gene, which is involved in the stimulation of the action of the *afsB* gene, was also detected. The production of actinorhodin by the transformant containing the *afsB* and C genes on a low copy number vector pIJ41 was three fold higher than that of transformants containing the *afsB* gene alone. Also, the recombinant plasmid, which was constructed by joining the high copy number vector, and *afsB* and *afsC* genes, was not introduced into *Streptomyves lividans*. It is likely that high expression of both the *afsB* and *afsC* genes caused the production of lethally large amounts of actinorhodin and undecylprodigiosin.

17.7 Production of New Antibiotics by Genetic Engineering

Whole genes for actinorhodin biosynthesis in *Streptomyces coelicolor* A3(2) have been isolated, and the location of each gene for biosynthesis on the DNA fragment cloned has been clarified (Figure 17.3). The introduction of whole genes or a part of a gene for actinorhodin biosynthesis into other *Streptomyces* strains that produce actinorhodin-related antibiotics was then attempted (Hopwood et al., 1985b). Some *Streptomyces* strains were not trans-

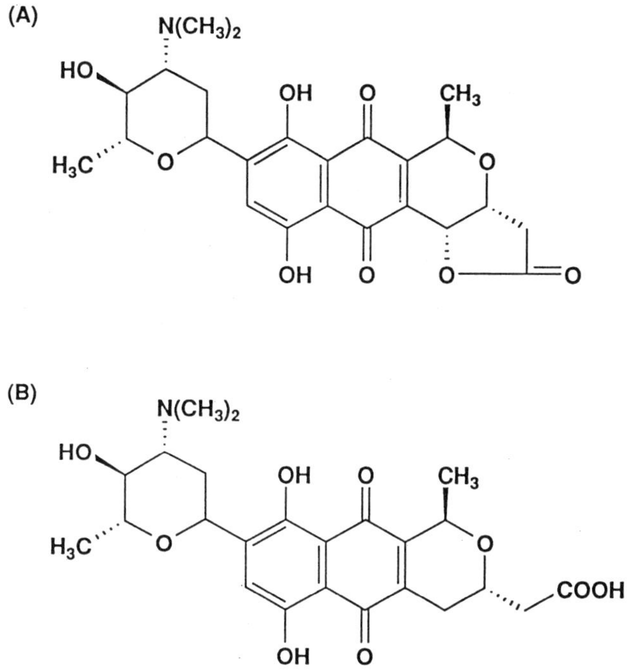

Figure 17.3 Structures of mederrhodins A and B.

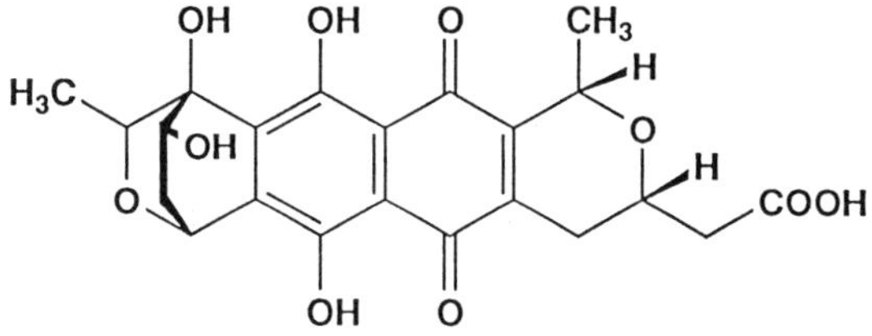

Figure 17.4 Structure of dihydrogranatirhodin.

formed by recombinant plasmids. However, two types of *Streptomyces* strains, kalafungin-producing *S. tanashiensis* Kala and medermycin-producing *Streptomyces* sp. AM7161 were affected. Production of new antibiotics was found in some transformants of medermycin-producing *Streptomyces* sp. AM7161 but not that of a kalafungin producer.

New antibiotics named mederrhodins A and B (Figure 17.4) were produced from the transformants that incorporated recombinant plasmids pIJ2301, pIJ2315 and pIJ2316 including *actV* (Omura et al., 1986). The transformants having recombinant plasmid pIJ2303, including whole genes for actinorhodin biosynthesis, did not produce mederrhodins A and B. However, they produced actinorhodin and medermycin. The production of mederrhodins A and B was quite stable without thiostrepton in the production medium, this being a selective marker for a plasmid vector. Among transformants carrying pIJ2301, pIJ2315 and pIJ2316, the productivity of mederrhodins A and B by the transformant carrying pIJ2315 was the highest. Three recombinant plasmids, pIJ2301, pIJ2315 and pIJ2316, included the transcriptional unit of *actV*. The *actV* gene would be transcribed from a vector promoter (pIJ922). This probably explains the stronger complementation of *actV* mutant by pIJ2315 than by pIJ2301 or pIJ2316 (Malpartida and Hopwood, 1986).

In the studies of granaticin-producing S. *violaceus-ruber*, the transformant with pIJ2303, which contains whole genes for actinorhodin biosynthesis, produced actinorhodin, granaticin, and a new antibiotic, granatirhodin. The configuration of granaticin was opposite to that of actinorhodin at C-1 and C-3, and granatirhodin has the actinorhodin configuration at C-3 and the granaticin configuration at C-1 (Hopwood et al., 1985b). The production of new antibiotics by genetic engineering has been tried using a 14-membered macrolide producer(McAlpaine et al., 1987). A part of the gene(s) for erythromycin biosynthesis by S. *erythreus* was obtained, and a recombinant plasmid containing a part of the erythromycin biosynthetic gene(s) was introduced into oleandomycin-producing S. *antibioticus*. The transformant produced a novel antibiotic, norerythromycin, which had a hybrid aglycon of erythromycin and oleandomycin.

17.8 Future Prospects

Modern genetic engineering has "opened the door" both to production of novel hybrid antibiotics and to greatly increased production of known types. It remains to be seen which of these will be of greatest utility and importance. Certainly, our ability to understand and manipulate the genes of antibiotic-producing organisms will lead to spectacular further developments, most of which can hardly be imagined at this time.

References

Bailey, C. R., Butler, M. J., Normansell, I. D., Rowlands, R. T. and Winstanley, D. J. 1984: Cloning a *Streptomyces clavuligerus* genetic locus involved in clavulanic acid biosynthesis. *Biotechnology* 2:808–811.

Chater, K. F. and Bruton, C. J. 1983: Mutational cloning in *Streptomyces* and the isolation of antibiotic production genes. *Gene* 26:67–78.

Chater, K. F. and Bruton, C. J. 1985: Resistance, regulatory and production genes for the antibiotic methylenomycin are clustered. *EMBO Journal.* 4:1893–1892.

Distler, J., Braun, C., Ebert, A. and Piepersberg, W. 1987: Gene cluster for streptomycin biosynthesis in *Streptomyces griseus*: Analysis of a central region including the major resistance gene. *Molecular & General Genetics* 208:204–210.

Feitelson, J. S. and Hopwood, D. A. 1983: Cloning of a *Streptomyces* gene for a O-methyltransferase involved in antibiotic biosynthesis. *Molecular & General Genetics* 190:394–398.

Feitelson, J. S., Malpartida, F. and Hopwood, D. A. 1985: Genetic and biochemical characterization of the red gene cluster of *Streptomyces coelicolor* A3(2). *Journal of General Microbiology* 131:2431–2441.

Fishman, S. E., Cox, K., Larson, J. L., Reynolds, P. A., Seno, E. T., Yeh, W.-K., Van Frank, R. and Hershberger, C. L. 1987: Cloning genes for the biosynthesis of a macrolide antibiotic. *Proceedings of the National Academy Science of the United States of America* 84:8248–8252.

Genthner, F. J., Hook, L. A. and Strohl, W. R. 1985: Determination of the molecular mass of bacterial genomic DNA and plasmid copy number by high-pressure liquid chromatography. *Applied and Enviromental Microbiology* 50:1007–1013.

Gil, J. A., Hopwood, D. A. 1983: Cloning and expression of a *p*- aminobenzoic acid synthetase gene of the candicidin-producing *Streptomyces griseus*. *Gene* 25:119–132.

Hopwood, D. A., Bibb, M. J., Chater, K. F., Kieser,T., Bruton, C. J., Kieser, H. M., Lydiate, D. J., Smith, C. P., Ward, J. M. and Schrempf, H. 1985a: *Genetic manipulations of Streptomyces: A LABORATORY MANUAL*. John Innes Foundation, Norwich, England.

Hopwood, D. A., Malpartida, F., Kieser, H. M., Ikeda, H., Duncan, J., Fujii, I., Rudd, B. A. M., Floss, H. G., Ōmura, S. 1985b: Production of 'hybrid' antibiotics by genetic engineering. *Nature* (London) 314:642–644.

Horinouchi, S., Hara, O. and Beppu, T. 1983: Cloning of a pleiotropic gene that positively controls biosynthesis of A- factor, actinorhodin, and prodigiosin in *Streptomyces coelicolor* A3(2) and *Streptomyces lividans*. *Journal of Bacteriology* 155:1238–1248.

Horinouchi, S. and Beppu, T. 1984a: Production in large quantities of actinorhodin and undecylprodigiosin induced by *afsB* in *Streptomyces lividans*. *Agricultural and Biological Chemistry* 48:2131–2133.

Horinouchi, S., Kumada, Y. and Beppu, T. 1984b: Unstable genetic determinant of A-factor biosynthesis in streptomycin-producing organisms: Cloning and characterization. *Journal of Bacteriology* 158:481–487.

Jones, G. H. and Hopwood, D. A. 1984: Molecular cloning and expression of the phenoxazinone synthase gene from *Streptomyces antibioticus*. *Journal of Biological Chemistry* 259:14151–14157.

Khokhlov, A. S., Tovarova, I. I., Borisova, L. N., Pliner, S. A., Shevchenko, L. A., Kornitskaya, E. Y., Ivkina, N. S. and Rapoport, I. A. 1967: A-factor responsible for the biosynthesis of streptomycin by a mutant strain of *Actinomyces streptomycini*. *Akademii Nauk Uzbekskoi SSR*. 177:232–235.

Malpartida, F.and Hopwood, D. A. 1984: Molecular cloning of the whole biosynthetic pathway of a *Streptomyces* antibiotic and its expression in a heterologous host. *Nature* (London) 309:462–464.

Malpartida, F. and Hopwood, D. A. 1986: Physical and genetic characterisation of the gene cluster for the antibiotic actinorhodin in *Streptomyces coelicolor* A3(2). *Molecular & General Genetics* 205:66–73.

Matsushima, P. and Baltz, R. H. 1985: Efficient plasmid transformation of *Streptomyces ambofaciens* and *Streptomyces fradiae* protoplasts. *Juornal of Bacteriology* 163:180–185.

McAlpaine, J. B., Tuan, J. S., Brown, D. P., Grebner, K. D., Whittern, D. N., Buko, A. and Katz, L. 1987: New antibiotics from genetically engineered actinomycetes. I. 2-norerythromycin, Isolation and structural determination. *Journal of Antibiotics* (Tokyo) 40:1115–1122.

McCann, P. A., Pogell, B. M. 1979: Pamamycin: A new antibiotic and stimulator of aerial mycelia formation. *Journal of Antibiotics* (Tokyo) 32:673–678.

Murakami, T., Anzai, H., Imai, S., Satoh, A., Nagaoka, K. and Thompson, C. J. 1986:

The bialaphos biosynthetic genes of *Streptomyces hygroscopicus*: Molecular cloning and characterization of the gene cluster. *Molecular & General Genetics* 205:42–50.

Ogawa, H., Imai, S., Satoh, A. and Kojima, M. 1983: An improved method for the preparation of *Streptomyces* and *Microspora* protoplasts. *Journal of Antibiotics* (Tokyo) 36:184–186.

Ohnuki, T., Imanaka, T. and Aiba, S. 1985: Self-cloning in *Streptomyces griseus* of an *str* gene cluster for streptomycin biosynthesis and streptomycin resistance. *Journal of Bacteriology* 164:85–94.

Ōmura, S., Ikeda, H., Malpartida, F., Kieser, H. M. and Hopwood, D. A. 1986: Production of a new hybrid antibiotics mederrhodins A and B by genetically engineered strain. *Antimicrobial Agents and Chemotherapy* 29:13–19.

Rodicio, M. R. and Chater, K. F. 1982: Small DNA-free liposomes stimulate transfection of *Streptomyces* protoplasts. *Journal of Bacteriology* 151:1078–1085.

Rudd, B. A. M. and Hopwood, D. A. 1979: Genetics of actinorhodin biosynthesis by *Streptomyces coelicolor* A3(2). *Journal General Microbiology* 114:35–43.

Samson, S. M., Belagaje, R., Blankenship, D. T., Chapman, J. L., Perry, D., Skatrud, P. L., Van Frank, R. M., Abraham, E. P., Baldwin, J. E., Queener, S. W. and Ingolia, T. D. 1985: Isolation sequence determination and expression in *Esherichia coli* of the isopenicillin N synthetase gene from *Cephalosporium acremonium*. *Nature* (London) 318:191–194.

Scribner, H. E., III, Tang, T. and Bradley, S. G. 1973: Production of a sporulation pigment by *Streptomyces venezuelae*. *Applied Microbiology* 25:873–879.

Stanzak, P., Matsushima, P., Baltz, R. H. and Rao, R. N. 1986: Cloning and expression in *Streptomyces lividans* of clustered erythromycin biosynthesis genes from *Streptomyces erythreus*. *Biotechnology* 4:229–232.

Index

Number
15B, 120, 129, 164, 166
1233A, 124, 164, 166, 298
6241-B, 226, 227, 228

A
A 60, 317
A 73 A, 296
A 47934, 21
A 58365, 120, 128, 129, 130
A-77003, 47, 52
A 82516, 122
A factor, 310, 332
AAD-609, 11, 21
AB-110-D, 293
abbeymycin, 4
acarbose, 122, 128
ACE, *see* angiotensin-converting enzyme
acetoacetyl-CoA thiolase, 163, 164
acetylpenicillide, 296
N-acetylthienamycin, 12, 20
Acholeplasma, 8, 20
acidophilic actinomycetes, 288
Acinetobacter lwoffii, 9
aclacinomycin, 107, 108

actinoidin, 11, 21
Actinomadura, 288, 290
actinomycetes
    habitat variation in soil, 281–286
    table of genera, number of isolates, 282
    table of genera, taxonomy, 299
    table of numbers in soil, 283
actinomycin, 81, 248
actinonin, 121
Actinoplanes, 290, 291
actinopyrone, 208, 209
actinorhodin, 331, 332
action mechanism, table of
    agricultural agents, 215
    antiparasitic agents, 75
    antiviral agents, 52
    targets on HIV replicative cycle, 56
aculeacin A, 33
aculeximycin, 248
acycloguanosine, 52
acyclovir, 47
acyl-CoA synthetase, 162, 164, 167
ADA, *see* adenosine deaminase
adechlorin, 125, 139, 140, 296
adecypenol, 92, 125, 139, 140, 296
adenine arabinoside, 52

337

adenosine aminohydrolase, *see* adenosine deaminase
adenosine deaminase, 128, 138–140
adenosine diphosphate (ADP), 198, 203
adiposins, 122
ADP, *see* adenosine diphosphate
affinity column, 21
agar diffusion method, *see also* paper disc method, 3, 12
agar medium, soil isolation, 287
agar plate assay, enzyme assay, 119
aggreceride, 200, 202, 296
aggreticin, 201, 202, 296
agricultural fungicide, table of, 215
ahpatinins, 121, 296, 317, 319
AIDS, *see* acquired immunodeficiency syndrome
akrobomycin, 93
aladapsin, 173, 181
alahopcin, 3
alanosin, 249
alboleutin, 296
aldose reductase, 133–135
aldostatin, 122, 135
alkaline medium, 320
alkaline phosphatase activity, 174
alkaline treatment of soil, 293
alkaloid, 264
  screening procedure, figure of, 268
alkalophilic actinomycetes, 288
allophane, 308
allosamidin, 122, 252
alphostatin, 125
altemicidin, 246, 249
aluminosilicate, 308
AM-2504, *see* dityromycin
AM-6201, *see* reductiomycin
amantadine, 47, 52
amastatin, 120, 177, 178
amauromine, 204, 205, 206
amino acid antagonist, 42
2'-amino-2'-deoxyadenosine, 296
3-amino-3-deoxy-D-glucose, 8, 21
aminoglycoside antibiotic, 22–23
  multiresistant actinomycetes to, 289
6-aminopenicillanic acid, 14, 15
aminopeptidase B, 128, 136–138

ammonium ion
  as regulatory factor, 306
  depressed fermentation, 317
  trapping agent, 308–309
amphidinolides, 295
amphomycin, 8, 20
amphotericin B, 32, 33, 53, 242
amycins, 267, 274
amylase inhibitor, 122
amylostatin, 122
anaerobic bacteria, 4
anaerobic condition, 3–4
ancovenin, 120, 129
andrimid, 295
angiotensin, 118, 119
angiotensin-converting enzyme, 118, 119, 128, 129, 130
Anguillula aceti, 70
animal model for antiviral agents, 51
animal tumor system, 84–87
anisomycin, 225
ankinomycin, 94
ansamycin antibiotic, 22
antagonism
  glycopeptide, 11
  with amino acid, 42
antagonized by L-glutamine, 233
anthelmintic antibiotics, 69–74
anthelmycin, 71
anthelvencins, 71
anthranylic acid, 248
antiamoebic antibiotic, 67
antibacterial activity, antitumor screening, 89–90
antibacterial agent, agricultural use, table of, 215
antibiotic
  for pretreatment of soil, 286
  table of, for selective isolation, 289
antibiotic TA, 11
antibody forming cell, 177
antichlamydial agent, 5
anticoccidial compound, 67, 69
  table of, 255
anticomplementary substance, 184–186
antifungal antibiotics, table of, 32
  agricultural use, 215
anti-HIV substances, 55–60

antihypertensive, 208
antiinflammatory screening, 203
antimicrobial activity, antitumor
   screening, 83
antimycin, 41, 175, 242, 248
antipain, 121, 266, 269
antiparasitic agents, table of, 64, 75
antiphage activity, antitumor
   screening, 83
antiprotozoal antibiotic, 65–69
antithrombotic drug, 200
antitrichomonal antibiotic, 67, 68
antitumor activity
   for immunomodulator, 173
   cell differentiation inducer, 106, 107
   enzyme inhibitor, 136, 137, 142
antitumor antibiotics, table of, 92, 93–95
antivermal antibiotics, 70
antiviral agents, table of, 46, 52, 53
6-APA, see 6-aminopenicillanic acid
APD, 125
aphidicolin, 53
API-2, 120
A-PK, see cAMP-dependent protein
   kinase
aplasmomycin, 285
acquired immunodeficiency syndrome,
   30, 55–59
arachidonic acid, 198
arglecin, 266, 271
aridicins, 11, 21
arildone, 52
armyworm, 247
arphamenines, 120, 137
arterial blood pressure, 207, 208
arugomycin, 93
asparenomycins, 10
aspergillomarasmine, 129, 130
Aspergillus niger, 38
aspermutarubrol, 276
aspiculamycin, 71, 246, 248
aspochracin, 246, 248
assasy method, see also screening
   method
   antiviral agent, table of, 48, 53
asukamycin, 296
AT-2433, 94
atherosclerosis, 128

atpenin, 296
augudin, 53
aurachin, 267, 275, 276
aurantinin, 296
aureothin, 242, 249
autochthonous tumors, 87
autoimmune disease, model mouse of, 191
autoradiography, 49
avermectin, 64, 72, 73, 75, 241, 243, 244, 248, 296
   table of activity, 245
avoparcin, 21
awamycin, 93, 296
axenomycins, 64, 71
azalomycin, 32, 274
azidothymidine, 47, 52, 56
azinomycins, 94
AZT, see azidothymidine
azukibean weevil, 247
azureomycin, 8, 20, 296

B
Bacillus, Dpm-requiring strain, 8, 20
Bacillus licheniformis, supersensitive, 9, 16
Bacillus subtilis, antitumor screening, 89, 90
   in synthetic medium, 89–90
   resistant mutant, 3
   sensitive mutant, 21
bacitracin production, 310
bacteria
   antitumor assay, table of, 83
   numbers in soil, 283, 284
bacterial cell wall inhibitors, table of, 7–11
Bacteroides, 4
BAYg5421, 122, 128
BE-10988, and BE-13793, 125
beauvericin, 242
benadrostin, 124
benanomicins, 53, 54, 55
benzidine staining method, 109
benzoisochromanequinone antibiotic, 202
benzylmalic acid, 120

BERDY data base, 6
bestatin, 92, 120, 128, 136, 137, 146, 177, 178
BHK 21 cell, 258
bialaphos, 225, 226, 227
bicyclic β-lactam, 9
bioactive compound by chemical screening, table of, 265–267
bioluminescence, 83
biosynthesis
 affecting factors, 304–312
 gene cluster of, 331
 regulatory factors, table of, 305–306
blasticidin S, 215, 216
bleb-forming assay, 141–142
bleomycin, 81, 242
blocked mutant, 321, 329
bloodstream clearance, 176
BMY-28160 and BMY-28251, 23
BMY-28438, 94
Botrytis cinerea, 219, 221
bredinin, 188, 189
brine shrimp, 250
bromovinyldeoxyuridine, 52
Brush inoculation method, 221
Bu-2743E, 120
bulge formation, 6, 15
bulgecins, 7, 15
Bursaphelenchus lignicolus, 64, 73, 74
butirosins, 22, 23

C
C-1027, 94
C1920, 93
C-19393, 7, 11
Caenorhabditis elegans, 64, 74
calcium 3-hydroxyquinoline-2-carboxylase, 275
calicheamicins, 94
calmodulin, 140
calphostins, 125, 142, 143
cAMP-dependent protein kinase, 140
candicidin production, 307
Candida albicans, 233
 morphological change, 38
Candida lipolytica, 164, 167
capoamycin, 93, 111
capreomycin, 23

carbapenem, 7, 9–11, 17
carbazomycins, 123
DD-carboxypeptidase, 19
carcinogen, 87
caricastatin, 121
carpetimycins, 7, 11
cationomycin, 315, 316
CD_4 analogs, 52, 56
celite, 312
cell culture, *see also* tissue culture
 anticoccidial assay, 257–258
 for tumor, 88–89
 for virus, table of, 48
 phytotoxic assay of herbicide, 231, 232
cell-free system, antitumor agent, 91
cell line
 leukemia, 105, 107, 109, 111, 141
 neuroblastoma, 110–111
 table of tumor cell, 82
 tumor, 86, 88
 virus, 48
cell-mediated immunity, 176
cell permeability, 274
cell proliferation, 174–175, 179–180
cell wall, cellulose containing, 228
cell wall inhibitor
 bacterial, 6–22
 fungal, 34–38
cellulose synthesis, 233
central nervous system tissue, 51
cephabacins, 7
cephalosporin, 7, 13, 14, 15, 16, 19
cephalosporinase, 13, 16
cephalotin, 16
cephamycin, 7, 12, 15, 19, 20
cephem, 7, 10, 11
cerulenin, 167, 296
cervinomycin, 296
cestodes, as test organism, 64
cGMP-dependent protein kinase, 140
chemical screening, antifungal antibiotics, 39
chemotactic method, figure of, 291
chemotactic property of actinomycetes, 290
chicamycin, 93
chimeramycins, 296, 322, 323

chitin medium, 287
chitin synthesis, 34, 35
chitinase inhibitor, 251
chitinovorins, 10
chloramphenicol, 215
chlorocaidicin, 11
8-chloroorbol, 296
2'-chloropentostatin, 125, 140
chlortetracycline production, 307
chochlioquinone, 253
cholesterol, 39, 128
  biosynthesis, 162, 164
  figure of biosynthesis, 163
choline esterase inhibitor, 241
chromostin, 124
chromoxymycin, 93
chrysomycins, 94
chymostatin, 120
cinnamic acid amide, 126
ciotrinin, 127
citrinin, 111
clavamycin, 37
clavulanic acid, 9, 14, 15, 17, 18
cloning genes for antibiotics synthesis, 329
clostomicins, 4, 296
Clostridium, 4
coccidiosis, 253
coformycin, 138, 139
colistin, 242
collagen, 198, 203
colony formation assay, 109
color change
  enzyme inhibitor, 139
  β-lactamase inhibitor, 18
  MTT assay for HIV, 57
  polyethers, 41
color reagent, 264
  table of, 265–267
coloradocin, 4
colorimetric assay, dye uptake method, 89
Comamonas terrigena, 9, 16
compactin, 123, 131, 132, 162, 163, 164
complement system, 184, 185
complestatin, 173, 186, 187
Con A, *see* concanavalin A
concanamycin, 173, 175

concanavalin A (Con A), 174
coriolin, 177
cosmomycin A, 111
CP-84657, 255, 256
CPE, *see* cytopathic effect
C-PK, *see* protein kinase C
culture condition, soil isolation, 287
CV-1, 5
cycloheximide, 41, 215, 225, 242
  production of, 309
cyclophellitol, 122
D-cycloserine, 20, 21
L-cycloserine, 6, 10
cyclophilin, 184, 187
cyclosporin A, 173, 175, 181, 183, 186, 187, 188, 190, 321
cyclothiazomycin, 121
cystargin, 293
cytopathic effect, 47, 48, 49, 52
cytosine arabinoside, 52
cytotoxicity
  antitumor screening, 82, 89
  antiviral compound, 49, 50

D

dactylarin, 64, 65, 66
dapiramicin, 219
daunomycin, 53
DC-86-M, 93
DC-89-A, 94
DC-92, 94, 95
DC-107, 95
deacetoxycephalosporin C, 10
deacetylcephalosporin C, 10
3'-O-decarbamoylirumamycin, 296
dehydrodicaffeic acid dilactone, 126
delayed-type hypersensitivity (DTH), 176–177
4'-deoxybutirosin, 22
deoxycoformycin, 92, 125, 128, 138, 139
deoxynojirimycin, 122
deoxyprepacifenol, 248
deoxyspergualin, 190
depsidomycin, 173, 175
10-descarbomethoxymarcellomycin, 108
destomycin, 241, 242, 244

dextransucrase, see glucosyltransferase
diacetyl reagent, 269
diabetes, 133, 134
diastovaricins, 111
diazaquinomycins, 124, 296, 317, 318, 319
4′,6-dichloroflavan, 47, 52
dienomycin, 266, 271
differanisol, 111
differenol, 111
differenolide, 267
dihydrocompactin, 132
6,7-dihydromethylspinazarin, 126
dihydromevinolin, 123, 132
dihydromonacolin L, 123, 132
2,5-dihydro-L-phenylalanine, 126
dihydrosarcomycin, 127
dihydroxerulin, 124
3,7-dihydroxytroporone, 94
diketocoriolin B, 177
3-(4,5-dimethylthiazole-2-yl)-2,5-diphenyl tetrazolium bromide, see MTT
dipeptidyl carboxypeptidase, see angiotensin-converting enzyme
1,3-diphenethyl urea, 265, 296
diplodialide, 123
diprotins, 120
directed biosynthesis, 321
distamycin A, 53, 54
distribution of microorganisms, figure of, 284
dityromycin, 265, 268, 270, 296
DNA polymerase, 46
DNA topoisomerase II, 91
dopastin, 126
dotriacolide, 125, 127
doxorubicin, 53, 81
Dragendorff's reagent, 264, 268
drug-free assay, 172
DTH, see delayed-type hypersensitivity
duocarmycin, 94, 95
duramycins, 124
dye staining, 82, 89
dye uptake method, 48, 89

E
E-64, 121
ebelactones, 127
echinocardin B, 32, 33

Ehrlich reagent, 271–274
Eimeria tenella, 64, 69, 254
elasamicins, 93
elasnin, 120, 296
elastatinal, 120
ELISA, see enzyme-linked immunosorbent assay
elloramycins, 266, 272, 273
enhancement of production, 309
enniatin, 39
enviroxime, 52
enzyme
　antitumor screening, table of, 92
　in parasites, 76
　specific for cancer cell, 91
　specific for virus, 49
enzyme and enzyme inhibitor, table of, 120–127
enzyme inhibitor
　as herbicide, 233
　as immunomodulator, table of, 177, 178
　for hybrid biosynthesis, 321
enzyme inhibitory activity
　to β-lactamase, 17
　to DD-carboxypeptidase, 19
enzyme-linked immunosorbent assay
　antibacterial compound, 5, 23
　antiviral compound, 49
epithienamycin, 7, 12, 20
erbstatin, 92, 125, 144, 145
ergosterol, 39
Erysiphe graminis, 221
erythroleukemia cells, 108
erythromycin
　derivatives by ELISA, 5
　producing organism, 334
　production, 311
Escherichia coli
　altered membrane, 83
　antibacterial assay, 4, 5, 15
　antitumor assay, 83
　spheloplast, 6, 9
　supersensitive, 6, 7, 8, 10, 11, 13, 15, 18
estatins, 121
esterastin, 127, 177, 178
ethericins, 267, 274, 275, 276
ethylenediamine disuccinic acid, 124

Index 343

2-ethyl-5(3-indolyl)-oxazole, 273
Eucoccidium dinophili, 69, 70
euglobin clot lysis test, 206–207
experimental infections, 180

F
F5-5 cell, 109–110
F-244, 124
factumycin, 296
false-positive, 161, 162, 168
fatty acid metabolism, 164–168
  figure of, 166
fatty acid synthetase, 162, 164, 168
ferric chloride reagent, 269
ferrithiocin, 267, 273, 274
fibrinolytic agent, 206, 207
fibrostatins, 127
FK-156, 173, 177, 181, 315, 316
FK-409, 205, 206
FK-506, 173, 176, 182–184, 190
FK-506-binding protein, 184, 190
FL-657, 109–110
5-fluorouracil, 137
FO-740, 296
focus reduction, 48
formadicins, 8
formycin, 139
5-formyluracil, 124
foroxymithine, 120, 129, 130
forphenicine, 125, 177, 178
forphenicinol, 92, 177
fosfazinomycin, 37, 38
fosfomycin, 10, 12
fosmidomycin, 8
FR-31564, FR-32863, and FR-33289, 8
FR-49175, 200, 203
FR-65814, 173, 176
FR-66979, 95
FR-900098, 8
FR-900130, 8, 20
FR-900137, FR-900148, and FR-900318, 8
FR-900405, FR-900406, 93
FR-900452, 200, 203
FR-900462, FR-900482, 94
FR-900483, 173, 180
FR-900494, 173, 180, 293
FR-900520 and FR-900523, 173, 176

FR-900840, 95
frenolicin B, 200, 255, 257, 296
Friend leukemia cell, 108–109
Friend leukemia virus, 88
fungi, numbers in soil, 283, 284
furaquinocins, 296
fusaric acid, 126
Fusobacterium, 4

G
galactostatin, 123
ganglionic cell line, 110–111
gene coding, HIV-1, 59
gene cluster of antibiotic biosynthesis, 331
genistein, 125, 126, 144, 145
gentamicin, 23
genus of actinomycetes, table of, 299
glidobactins, 94
globomycin, 9
globopeptin, 37, 296, 317
β-1,3-glucan synthesis, 34, 35
glucoallosamidins, 122
glucopiericidinols, 296
glucose as regulatory factor, 306
α-glucosidase, 128
glucosyl S-GI, 122
glucosyltransferase, 131–133
glutamine antagonist, 233
glutamine synthetase, 168
glycopeptide antibiotic, 8, 11, 21
glycoprotein, 60, 108
glyo-I, 92, 127
glyo-II, 92
gostatin, 127
G-PK, *see* cGMP-dependent protein kinase
gramicidin S, 242
granaticin producing organism, 334
granatirhodin, 334
greening-inhibitor, 230
griseofulvin, 32, 33, 215
griseolic acid, 125
GTase, *see* glucosyltransferase
guanine-7-oxide, 93

H
habitat of actinomycetes, 284
Haemonchus contortus, 64, 71
haim-I and II, 122

heart rate measurement, 207, 208
heat hemolysis of erythrocyte, 203
heat treatment of soil, 290, 294
hemadsorption, 48
hemolytic activity, 38, 39, 40
heptidilic acid, 122
herbicidal activity, 41
herbicidal antibiotic, 113
herbicidal metabolite, table of, 226
herbicidin, 225, 226, 227
herbimycin, 112, 113, 225, 226, 227, 297
herpes simplex virus, 48, 51, 52, 53
herquline, 265, 269, 270, 297
himastatin, 95
histargin, 120
hitachimycin, 297
HIV, see human immunodeficiency virus, 48, 52, 55–60
HIV replicative cycle, table of, 56
HIV-1 protease assay, 59
HL-60 cells, see human promyelocytic leukemia cell
HMG-CoA reductase, 128–131, 146, 162, 163
HMG-CoA synthase, 162, 164
HMPGG, see 7-hydro-8-methylpreroylglutamylglutamic acid
homoalanosine, 226, 228
HSV, see herpes simplex virus
human immunodeficiency virus, 48, 52, 55–60
human promyelocytic leukemia cell, 107, 108
hybrid biosynthesis, 321, 323
hybridization technique, 49
hydantocidin, 226, 227, 228
7-hydro-8-methlpreroylglutamylglutamic acid (HMPGG), 24, 297
6-(3-hydroxybutyl)-7-O-methylspinochrome B, 126
hydroxychlorothricin, 94
hydroxydaidzein, 297
8-hydroxygenistein, 126
13-hydroxyglucopiericidin A, 297
7-hydroxyguanine, 93
3-hydroxy-3-methylglutaryl coenzyme A, see HMG-CoA

hydroxypepstatin, 121
p-hydroxyphenylacetaloxime, 123
L-β-(5-hydroxy-2-pyridyl)-alanine, 271
L-β-(3-hydroxyureido)-alanine, 271
hygromycin B, 53, 54, 241, 244
hypercholesterolemia, 128, 131, 146
hypersensitive mutant, see also supersensitive mutant
  to aminoglycoside antibiotics, 22–23
  to β-lactamase antibiotics, 7, 9, 13, 16
hypersensitivity, immunomodulator, 176–177
hypertension, 118
hypertensive system, 119
hypocholestemic agent, 162, 163
hypotensive system, 119

I
I-6123, 126
identification of known antibiotics, 5–6, 40–41
IgM, see immunoglobulin M antibody
ikarugamycin, 64, 66
IL, see interleukin
immunoglobulin M antibody, 172
immunomodulator as enzyme inhibitor, 136
  table of, 177, 178
immunophilins, 184, 190
immunopotentiator, 177, 178–181
immunosuppressive factor, 178–179
immunosuppressor, 176, 177, 181–186
inactivation by/with β-lactamase, 7, 10, 15, 16
induction of β-lactamase, 18
influenza virus, 48, 50, 51, 52
inhibition of
  antibiotic production, 307
  cellulose synthesis, 233
  cytopathic effect, 47
  dimorphic conversion, 38
  DNA topoisomerases, 91
  hemadsorption, 48
  HIV cytopathic effect, 57
  β-lactamase, 9, 17, 18
  life cycle of virus, 52

IL-2 production, 181
reverse transcriptase, 56
starch synthesis, 231
syncytium formation of HIV, 58
virus specific enzyme, 49
inhibitor
  ACE, table of, 129
  bacterial cell wall synthesis, 6–22
  chitin synthesis, 34
  enzyme, table of, 120–127
  fatty acid synthesis in bacteria, 21
  fungal cell wall synthesis, 34
  greening, 230
  α-mannosidase, 180
  platelet aggregation, 198
  specific for cancer cell, 92
inorganic phosphate, 306, 314
inostamycin, 92
insect, table of, 238–239
insecticide, by chemical synthesis, 240–241
insecticide-resistant organism, 251
interferon, 56
interleukin-2 (IL-2), 181–182, 190
intracellular factors, table of, 310
iododeoxyuridine, 52
irumamycin, 219, 297
irumanolide, 297
isoflavonoids, 123
isosulfazecin, 7, 15
ivermectin, 241
ivermectin binding assay, 253
izumenolide, 127
izupeptins, 8, 20, 297

J
jietacins, 64, 71, 72, 249, 297, 317, 319
juglorin, 127

K
K-4, K-13, and K-26, 120, 128, 129, 130
K-76, 173, 186, 187
K-252, 125, 142
K-254, and K-259, 126
KA-107, 17, 127, 297

kalafungin, 200, 333
kallikrein–kinin system, 118, 119
kanamycin, 53
kaolin, 312
kapurimycins, 95
karabemycin, 297
kasugamycin, 215, 216, 217, 220
kazusamycin, 93, 297
KD16-U1, 267
kerriamycin, 93
ketalin, 272, 273
KF77-AG6, 266
KF8940, 123
kibdelins, 11, 21
kifunensin, 173, 180
kijimycin, 255, 257
kinamycins, 297
kininase II, see angiotensin-converting enzyme
Kitasatosporia, 290, 291, 292, 293
Klebsiella aerogenes, 17
Klebsiella pneumoniae
  sensitive strain, 18, 23
  resistant strain, 22
KM-8, 297
KMC data base, 6
KO-3599, 297
Kohsaka's method, 220
kojic acid, 126
koningics, 122
KS-501, KS-502, KS-504, and KS-619, 126
KT5720, and KT5822, 142

L
L-155,175, 64, 71
L-681,110, 126
L-681,176, 120, 129, 130
LA-1, 297
lactacystin, 112, 297
β-lactam antibiotic
  antibacterial, 7, 12, 13, 17
  antifungal, 37, 40
β-lactamase, 7, 9–11, 16, 17, 18, 19
lactocillan, 106
lactoquinomycin, 93
lankacidin production, 309
lasalocid, 254, 255, 257

lavanducyanin, 95
lavendustin A, 92
leaf disk method, 221
lecanoric acid, 126
leinamycin, 95
lentinan, 173
leptomycin, 37, 38
leucanicidin, 246, 249
leucine aminopeptidase, 137
leucomycin, 297
leuhistin, 120
leukemia cell line, 105
leupeptin, 120
limocrocin, 124
lipid metabolism, 162, 164
lipopeptin, 35, 36
lipstatin, 124
LL-AM-347, 21
LL-D49194, 95
LL-F28249, 64, 71, 72
locomotive activity, 106
lovastatin, 128, 131, 132, 146
luminamicin, 4, 297
lustromycin, 4, 297
luteosporin, 124
lysozyme activity, 105

M

M119 complex, 288
maduramicin, 254, 255, 256
magnesium carbonate, and phosphate, 308
mannan synthesis, 34, 35
mannostatins, 123
α-MAPI, 121
marcellomycin, 107, 108
marine microorganism, 312
marine sediments, 285
MC-696-SY, 127
mechanism of action, see action mechanism
mechanism of regulation, table of, 305–306
medermycin, 200, 202
mederrhodin A, 297, 333
medium
    affecting antibiotic biosynthesis, 304
    control of fermentation, 312
    for actinomycetes, table of, 288
melanocidin, 177, 178
metabolites 2, 3, and 4, 7
metal ions, 307
    for antibiotic production, table of, 308
methisazone, 52
7-methoxy-deacetylcephalosporin C, 20
methylallosamidin, 122
2-methyl-4-quinazolinone, 124
methylrosaniline staining, 163
methylspinazarin, 126
7-O-methylspinochrome B, 126
mevalonate biosynthesis, inhibition, 162–164
mevinolin, 123, 128, 131, 132, 162
M-GCI, 123
MH237-CF8, 113
microbial conversion, 320
Microbispora, 288
Micrococcus flavus, 3
Micromonospora, 286, 288, 293
Microtetraspora, 288
miharamycin, 219
milbemycin, 241, 243, 244, 249
mildiomycin, 215, 216, 218, 221
mishigamycin, 297
mite, table of, 238–239
mitogen of T lymphocyte, 174, 175
mitomycin C, 81, 137
mixed culture, 320
mixed lymphocyte reaction, 176, 182–183
MK-I, 121
M1 cell, see mouse myeloid leukemia cell
ML-236, 123, 132
MLR, see mixed lymphocyte reactions
MM 4550, 9, 17, 127
MM 13902, and MM 17880, 9, 17
MM 22380, MM 22381, MM 22382, and MM 22383, 9, 18
MM 27696, and MM 42842, 9
mode of action, see action mechanism
monacolin, 123, 128, 131, 132, 162, 164
monensin, 254, 255, 256

monobactam, 7, 9, 10, 16
monocyclic β-lactam, 7, 8, 11, 13, 16
moroyamycin, 242
morphological change
   bacterial cell, 6, 15, 18
   cell differentiation inducers, 106, 113
   fungal cell, 37–38
   tumor cell, 82, 87, 88–89
   Vero cell, 164, 165
mosquito, 245
mouse myeloid leukemia cell (M1 cell), 105, 106, 107, 111
MRL/1pr mice, 191
MS-3, 92
MTT assay, 57, 58, 89, 174
Mucor, 38
muraceins, 120, 129, 130
murasmine, 120
musettamycin, 107, 108
mutant
   antibiotic resistant, 2, 3
   supersensitive to antibiotic, 3
mutastein, 122, 128, 133, 146
mutational cloning, 329–330
MY3-469, 123
MY12-62, 123
mycelianamide, 126
mycenon, 127
Mycoplasma, 20
mycosis, table of, 30
myrocin C, 95
myxalamid, 276
myxobacteria, 294
myxochelin, 266, 269, 271
myxovirus, 48

N

NA-337 A, 264, 265, 297
nanaomycin, 201, 297, 317, 318, 319
naphthomevalin, 267
naphthoquinomycins, 22
narasin, 254
nasal tissue, 50
NB-1, *see* neuroblastoma cell, 111
nelanocidins, 124
nematocidal antibiotics, 69, 70, 73

nematode, as test organism, 64, 73, 251
Nematospiroides dubius, 64, 72, 74
neoaureothin, 242, 249
neopeptins, 35, 36, 219
neopolyoxins, 35, 36, 219
neorustmicin, 219
3,3'-neotrehalosadiamine, 23
neoxaline, 265, 269, 270, 297
neuraminin, 122
neuroblastoma cells, 110–111
new Ifv, 126
nigrifactin, 265, 270
nikkomycin, 39, 41
ninhydrin reagent, 269
nitrocefin, 18
nitropeptin, 219
Nocardia, 288, 293
nocardicins, 8, 13, 14, 15, 315, 316
nojirimycin B, 122
norerythromycin, 334
notonesomycin, 219
novobiocin, 53, 215, 290, 292
nucleic acid hybridization assay, 49
nucleoticidin, 124, 177–178
nutritional factor, 304–308
nystatin, 32, 33

O

O-2867, 297
OA-6129, 9, 17
octylpentanedoic acid, 124
OF-4949, 120, 137
oganomycins, 11
OH-1049R, 95
OH-2223, 297
okaramine B, 246, 248
oleandomycin, producing organism, 334
oligopeptide antibiotic, 3
oligopeptide transport deficient mutant, 3
oligosaccharide, 108
olivanic acid, 9, 17, 19
OM-173, 297
OM-4842, 201, 202
onchocerciasis, 73
oncogene, 142

oncovirus, 87
oosponol, 126
opportunistic infection, 4
organ culture, see also cell culture for antiviral screening, 50
organic solvent, for pretreatment of soil, 286
orobol, 125, 126, 144, 145
OS-3256B, 297
OS-1000, 298
osmotic pressure, 312
oudenone, 126
oxetanocin, 53, 54, 315, 316
oxetin, 127, 226, 227, 228, 298
oxicenone, 298
oxytetracycline, 215

P

P-1894B, 127
PAF, see platelet activating factor
paims, 122
paper disc method, see also agar diffusion method, 32, 34, 219
pannorin, 123
panosialin, 122
paraffin bait technique, 293–294
parasites, table of test organisms, 64
parvodicin, 11, 21
pathogen and plant disease, table of, 214
pathogenic fungi, table of, 30
patulin, 242
PB-5266, 10
PD116,152 and PD124,895, 93
PDE, 125
pellet formation, 309, 311, 312
Pellicularia sasaki, 225
penam, 8
penicillin G, 13, 14, 21
penicillin N, 7, 14, 19, 20
penicillin production, 311
penicillinase, 8, 13, 16
pentalenolactone, 298
pepstanone, 121
pepstatin, 53, 54, 121
peptidoglycan biosynthesis, 19
peptidyl prolyl cis-trans isomerase, 187

Peptococcus, 4
pereniporin, 230
permeation-damaged cell, 274
pesticidal antibiotic, table of, 242, 248–249
pesticide, 240, 241–244
PFC, see plaque-forming cell
PHA, see phytohemagglutinin
phage, antitumor assay, 83
phagocytic activity, 105
phenacein, 120, 128, 129, 130
phenazinomycin, 95, 298
phenicin, 123
phenomycin, 92
phenopicolinic acid, 126
phosphate
  high concentration in medium, 312
  ion-depressed fermentation, 317
  ion-trapping agent, 308, 309
phosalacine, 127, 168, 226, 227, 228, 233, 293, 298
L-phosphinothricyl-alanyl-alanine, 3
phosphinothricine, 168, 225
phospholine, 95
phospholipase on platelet membrane, 198
phosphonic acid, 8
phosphonoformic acid, 52
phosphoramidon, 121, 266, 273
Photobacterium leignathi, 83
phthomevalin, 274
phthoramycin, 298
phthoxazolin, 226, 227, 228, 233, 298, 317
physiological active compound, see also bioactive compound, 265
phytohemagglutinin (PHA), 174
Phytophthora parasitica, 228, 233
phytotoxic compound, 231
piericidin, 242, 246, 248, 249
pimaricin, 32, 33
pimprinethine, 273
pimprinine, 126, 205
placetins, 202
plant disease, 213
  and pathogen, table of, 214
plaque-forming cell, 172
plaque reduction assay, 47, 48, 52
plastatin, 124

platelet activating factor (PAF), 198
platelet aggregation inhibitor, 198–204
plipastatins, 124
pluracidomycins, 11
PMN, *see* polymorphonuclear leucocyte
polyene antibiotic, 39, 40
polyether ionophor, 41, 254, 255
poly-L-malic acid, 121
polymorphonuclear leucocyte (PMN), 181
polymyxin B, 242
polyoxin, 34, 36, 41, 215, 216, 217
portimicin, 256
poststatin, 120
pot test method, 219
powdery mildew, 218, 221
pradimicin, 52, 53, 54, 55
pravastatin, 131, 146, 162
pretreatment of soil, for selective isolation, 286
probestin, 120
prodigiosin 25C, 173
prohisin, 121
propanocin, 219
prophage, 83
propioxantins, 121, 293
protease, 46, 59
protein kinase, 46
protein kinase C, 140–142, 269
protein phosphatase inhibitor, 142
Proteus vulgaris, 7
protomycin, 64, 66, 67
protoplast, 327
protozoa, as test organism, 64, 65
protylonolide, 298
prumycin, 298
PS-5, PS-6, PS-7, and PS-8, 9, 17
pseudomembranous colitis, 4, 17
Pseudomonas aeruginosa, hypersensitive, 7–9, 11, 13, 15, 21, 22
pseudospheroplast, 37, 38
psi-tectorigenin, 126
Puccinia recondita, 219
purpactins, 123, 298
Pyricularia oryzae, 35, 215, 217, 219, 220

pyridindolol, 123
pyrindamycins, 94
pyrindicin, 264, 265, 270, 298
pyrrolnitrin, 32, 33
pyrrolnitrin production, 312, 314
pyrromycin, 107, 108
pyrrothine antibiotic, 203

Q
quinoline-2-methanol, 265, 268, 298

R
radioactive assay method
 for antibacterial agent, 20, 22
 for antitumor agent, 82
 for antiviral agent, 49
radiochemical assay, 49
radioimmune assay, 5
rapamycin, 189
rare actinomycetes, 284, 287, 289
 figure of isolation, 287
 selective isolation, 289–294
rebeccamycin, 94
recombinant virus DNA, 49
red blood cell, 48
reductiomycin, 265, 268
regulation of production, by gene, 332
regulatory factors, table of, 305–306
renin, 118, 119
resistant mutant, bacterial, 16–17
resistant strain, fungal, 40
resorthiomycin, 95
retinoid, 114
reticulol, 125
retrostatin, 124
reversal of antifungal activity, 41
reverse transcriptase assay, 59
revistin, 124
rhino virus, 48
Rhizoctonia solani, 219
rhizoxin, 93
RI-331, 42
ribavirin, 52
ribocitrin, 122, 133
riboflavin, 123
rice plant-infusion method, 220
rifampicin, 52

rimantadine, 52
ristocetins, 21
RK-286, 125, 142
RNA polymerase, 46
Ro09-0766, -0767, and -0768, 122
Rous sarcoma virus
  cell differentiation inducer, 112
  herbimycin, 225
  tumor, 90
RSV, 48
rustmicin, 219
Rydon-Smith reagent, 269

S

S-15-1, 53
S-AI, 122
Saccharomyces cerevisiae
  antifungal assay, 35
  antitumor assay, 83
SAGP, 95
saintopin, 91, 92
salinomycin, 254, 255, 257
salts in medium, 317
  table of, 314
Sakaguchi's reagent, 269
sakyomicin A, 53, 54
sandramycin, 95
saphenic acid methyl ether, 267, 274, 275
sarcomycin, 127
Scenedesmus obliquus, 230
Sch 38519, 201, 203
screening method, see also assay method, and screening system
screening method, table of, or figure of
  antiparasitic antibiotics, 64
  antitumor compounds, 82, 83, 85, 92
  antiviral agents, 51
  bacterial cell wall inhibitors, 7–11
  cell differentiation inducer, 111
  immunomodulator, 173
  nematocidal substance, 74
  synergy with polyenes, 39
setamycin, 64, 66, 67, 293, 298
setomimycin, 298
SF-1239, 226

SF-1293, 3, 4
SF-1623, 10, 19
SF-2050, and SF-2103, 10
SF-2140, 53
SF-2312, and SF-2339, 3, 4
SF-2494, 226, 227, 228
S-GI, 122, 123
siastatins, 122
sibanomycin, 94
siccanin, 32, 33, 242
simaomicin, 255, 257
simvastatin, 131, 146, 162
SM 196, 267, 274
S-MPI, 121
sohbumycin, 93, 298
soil microorganisms, number of, table of, 283
sorbistin, 22, 23
spheroplast, 6–8, 13, 20
sphydrofurans, 266, 272, 273
S-PI, 121
spicamycin, 107
spiramycin, 5
spleen cell proliferation, 174–175, 179
sporamycin, 298
sporulation, 332
spot inoculation method, 220
SQ 26,180, and SQ 27,860, 9
SQ 26,445, 5, 16
SQ 26,517, SQ26,700, SQ26,812, SQ 26,823, SQ 26,875, SQ 26,970, SQ 28,332, SQ 28,502, SQ 28,503, SQ 28,516, and SQ 28,517, 10
S-SI, 120
Src gene expression, 113
staccopin, 121
staining reagent, 274
Staphylococcus aureus, 13, 20
  resistant mutant, 3, 11
staurosporine, 111, 125, 141, 142, 202, 265, 268, 269, 270, 298
stigmatellin, 276
strepin, 121
streptazoline, 266, 272
Streptococcus mutans, enzyme from, 131–132
Streptomyces, 286

streptomycin, 215, 332
Streptosporangium, 294
streptothricin group antibiotics, 23
streptovirudins, 53
streptovitacin, 242
5′-O-sulfamoyltubercidin, 228
sulfazecin, 7, 9, 15, 16, 20
superfusion method, 204
supersensitive mutant, *see also*
        hypersensitive, 3, 7, 8, 9, 13, 15, 18
survival effect, 87
swainsonine, 173, 179–180
symbiosis, 295
synergistic activity with
  cycloserine, 8, 20
  β-lactams, 4, 15
  polyene antibiotic, 39

T
T cell, 55
talopeptin, 121
takaokamycin, 298
TAN-868 A, 3
TAN-931, 124
target directed screening, antibacterial, 6–23
target virus, table of, 52
tautomycin, 142, 143, 219
temperature effect, 311
terferol, 125
terpentecin, 293
tetranactin, 241, 243, 244, 249
tetrazolium reagent, 271–274
thaimycins, 64, 71
thermophillic actinomycetes, 288
thiazocins, 122
thielavins, 123
thielocins, 124
thienamycin, 7, 12, 13
thiolactomycin, 21
thiolactone, 11
thiolstatins, 121
thiolutin, 201, 204
thiosarin, 201, 204
thiotetromycin, 4, 298, 317, 319
thrazarine, 94
thrombin, 199

thymidine kinase, 46
thymidylate synthetase, 168
tirandalydigin, 4
tirandamycin, 298
tissue culture, *see also* cell culture
  virus, 50, 51
TM-64, 264, 265, 288, 298
topostins, 125
toyocamycin, 125, 225
tracheal tissue, 50
transformation
  and tansfection, 328
  fibroblast, 90
  MTT assay, 174
transplantable tumor, 84–89
trehalase inhibitor, 252
trehalostatin, 123, 252
trehazoline, 123
trestatins, 122
triacsins, 124, 167, 168, 298, 317, 318, 319
1,2,4-triazole-3-alanine, 298
Trichomonas foetus, 64, 67, 68
trichomycin, 32, 33, 39
trichostatin, 67, 111
trienomycins, 93, 298
trifluorothymidine, 52
tripeptide antibiotic, resistant to, 3
trypacidin, 64, 65, 66
trypan blue exclusion test, 49, 50
tumor cells, table of, 82
tumor transplantation, 86–87
tunicamycin, 293
two-spotted spider mite, 250
tylosin production, 307
tyrosine kinase, 142–145
tyrosine-specific protein kinase, 142–145
tyrostatin, 121, 293

U
U-68,204, 11
ubenimex, *see* bestatin
UCN-01, and UNC-02, 125, 141, 142
undecylprodigiosin, 332
urdamycin, 266, 272
UV absorption spectrum, polyene, 40, 41

V

valanimycin, 93
validamycin, 215, 216, 217, 220
validoxylamine A, 252
valielactone, 127
valinomycin, 248
Valsa ceratosperma, 219
vancomycin, 21
vanoxonin, 124
variotin, 32, 33
vasodilator, 204–206, 208–209
vermiculine, 64, 65, 66
vertical distribution of soil microorganism, figure of, 284
Vero cells, 163, 164, 168
Vibrio percolans, 7
vidarabine, 47
vineomycins, 127, 298
vinigrol, 208
virantomycin, 53, 54, 298
virus, malignant tumor, 88, 90
virus and antiviral agents, table of, 52, 53
virus and viral diseases, table of, 46
virustomycin, 298
viscosity of culture fluid, 311
VM44857, VM44864, and VM44866, 64, 71, 72

W

WB-3559, 207
WF-2421, 122
WF-3405, 94
WF-3681, 122, 135
WF-5239, 200
WF-10129, 120, 129, 130
Wood reagent, 269
WS-1228, 204, 205
WS-1358, 127
WS-5995, 64, 66, 67, 255, 256
WS-9659, 123
WS-30581, 204, 205

X

Xanthomonas oryzae, 41
xanthomycin-like, 64

Z

zeolite, 308, 309, 312
Z-laureatin, 246, 248

Brock/Springer Series in Contemporary Bioscience

(Continued from page ii)

John H. Andrews
COMPARATIVE ECOLOGY OF MICROORGANISMS
AND MACROORGANISMS

Ryszard J. Chróst (Editor)
MICROBIAL ENZYMES IN AQUATIC ENVIRONMENTS

Alan G. Williams and Geoffrey S. Coleman
THE RUMEN PROTOZOA

John H. Andrews and Susan S. Hirano (Editors)
MICROBIAL ECOLOGY OF LEAVES

J.M. Odom and Rivers Singleton, Jr.
THE SULFATE-REDUCING BACTERIA: CONTEMPORARY
PERSPECTIVES

Satoshi Ōmura (Editor)
THE SEARCH FOR BIOACTIVE COMPOUNDS
FROM MICROORGANISMS

Madeleine Sebald (Editor)
GENETICS AND MOLECULAR BIOLOGY OF ANAEROBIC BACTERIA